INFINITESIMAL METHODS OF MATHEMATICAL ANALYSIS

'Mathematics possesses not only truth, but supreme beauty - a beauty cold and austere like that of sculpture, and capable of stern perfection, such as only great art can show,'

Bertrand Russell (1872-1970) in *The Principles of Mathematics.*

HORWOOD

ABOUT THE AUTHOR

Born in the region of Cheires, Portugal, José Sousa Pinto originally qualified as a chemical engineer at the University of Porto. An increasing interest in, and aptitude for, mathematics led him to change direction culminating in postgraduate study at the Cranfield Institute of Technology (now Cranfield University) where he obtained his M.Sc. in mathematics and subsequently a Ph.D for a thesis in applied functional analysis. He returned to Portugal first as a mathematics lecturer in the University of Coimbra and then as Associate Professor of Mathematics at the University of Aveiro, where he remained until his untimely decease in 2000.

Sousa Pinto's research activity was primarily centred on aspects of the theory of distributions and other generalised functions. He published a number of papers in this field, and was particularly concerned with applications of generalised functions to sampling theorems and systems theory. Later his attention was drawn to the rapidly expanding subject of nonstandard analysis which had been introduced in the late 1960s by Abraham Robinson, and he became increasingly concerned to explore the application of nonstandard methods to the study of generalised functions. He was largely responsible for the organisation of the highly successful *International Colloquium of Nonstandard Mathematics* held at the University of Aveiro in 1994, and for the subsequent developmemt of research in that field at Aveiro. His own personal contribution to that research has been a series of papers on nonstandard theories of Schwartz distributions, ultradistributions and Sato hyperfunctions, and he was working most recently on the hyperfinite representation of generalised functions following the approach initiated by Kinoshita. He was co-author with R.F.Hoskins of a comprehensive study of generalised functions entitled *Distributions, Ultradistributions and Other Generalised Functions*, which was published in 1999 by Ellis Horwood.

INFINITESIMAL METHODS OF MATHEMATICAL ANALYSIS

J.SOUSA PINTO, MSc,PhD
Associate Professor of Mathematics
University of Aveiro, Portugal

Translator
R.F.HOSKINS
sometime Professor of Applicable Mathematics
Cranfield University, Bedford

Horwood Publishing
Chichester, West Sussex

HORWOOD PUBLISHING LIMITED
International Publishers in Science and Technology
Coll House, Westergate, Chichester
West Sussex PO20 3QL England

First published in 2004

British Library Cataloguing in Publication Data
A catalogue record of this book is available from the British Library

ISBN: 1-898563-99-3

Translator's Preface

This text is a translation of the book *Métodos Infinitesimais de Análise Matemática* by José Sousa Pinto of the University of Aveiro, Portugal, who died in August 2000 after a prolonged and debilitating illness. His interest in nonstandard methods, particularly in their application to the study of generalised functions, was long standing and he will be especially remembered for his part in the organisation of the highly successful *International Colloquium of Nonstandard Mathematics* held at Aveiro in 1994. A most modest and unassuming mathematician, his contributions to NSA are less well known than their value deserves. This translation will make his work more widely available and stand as a tribute and memorial to him. From a personal point of view I would also wish to take this opportunity to acknowledge the value and pleasure I have had in working with him over many years.

The book is intended primarily for graduate level students or for final year undergraduates with a general background in calculus, and offers an introduction to the use of modern infinitesimal methods in both elementary and advanced real analysis. The differential and integral calculus is generally considered to have been conceived and developed in the 17th and 18th centuries, based largely on the work of Newton and Leibniz. In its early treatment much use was made of the concept of infinitesimal, or "infinitely small" quantities, but later on this was increasingly regarded by mathematicians as ill-defined, unreliable and even logically inconsistent. By the 19th century a rigorous foundation for the calculus was devised by Weierstrass, the infinitesimal being totally excluded, and the subject area being more generally described as **analysis** (more specifically, **mathematical analysis**). The resulting "Weierstrassian" form of analysis has since become firmly established as the orthodox or "standard" way in which the fundamental theory of the calculus should be treated, and until the last quarter of the 20th century there was no viable alternative formulation of the theory.

However, the concept of the infinitesimal never really disappeared and it continued to be employed, loosely but often quite successfully, by physicists and by engineers and even by many applied mathematicians. Moreover, during the 20th century various new theories of infinitesimals were developed and were seriously re-examined by (pure) mathematicians. The boldest and most comprehensive of these stemmed from the fundamental researches of the mathematical logician Abraham Robinson in the 1960s. His work actually involved much more than the re-instatement of the infinitesimal (although this is its most immediately apparent feature). It is perhaps somewhat unfortunate that Robinson chose the name **Nonstandard Analysis** for his new theory, presumably to emphasize the contrast with the orthodox **standard** treatment based on Weierstrassian lines. For good or ill the name has stuck, and Nonstandard Analysis (or NSA) is now a well-established title for a specific area of mathematics.

Despite its power as a mathematical tool, NSA in its original form is not easy to understand, since it is necessary to become acquainted with a complex and recondite logical foundation. Accordingly various attempts have been made to present the

Robinson theory in a more widely accessible form, suitable for teaching to undergraduate mathematical students and also to physicists and engineers whose knowledge of mathematical techniques may be extensive but essentially pragmatic. Sousa Pinto's book is an important example of such an approach to NSA. Applications of NSA to elementary calculus, topology of the real line and real variable theory are discussed in order to illustrate the power and versatility of the infinitesimal methods which the theory makes available. But the main purpose of the text is to describe nonstandard treatments of more advanced material, in particular of measure theory, integration and generalised functions. To do this Sousa Pinto introduces the so-called "hyperfinite line" and develops a hyperfinite representation of Schwartz distributions, initiated by Kinoshita. The final chapter applies this representation to the study of generalised Fourier transforms and harmonic analysis,

I have tried to retain throughout the character of Sousa Pinto's style, but have had on occasion to make some minor alterations of the original text and sometimes of his notation, for the sake of clarity and readability. In particular the end of each proof is indicated by the use of the symbol •. The production of this English version has been greatly assisted by the comments and constructive criticism of Professor Vítor Neves, of the University of Aveiro, although I must claim full responsibility for such inaccuracies and infelicities which still remain.

R.F.Hoskins

Bedford, 2003

Preface to original Portuguese edition

Nonstandard Analysis was the title chosen by Abraham Robinson (1918-1974) for the book published in 1966 in the North-Holland series *Studies in Logic and Foundations of Mathematics.* For the first time in modern times the rehabilitation of the concept of *infinitesimal* was established rigorously in this book, using modern mathematical logic, in particular model theory.

Nonstandard Analysis might better be called Infinitesimal Analysis: however there is already a classical connotation for the latter which would make it difficult to use in the required context. For some mathematicians the choice of name is not a happy one since the term "nonstandard" suggests opposition to "standard" classical analysis. This is not the case. Nonstandard Analysis simply offers to the mathematical analyst a greater variety of mathematical objects with which to work.

At first the main interest in NSA stemmed from its possible application to the study of elementary analysis. The real number system $\mathbb{R}$ can be extended to the so-called hyperreal number system ${}^*\mathbb{R}$ which contains both infinite numbers and infinitesimals as well as (copies of) the ordinary real numbers themselves: further, every function $f : \mathbb{R} \to \mathbb{R}$ admits a canonical extension to a nonstandard function ${}^*f : {}^*\mathbb{R} \to {}^*\mathbb{R}$. Within this extended system it is possible to make an explicit Leibnitzian formulation of elementary calculus by exploiting the infinitesimal structure of ${}^*\mathbb{R}$. But this is not the most important or the most significant application of NSA: on the contrary, NSA has been shown to be an important technique in a great number of advanced mathematical areas, both pure and applied.

The idea of writing this book on NSA had its origin in the notes made for a course on Infinitesimal Analysis in the Master's degree offered by the Department of Mathematics in the University of Aveiro in the academic year 1997/98. The intention was to present an exposition of NSA and its applications at two distinct levels: for elementary analysis on the one hand, and for some areas of higher mathematical analysis on the other.

In Chapter 1 we give a preliminary account of the concept of tha infinitesimal in order to show the reader something of its past history. This is by no means a deep study or a complete history of what is a rich and complex field: it is simply intended to place the reader in the period, showing that NSA is to a certain extent a return to classical problems which have been considered by many highly respected figures in the history of mathematics.

Before launching into NSA proper we consider some other possible elementary treatments of infinitesimals in Chapter 2. We examine some relatively simple algebraic structures involving infinitesimals and show that their relevance to mathematical analysis is interesting but comparatively limited. Such treatments of infinitesimals are described as "standard" to emphasise that they are essentially different in character from NSA and from what has since been achieved by the use of genuine nonstandard methods. This chapter has no further interaction with the remainder of the book and, if desired, may be omitted without significant loss of understanding.

The development of NSA itself is conveniently presented in three essential stages. The first stage, which we describe as "Elementary NSA" forms the content of Chapters 3,4,and 5. Chapter 3 gives an account of the basic tools of nonstandard analysis: hyperreal numbers, canonical extrensions of sets of real numbers and of real functions of a real variable, and a simple version of the Transfer Principle allowing transfer from $\mathbb{R}$ to $^*\mathbb{R}$ (and vice versa) of certain properties belonging to a well determined class. This material is used in Chapter 4 for the study of some familiar aspects of elementary real analysis. In Chapter 5 we go on to consider an important issue in the proper domain of NSA, namely that of the so-called internal sets and functions, in order to show how we can use them to represent some relatively complicated standard objects in much simpler form.

The second stage of the development of NSA forms the subject matter of Chapter 6. In this we need to have at our diposal a sufficiently fundamental basis for nonstandard analysis. Accordingly we give some idea of the questions which lie at the foundation of a rigorous formulation of NSA, without however going into the matter too deeply or extensively. The interested reader should consult the more specialised literature, particularly that which deals with model theory. For the mathematical analyst it is usually enough to know that the objects with which he works do have a secure foundation, without having to enquire too deeply into that foundation.

Finally, in Chapters 7, 8 and 9 we describe some applications of NSA to more advanced topics of analysis. For the study of these particular areas - measure, generalised functions, Fourier analysis - we work in terms of the so-called hyperfinite line, a certain discrete set of hyperreal numbers which can be briefly summarised as follows. Let κ be an even infinite positive number. Then the hyperfinite line is composed of the set of points ranging from $\kappa/2$ (inclusive) to $\kappa/2$ (exclusive) in infinitesimal steps of magnitude $1/\kappa$. Such a set, although of infinite cardinality, possesses the characteristic properties of finite sets, a fact which turns out to be especially valuable for the treatment of analysis: it effects in a way a synthesis of the continuous and the discrete.

The book contains two appendices. The first discusses some of the key stages in the history of the development of the concept of *function*. In the second there is a simplified proof of the Theorem of Lŏs, dedicated to those readers who would like to know what it does but who do not wish to pursue the subject in depth.

Throughout the book. but more especially in the first part, we have given various exercises designed to aid comprehension.

Acknowledgments The author would like to make two personal acknowledgment. First to Professor R.F Hoskins for the support which he has given me throughout all the work on this material; and second with much pleasure to Dr. Antonio Batel for all his help in preparing the text and his unremitting patience in so doing.

Finally my thanks to Fundaciao Calouste Gulbenkian and particularly to its Education Section for the welcome given to this project.

J.Sousa Pinto

University of Aveiro, Portugal, 2000

Notation

$\mathbf{C}$	Field of complex numbers
$\mathcal{C}$	Space of continuous functions
$\mathcal{C}^p$	Space of p-times continuously differentiable functions
$\mathcal{C}^\infty$	Space of infinitely differentiable functions
$\mathcal{C}_\infty$	Space of finite order Silva distributions
$\mathcal{D}$	Space of infinitely differentiable functions of compact support
$\mathcal{D}_K$	Space of infinitely differentiable functions with support in K
$\mathcal{D}'$	Space of distributions
$\mathcal{D}'_{fin}$	Space of finite order distributions
$\mathbf{D}_\infty$	Module of finite order pre-distributions
$\mathbf{D}_\Pi$	Module of global distributions
$\delta(x)$	Dirac delta function
$\Delta_0(x)$	Nonstandard delta function
$\mathbf{F}_\Pi$	Set of internal functions on the hyperfinite real line Π
$\mathcal{F}[f]$	Fourier transform of f
$\mathbf{H}(x)$	Heaviside unit step function
$\mathbb{N}$	Set of natural numbers (positive integers)
$\mathbb{N}_0$	Set of non-negative integers
$^*\mathbb{N}$	Set of hypernatural numbers
$^*\mathbb{N}_\infty$	Set of infinite hypernatural numbers
Π	Hyperfinite real line
Π_b	Finite part of the hyperfinite real line
$\mathbf{Q}$	Field of rational numbers
$^*\mathbf{Q}$	Field of hyperrational numbers
$\mathbb{R}$	Field of real numbers
$^*\mathbb{R}$	Field of hyperreal numbers
$^*\mathbb{R}_b$	Set of finite hyperreal numbers
$\mathcal{U}_\mathbf{N}$	Ultrafilter over $\mathbb{N}$
$\mathcal{V}(X)$	Superstructure over the set X
$\mathcal{Z}$	Set of all integers
$^*\mathcal{Z}$	Set of all hyperintegers

Contents

Chapter 1

Calculus and Infinitesimals

"It is singular that nobody objects to $\sqrt{-1}$ as involving any contradiction, nor, since Cantor, are infinitely great quantities objected to, but still the antique prejudice against infinitely small quantities remains."

C.S.Peirce in *The New Elements of Mathematics.*

The use of infinitely large and infinitely small numbers in mathematics has a long history: the concept of the infinitesimal has actually been around for at least 23 centuries! Until the appearance of Nonstandard Analysis (NSA) the rational basis of such numbers was not well founded, and they were treated with suspicion. In consequence the existence of infinitesimals, in particular, was almost invariably seen as controversial, although their use as a practical tool by physicists and engineers, for example, never ceased.

The term "calculus", in its broadest sense, can mean any system whatever of rules and symbols which allows the solution of a given class of mathematical problems to be reduced to a species of routine algorithms. In normal usage, however, the calculus is understood to refer to the familiar specific disciplines of the differential and integral calculus, considered by many as the most significant landmark in the history of mathematical thought. The importance of infinitely large and infinitely small numbers lies ultimately in their effectiveness in the discovery and development of the calculus in this specific sense. Before the 17th century the calculus could be considered to consist of two distinct branches, each having its own special significance. The first, dealing with the determination of the *tangents* to plane curves, may be identified with modern differential calculus; the second, concerned with the determination of *plane areas* bounded by curves, with the origin of the integral calculus of the present day. These problems (of determining tangents and calculating areas), although solved classically in a surprisingly wide number of situations, were not up to that time considered in full generality, but only in certain very particular cases. Moreover it was not then obvious to contemporary scientists that these two types of problem were intimately connected, let alone that they were, in a certain sense, mutually inverse.

The work of Isaac Newton (1642-1727) and Gottfried W. Leibniz (1646-1716) allowed the calculus to be understood in its first general formulation and unified the two branches by means of the fundamental theorem of the calculus. Last, but not least, the first applications were made not only to mathematics itself but also to other sciences, particularly to physics and astronomy. Newton and Leibniz achieved their discoveries independently of one another. Although each of them obtained the same types of results, their methods were completely distinct. Newton thought more in terms of what eventually evolved into the modern theory of limits. Leibniz founded his reasoning in terms of *infinitesimals* (understood as static quantities), thereby giving the calculus a more algebraic form, and incidentally conferring on the subject the appropriate name of the **Infinitesimal Calculus.**

As will be seen, Nonstandard Analysis involves in some sense a kind of return to Leibniz, and there are, in effect, two reasons for regarding him as the founder of the essential ideas of what we now call NSA. In the first place Leibniz openly advocated the use of infinite numbers and infinitesimals in the development of the calculus. In the second place he was a precursor of modern mathematical logic, and therefore of mathematical tools powerful enough to justify the use of these quantities. Such tools are provided by the **theory of models** which has been described[1] as "analysis of the relations existing between a concrete mathematical structure and its theory, in the formal sense of that term". This was a development of mathematical logic in the 20th century which came to be seen as crucial for the foundation of the notion of the infinitesimal. In fact its application to nonstandard analysis goes much deeper than this, and in its turn, NSA itself is in no way confined to the re-creation or re-interpretation of Leibnizian Infinitesimal Calculus.

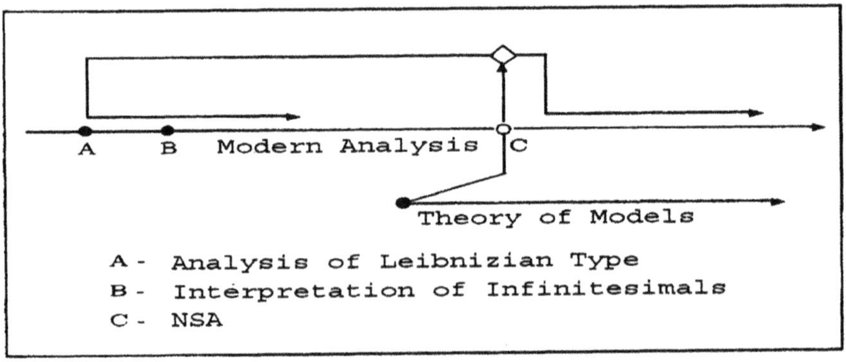

The above diagram (taken from the Encyclopedia EINAUDI, op. cit.) offers a schematic idea of the recent history of the infinitesimal, starting from the 17th century.

[1]See "Infinitesimal", by Jean Petitot, *Encyclopedia EINAUDI*, vol 4, pp209-285,

1.1 The Origins of the Calculus

"[...] It can be contended that Leibnitz's way of writing the calculus approaches the poetic. One can be borne up and carried along purely by his symbolism, while his symbols themselves may appear to take a life of their own. Mathematics and poetry are different, but they are not so far apart as one might think.
W.M.Priestley in [35].

Naturally enough the calculus developed by Newton and Leibniz, as described in the article [26] of I.Kleiner, did not have the form in which it is known today. In particular it was not a *calculus of functions.* The objects of study in the 17th century were essentially geometric: that is to say they were curves - circles, ellipses, parabolas, spirals etc. The appearance of the calculus in the 17th century was, accordingly, motivated by geometric (and kinematic) considerations. Workers were interested in the development of methods of solving problems such as the determination of areas under curves, of finding tangents to curves, or of the velocities of points moving along curves. The variables associated with a curve were similarly geometric - abscissas, ordinates, subtangents, subnormals, radii of curvature, etc. The notion of function, as an object of mathematical study, did not exist even though in the (geometric) concept of curve there would nevertheless have been a certain basic element of functionality.

1.1.1 Determination of tangents

Let Γ be a plane curve with equation $y = f(x)$ which passes through the point P_0 with coordinates (x_0, y_0): we could define the tangent to Γ at P_0 as being the straight line $\tau_f(x_0)$ which most closely approximates Γ in a suitable neighbourhood of P_0. This notion can be conceived intuitively in two distinct ways: *dynamic* (Newtonian) and *static* (Leibnizian). In the first, the straight line $\tau_f(x_0)$ is the *limit* of the secants P_0P of Γ, where P is a point which is *displaced* along Γ from P_0. The complete theoretical treatment of this method of approximation to the problem requires the formalisation of the notion of limit, which involves travelling along the historical path from Newton to Weierstrass, passing through contributions from d'Alembert and Cauchy (and others) along the way.

In the static geometric conception adopted by Leibniz the tangent $\tau_f(x_0)$ is a straight line which cuts Γ in two *infinitely close* points belonging to an *infinitesimal neighbourhood* of P_0. Equivalently $\tau_f(x_0)$ is a straight line which, in an infinitesimal neighbourhood of P_0 and up to an infinitesimal, may be identified with the curve Γ. Moreover, there exists no other straight line with the *local* property of passing through P_0 and lying between Γ and $\tau_f(x_0)$.

Let us consider for example the problem of finding the tangent to the parabola Γ with equation $y = x^2$ at the point $P_0 \equiv (1, 1)$. Note that this can be treated (informally) without making explicit use of the theory of limits (which, as already remarked, of course did not exist in the early days of the calculus).

The Cartesian equation of the straight line which passes through the two points of the real co-ordinate plane $P_1 \equiv (x_1, y_1)$ and $P_2 \equiv (x_2, y_2)$, with $x_1 \neq x_2$, is of the form

$$y - y_1 = m(x - x_1)$$

where m, the slope, is given by

$$m = \frac{y_1 - y_2}{x_1 - x_2}.$$

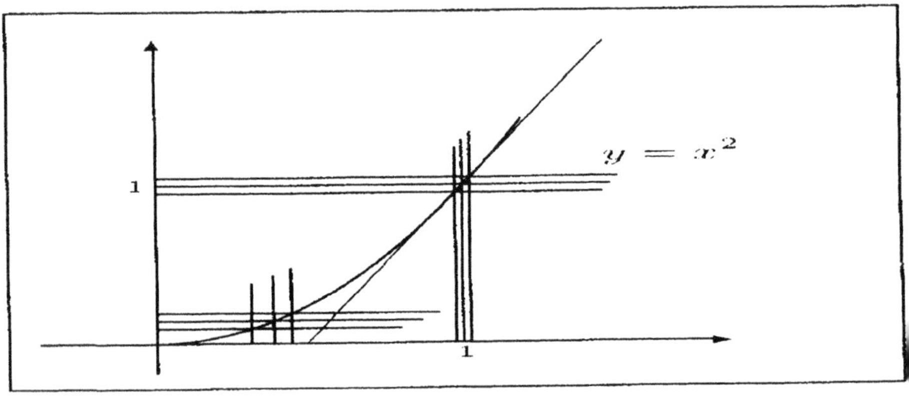

Figure 1.1: Graph of the Parabola $y = x^2, \quad x \geq 0$

In the problem of finding the tangent, however, this is not exactly what is given. We know only one point, the information regarding the other point being replaced by the condition of tangency. That is to say, in the present case we know just the one point, $(1, 1)$, and we have at our disposal the supplementary information that the straight line through this point is supposed to be tangential to the parabola there. To solve this problem "in the manner of Leibniz", let us assume that there exists an *infinitesimal* number (not null), with which we can operate "normally". That is to say, let us suppose that there exists a number which is (strictly) positive, but which is smaller than any positive real number, and which satisfies the usual elementary rules of algebra. (Of course no such number actually exists in $\mathbb{R}$.)

Let us therefore consider, on the curve $y = x^2$, two (distinct) points which are *infinitely close*: that is, points whose distance apart, although positive, will be less than any positive real number, and whose coordinates are $(1, 1^2)$ and $(1+\epsilon, (1+\epsilon)^2)$. The slope of the chord passing through these two points is given by the quotient

$$\frac{dy}{dx} = \frac{(1+\epsilon)^2 - 1^2}{\epsilon} = 2 + \epsilon$$

where, as Leibniz did in the first place a little more than 300 years ago, we denote by dx the infinitesimal increment of the independent variable and by dy the corresponding increment of the dependent variable. In the same way, the slope of the chord passing through the points $(1, 1)$ and $(1-\epsilon, (1-\epsilon)^2)$ is given by

$$\frac{dy}{dx} = \frac{1^2 - (1-\epsilon)^2}{\epsilon} = 2 - \epsilon \ .$$

What then will be the slope of the tangent to the parabola $y = x^2$ at the point $(1, 1)$? If this is a question posed in a universe in which only real numbers play a part then it must ultimately have an answer which is itself free from any quantities which are not real. Now calculations carried out in an enlarged number system (supposed to contain infinitesimal quantities) show that the (real) slope of the tangent to the parabola at the point $(1, 1)$ must be between $2 - \epsilon$ and $2 + \epsilon$. The *unique* real number lying between these two elements is 2 and, consequently, the slope of the tangent to the parabola $y = x^2$ at the point $(1, 1)$ is equal to 2. Accordingly

$$y = 2x - 1$$

will be the equation of the tangent $\tau_{x^2}(1)$ to Γ at the point $(1, 1)$. We can see that the solution obtained, namely the straight line with equation $y = 2x - 1$, satisfies what was required; that is that it coincides with Γ *in an infinitesimal neighbourhood of* P_0, *at least up to an infinitesimal.* For $x = 1$ we have $y = 1$ on the parabola and $y = 1$ on the tangent line $\tau_{x^2}(1)$; for $x = 1 + \epsilon$ we have $y = 1 + 2\epsilon + \epsilon^2$ on the parabola and $y = 2(1 + \epsilon) - 1$ on the tangent line $\tau_{x^2}(1)$. The distance between these two points is

$$|1 + 2\epsilon + \epsilon^2 - (1 + 2\epsilon)| = \epsilon^2$$

which is an infinitesimal quantity, thus confirming the assertion made above. This is the unique real line which passes through P_0 and which, up to an infinitesimal, passes through every point of Γ *infinitely close* to P_0.

Note 1.1 In the 17th Century the argument would probably have been shortened to the following

$$\frac{dy}{dx} = \frac{(1 + dx)^2 - 1^2}{dx} = 2 + dx = 2 \quad (!)$$

where, contradictorily, dx is different from zero in the first stage (so that we can divide by dx), but is allowed in the second stage to be equal to zero (in order to obtain the final result, which is a real number). The infinitesimal increment dx thus appears as a semantically inconsistent concept (dx is both equal and not equal to zero).

1.1.2 The calculation of plane areas

Let Γ be the curve with equation $y = f(x)$ represented in the following figure and denote by A the area of the plane region bounded laterally by the lines $x = a$ and $x = b$, where $a \leq x \leq b$, and bounded above and below respectively by the lines with equations $y = f(x)$ and $y = 0$. In principle at least the calculation of the area A can be carried out very simply: since the area of a rectangle is easily obtained (base $\times$ altitude), we replace A by another plane figure consisting of an *infinite number* of juxtaposed rectangles each of *infinitesimal* base, as suggested in the figure. (It is understood that by *infinite number* is meant a number larger then every natural number, a number whose existence, again *as conceived by Leibniz* is supposed to be assured.) More specifically, if ν is such an infinite positive whole number, then taking for the common base of the rectangles the *infinitesimal length*

$$dx = \frac{(b - a)}{\nu}$$

and considering for each one an altitude equal to $f(a + jdx)$ for the corresponding $j = 0, 1, \ldots, \nu - 1$, we obtain the (infinite) sum

$$f(a)dx + f(a + dx)dx + \ldots + f(a + (\nu - 1)dx)dx$$

$$\equiv \sum_{j=0}^{\nu-1} f(a + jdx)dx$$

formed by the ν portions.

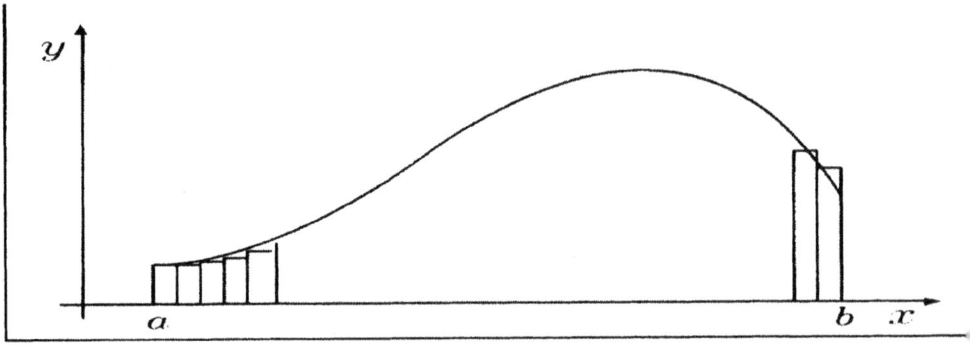

Figure 1.2: Approximation by infinitesimal rectangles.

From the inequality

$$\sum_{j=0}^{\nu-1} f(a + jdx)dx \leq \left\{\max_{a \leq x \leq b} f(x)\right\} \sum_{j=0}^{\nu-1} dx \equiv (b - a) \left\{\max_{a \leq x \leq b} f(x)\right\}$$

it follows that this sum is finite, and it can be expressed in the form $A + \delta$, where δ is an *infinitesimal number* and A is the real number whose value it is required to calculate. Using the modern notation for the integral (also devised by Leibniz) we can write

$$A = \int_a^b f(x)dx \quad .$$

Example 1.2 Let $f(x) = x^2$, $0 \leq x \leq 1$ and $dx = \nu^{-1}$ where ν is an "infinite positive whole number". In this case

$$\sum_{j=0}^{\nu-1} f(0 + j/\nu)\frac{1}{\nu} = \frac{1}{\nu}\left\{0^2 + \left(\frac{1}{\nu}\right)^2 + \left(\frac{2}{\nu}\right)^2 + \ldots + \left(\frac{\nu - 1}{\nu}\right)^2\right\}$$

$$= \frac{1}{\nu}\left\{\frac{1}{\nu^2} + \frac{2^2}{\nu^2} + \ldots + \frac{(\nu-1)^2}{\nu^2}\right\}$$

$$= \frac{1}{\nu^3}\sum_{j=1}^{\nu-1} j^2 = \frac{1}{\nu^3}\frac{\nu(\nu-1)(2\nu-1)}{6}$$

$$= \frac{1}{3} - \frac{1}{6\nu}\left(3 - \frac{1}{\nu}\right)$$

whence, taking account of the fact that $\frac{1}{6\nu}(3-\frac{1}{\nu})$ is an infinitesimal number, it follows that

$$A = \int_0^1 x^2 dx = \frac{1}{3} \quad .$$

That is to say, the area under the arc of the parabola $y = x^2$ from $x = 0$ to $x = 1$ is $\frac{1}{3}$.

1.2 Infinitesimals in Historical Perspective

"C'est par la logique qu'on demonstre, c'est par l'intuition scientifique qu'on invente."

Henri Poincaré in *Science et Methode*

1.2.1 In Archimedes' time

Remarkably, many of the problems of finding tangents and calculating areas were extensively explored in Greek antiquity, and with great success considering the mathematical tools available at the time. In particular, we may point to the work carried out by the great Archimedes. He was born in 287BC, and killed by a soldier in 212BC while defending Syracuse when it was besieged by the Romans. Archimedes studied the tangent problem in various situations, some of them by no means trivial, as in the case of the spiral[2] which bears his name. But even more remarkably Archimedes found the areas of many regions bounded by curves, including for example that of a segment of a parabola as shown in Figure 1.3 below.

According to some historians, many of the concepts of the modern calculus were already present in the mind of this man, and his premature death may well have delayed the complete development of the subject for some 1800 years.[2] It is often remarked in the literature that the proofs of Archimedes are based on the so-called *method of exhaustion*, whose form anticipates that of modern proofs based on the notion of limit (sometimes referred to as $\epsilon - \delta$ proofs).

Following the later rigorous formulation developed by Weierstrass, this type of reasoning became accepted as the standard of rigour to be followed in the calculus.

[2]In polar coordinates this has the equation $\rho = a\theta,$, $\theta \geq 0$.

In the beginning of the 20th Century, however, a previously unknown treatise by Archimedes called *"The Method"* was discovered in which he at last admitted also using infinitesimals in his works, *not for proving results, but for discovering them.*

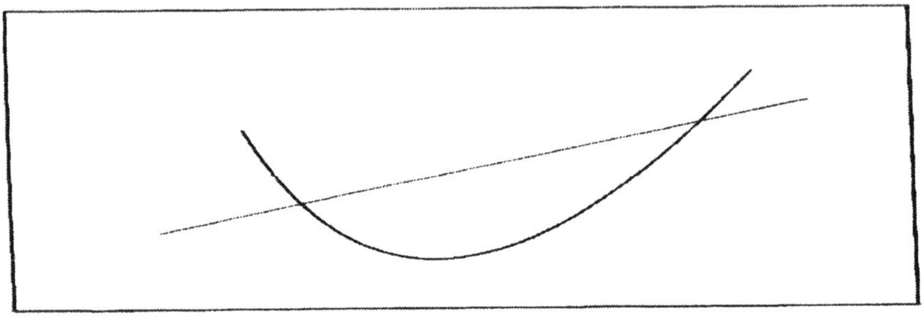

Figure 1.3: Segment of a parabola

1.2.2 The 17th, 18th and 19th centuries

The idea of the infinitesimal (and of the infinite number) cannot be realised in a universe constructed on the basis of the set $\mathbb{R}$ of the real numbers. Leibniz, however, openly proposed for the study of the infinitesimal calculus the adoption of a system of numbers more extensive than that of the real numbers which would include, over and above them, *"ideal"* infinite and infinitesimal numbers but which nevertheless would continue to satisfy the usual laws of the ordinary numbers. These two objectives, formulated just like this, were of course mutally contradictory!

Nevertheless, the use of infinite numbers and of infinitesimals persisted throughout all of the 18th century and part of the following one. Euler, the Bernoullis, Lagrange, D'Alembert, Bolzano and Cauchy obtained excellent results using infinite numbers and infinitesimals, without establishing any fundamental logical grounds for their existence. Leonhard Euler (1707-1783), one of the most prolific mathematicians of all time, offers a prime example of this practice, and it is worth while to consider briefly the way in which he obtained some of his important results.

The infinitesimal calculus of L.Euler In 1748 Euler published the first volume of his three-volume work *Introductio Analysin Infinitorum* [8], rated as one of the most important in the whole history of the calculus. In what follows we consider some specific examples of Euler's methods so as to illustrate the use he made of infinitesimals and infinite numbers.

1. In Chapter 7 Euler introduces the number e of Napier as follows: Since we have

$a^0 = 1 \quad (a > 1)$ then, for a value ϵ which is infinitely small,

$$a^\epsilon = 1 + k\epsilon \quad .$$

If x denotes a positive real number, x/ϵ is, in the proper sense of Euler, infinitely large and may be identified with a specific infinite number ν. Thus,

$$a^x = a^{\nu\epsilon} = (a^\epsilon)^\nu = (1 + k\epsilon)^\nu.$$

Now applying Newton's binomial formula we get,

$$a^x = \left(1 + \frac{kx}{\nu}\right)^\nu \tag{1.1}$$

$$= 1 + \nu\left(\frac{kx}{\nu}\right) + \frac{\nu(\nu-1)}{2!}\left(\frac{kx}{\nu}\right)^2 + \ldots + \frac{\nu(\nu-1)\ldots(\nu-n+1)}{n!}\left(\frac{kx}{\nu}\right)^n + \ldots$$

$$= 1 + \left(\frac{k}{1!}\right)x + \frac{(\nu-1)}{\nu}\left(\frac{k^2}{2!}\right)x^2 + \ldots + \frac{\nu-1}{\nu}\frac{\nu-2}{\nu}\ldots\frac{\nu-n+1}{\nu}\left(\frac{k^n}{n!}\right)x^n + \ldots$$

Since ν is infinitely large, Euler without more ado, writes

$$1 = \frac{\nu-1}{\nu} = \frac{\nu-2}{\nu} = \ldots\frac{\nu-j}{\nu} = \ldots$$

from which, after substituting, he obtains the series expansion

$$a^x = 1 + \frac{x}{1!}k + \frac{x^2}{2!}k^2 + \ldots + \frac{x^n}{n!}k^n + \ldots$$

Setting $x = 1$, the number e can be defined as being the *value of a for which the constant k is equal to 1* or, equivalently

$$e = 1 + \frac{1}{1!} + \frac{1}{2!} + \ldots + \frac{1}{n!} + \ldots$$

From (1.1) we get

$$e^x = \left(1 + \frac{x}{\nu}\right)^\nu = 1 + \frac{x}{1!} + \frac{x^2}{2!} + \ldots + \frac{x^n}{n!} + \ldots$$

and in particular for $x = 1$

$$e = \left(1 + \frac{1}{\nu}\right)^\nu = 1 + \frac{1}{1!} + \frac{1}{2!} + \ldots + \frac{1}{n!} + \ldots$$

where, as remarked, ν is an infinite number.

2. The derivation of the MacLaurin series for the trigonometric function $\cos v$, where $v \in \mathcal{R}$, is made as follows: Starting from the formulas of De Moivre,

$$(\cos z + i \sin z)^n = \cos nz + i \sin nz$$

$$(\cos z - i \sin z)^n = \cos nz - i \sin nz,$$

we obtain the expression

$$\cos nz = \frac{1}{2}\{(\cos z + i\sin z)^n + (\cos z - i\sin z)^n\}$$

$$= \cos^n z - \frac{n(n-1)}{1.2}\cos^{n-2} z \sin^2 z + \frac{n(n-1)(n-2)(n-3)}{1.2.3.4}\cos^{n-4} z \sin^4 z + \ldots,$$

Leibniz applied this to the case when n *is an infinite number and* z *an infinitesimal* such that the product nz is a finite (real) number v. Having observed that, under these conditions, one has

$$\sin z = z, \qquad \cos z = 1 \tag{1.2}$$

$$nz = v, \quad (n-1)z = v, \quad (n-2)z = v, \quad \ldots$$

by substitution he obtained

$$\cos v = 1 - \frac{v^2}{2!} + \frac{v^4}{4!} - \ldots$$

which is the well known formula for the development of the Maclaurin series of the function $\cos v$, $v \in \mathbb{R}$.

3. The next example, one of the most significant results of Euler's inspired mathematical intuition, is an interpretation of how he might have determined for the first time in history that, as we now know, the sum of the series

$$\sum_{n=1}^{\infty} \frac{1}{n^2} \tag{1.3}$$

is $\frac{\pi^2}{6}$. Bearing in mind the state of knowledge at this time, who could then have guessed that the ever fascinating transcendental number π would somehow have been involved in the sum of this series? The way in which Euler arrived at this result in [8], can be summarised as follows:

i. After having discussed the decomposition of polynomials with real coefficients into irreducible factors over $\mathbb{R}$, he applied the results obtained to certain entire transcendental functions, treating them as if they were, so to speak, polynomials of *infinite* degree. He factored

$$\frac{\sin x}{x} = \left(1 - \frac{x^2}{\pi^2}\right)\left(1 - \frac{x^2}{4\pi^2}\right)\left(1 - \frac{x^2}{9\pi^2}\right)\cdots$$

thereby showing that the polynomial of infinite degree

$$1 - \frac{1}{3!}x^2 + \frac{1}{5!}x^4 - \frac{1}{7!}x^6 + \ldots$$

possesses an infinity of roots, namely $\pm\pi, \pm 2\pi, \pm 3\pi, \ldots$ (where each pair of symmetric roots is associated with one of the factors into which the function has been decomposed).

From the above relations he then deduced the following identity between two polynomials of infinite degree:

$$1 - \frac{x^2}{3!} + \frac{x^4}{5!} - \frac{x^6}{7!} + \ldots = \left(1 - \frac{x^2}{\pi^2}\right)\left(1 - \frac{x^2}{4\pi^2}\right)\left(1 - \frac{x^2}{9\pi^2}\right)\cdots \tag{1.4}$$

ii. Euler applied to (1.4) the formulas of Viète relating the coefficients of a polynomial to its roots. In the present case considering the polynomial (with only even powers)

$$1 + a_1x^2 + a_2x^4 + \ldots + a_nx^{2n},$$

which admits the roots $\pm\alpha_1, \pm\alpha_2, \ldots, \pm\alpha_n$, we have

$$\begin{aligned} 1 + a_1x^2 + a_2x^4 + \ldots + a_nx^{2n} &= a_n(x^2 - \alpha_1^2)(x^2 - \alpha_2^2)\ldots(x^2 - \alpha_n^2) \\ &= (-1)^n a_n\alpha_1^2\alpha_2^2\ldots\alpha_n^2\left(1 - \frac{x^2}{\alpha_1^2}\right)\left(1 - \frac{x^2}{\alpha_2^2}\right)\cdots\left(1 - \frac{x^2}{\alpha_n^2}\right) \\ &= (-1)^n a_n\alpha_1^2\alpha_2^2\ldots\alpha_n^2\left\{1 - \left(\frac{1}{\alpha_1^2} + \frac{1}{\alpha_2^2} + \ldots + \frac{1}{\alpha_n^2}\right)x^2 + \right. \\ &\left. \left(\frac{1}{\alpha_1^2\alpha_2^2} + \ldots + \frac{1}{\alpha_{n-1}^2\alpha_n^2}\right)x^4 + \ldots + \frac{(-1)^n}{\alpha_1^2\alpha_2^2\ldots\alpha_n^2}x^{2n}\right\} \end{aligned}$$

from which we can derive various relations, among which are

$$(-1)^n a_n\alpha_1^2\alpha_2^2\ldots\alpha_n^2 = 1 \tag{1.5}$$

and, further,

$$\frac{1}{\alpha_1^2} + \frac{1}{\alpha_2^2} + \ldots \frac{1}{\alpha_n^2} = -a_1. \tag{1.6}$$

Euler applied equality (1.6), deduced for polynomials, to the equation (1.4) since this also establishes a formal relation between *two polynomials* although now ones of *infinite degree.* Taking into account that $\alpha_j = j\pi, j = 1, 2, \ldots$, he thus obtained the result

$$\frac{1}{\pi^2} + \frac{1}{(2\pi)^2} + \ldots + \frac{1}{(j\pi)^2} + \ldots = -\left(-\frac{1}{3!}\right)$$

from which he concluded the following

$$\frac{1}{1^2} + \frac{1}{2^2} + \ldots + \frac{1}{j^2} + \ldots = \frac{\pi^2}{6}.$$

Note 1.3 This type of reasoning allows the calculation of the sums of many other infinite series. Some further examples, among many others which could have been chosen, are

$$\sum_{j=1}^{\infty} \frac{1}{(j\pi)^4} = \frac{\pi^4}{90}$$

$$\sum_{j=1}^{\infty} \frac{1}{(2j-1)^2} = \frac{\pi^2}{8} \qquad \text{etc.}$$

No such results have been obtained, however, for the sums of series of reciprocals of odd powers of natural numbers.

1.2.3 The rigour of mathematical analysis

In spite of its success in practice, the notion of the infinitesimal was never properly clarified, and its unrestrained use gave rise to serious inconsistencies and difficulties within the calculus which could not then be overcome. In the 19th century Karl Weierstrass (1815-1897) finally achieved a complete and rigorous formulation of foundations for the calculus based on the present definition of *limit*. We now say, following Weierstrass, that a real function $f(x)$ defined on $\mathbb{R}$ *tends towards* $b \in \mathbb{R}$ *when* x *tends to* $a \in \mathbb{R}$ if and only if it satisfies the condition[3]

$$\forall_{r>0}\exists_{s>0}\forall_{x\in\mathbb{R}}[|x-a|<s \Rightarrow |f(x)-b|<r]. \tag{1.7}$$

Since then all such alien elements as infinitesimals have been excluded from the set $\mathbb{R}$ in most texts on mathematical analysis, although references to infinitesimals have persisted until the present day in texts on other scientific disciplines, such as physics, which make large use of the calculus.

From now on we shall denote the notion of *infinitely close* by the symbol $\approx$. As we shall see later on, in a system in which such a notion can be rigorously formulated, the Weierstrassian significance of the expression

$$\lim_{x\to a} f(x) = b$$

given in (1.7) is equivalent to the following condition[4]

$$\forall_x[x \approx a \Rightarrow f(x) \approx b] \tag{1.8}$$

where x now varies over a number system more extensive than $\mathbb{R}$ in that it contains infinitesimal numbers.

In the 19th century condition (1.7) was responsible for an enormous advance in the rigorous foundation of mathematical analysis. However, as is all too well hnown, it involves a condition often found difficult to understand and one which, in practice, can become exceedingly technical. But the reward was considerable! Mathematical analysis was made totally free from the intrusion of unfamiliar and questionable objects into the orthodox universe constructed on the basis of the set $\mathbb{R}$ of the real numbers.

By contrast (1.8) is much more intuitive: in fact this expression is the symbolic translation of our ingenuous conception of the idea of *limit*. But it makes no sense in $\mathbb{R}$: in order to use it one needs somehow to admit the existence of infinitesimals and other related concepts.

1.2.4 End of the 20th century: nonstandard analysis

Nonstandard Analysis (NSA) was discovered by the mathematician Abraham Robinson (1918-1974) in the middle of the 20th century, at least forty years ago[5] Its *official*

---

[3]Some French authors formally denote this condition by the term *contravariant*.

[4]*"for all x "infinitely close"to a, the value of $f(x)$ is "infinitely close"to $f(a)$"*

[5]Robinson was born on 6th October 1918 in Waldenburg, Germany. With the advent of German Nazism Robinson, together with his family, emigrated to Palestine and entered the Hebraic University of Jerusalem as a student of philosophy and mathematics. It was Professor A. Fraenkel who led him to concentrate his studies in mathematical logic and set theory.

appearance can perhaps be placed at the time of publication of the article

Nonstandard Analysis
Proc. Roy. Acad, Sci. Amsterdam (A), 64, (1961), pp.437-440.

in which Robinson, for the first time, used a nonstandard model of the real number line[6] to elaborate a developmemt of *Infinitesimal Calculus* which followed very closely the style of its creators in the 17th century, especially that of Leibniz. Later, in his celebrated text book

Nonstandard Analysis
North-Holland, 1966

Robinson showed that one could apply the methods of nonstandard analysis to many distinct areas of Mathematics with advantage. However, it was the study of the development of elementary calculus which first generated the greatest interest of the scientific community in NSA, although perhaps this was not to be the most significant of its applications nor that of the richest consequence.

[6]described explicitly in subsequent chapters.

Chapter 2

"Standard" Infinitesimal Methods

2.1 Introduction

20th century mathematical analysis, soundly based on Weierstrassian principles, has no need of infinitesimals or any other such entities foreign to the real number system. Nevertheless, both before and after Robinson's creation of NSA, other systems of numbers containing both infinite and infinitesimal elements have appeared. We shall examine some of these systems here, before introducing NSA itself, in order to give the reader a first contact with some of the relevant ideas which have been seriously studied throughout the first three quarters of the 20th century.

The **Axiom of Archimedes** states that given two positive numbers a, b, with $a < b$, it is always possible to exceed the second by successive addition of terms equal to the first. To put this another way, there always exists a natural number n for which the following relation is satisfied:

$$a + a + \ldots + a \quad (n \quad \text{times}) \quad > b.$$

A number system which satisfies this Archimedian property cannot contain infinitesimals. Nevertheless, as remarked earlier, infinitesimals had an important and crucial role to play in the creation and development of the Calculus, especially during the 17th and 18th centuries. Now the theories of infinitesimals which are described below do not have either the depth or the power of nonstandard analysis. But, on the other hand, they are simpler to construct and they have a more immediate, albeit much more limited, application. Accordingly, the title for this chapter has been deliberately chosen to make it plain that the methods described are very different from those which NSA can offer.

In the articles [11] and [51] D.Fisher and D.Tall respectively construct by very simple and fundamentally algebraic processes systems of numbers which extend $\mathbb{R}$ and which contain infinitesimals. These systems can be used to explore certain aspects of the calculus, to a limited extent, and they will be described in what follows in terms of operations on certain upper triangular matrices of a specific type. There are several advantages to this approach. First, the arithmetic of the new numbers is defined in a natural way in terms of well understood matrix operations. Secondly, it is then possible to pass from one such system to others by simply extending the size of the matrices concerned. As a result the systems of Fisher and Tall appear in a logical sequence leading from the one to the other.

2.2 Rings of ι-numbers

2.2.1 ι-Numbers of the first order

Consider the set $\mathcal{M}(\mathbb{R}^2) \subset \mathbb{R}^{2\times 2}$ comprising all square matrices of the form

$$M(a,b) = \begin{pmatrix} a & b \\ 0 & a \end{pmatrix}$$

Equipped with the operations of addition and multiplication for square matrices and of multiplication by a scalar, $\mathcal{M}(\mathbb{R}^2)$ is a commutative algebra with unit element. This algebra can be given a total ordering, defined for any two matrices $M(a,b), M(c,d)$ first by a strict order relation $<$, given by

$$M(a,b) < M(c,d) \text{ if and only if } [a < c \text{ or } [a = c \text{ and } b < d]]$$

and then by a total order relation, denoted by $\leq$, defined by

$$M(a,b) \leq M(c,d) \text{ if and only if}$$

$$[M(a,b) = M(c,d) \text{ or } M(a,b) < M(c,d)].$$

The set $\mathbb{R}$, with its ordered field structure, can be embedded in $\mathcal{M}(\mathbb{R}^2)$ by means of the mapping

$$\phi : \mathbb{R} \to \mathcal{M}(\mathbb{R}^2)$$

$$a \mapsto \phi(a) = M(a,0) = \begin{pmatrix} a & 0 \\ 0 & a \end{pmatrix}.$$

The structure $\mathbb{R}[\iota] \equiv (\mathcal{M}(\mathbb{R}^2), +, \times, \leq)$ thus constitutes a proper extension of $\mathbb{R} \equiv (\mathbb{R}, +, ., \leq)$. Identifying each number $a \in \mathbb{R}$ with its image $\phi(a) \in \mathcal{R}[\iota]$ allows us to write $\mathbb{R} \subset \mathbb{R}[\iota]$ in this sense. To verify that this is a proper extension consider, for example, the element

$$\iota \equiv \begin{pmatrix} 0 & 1 \\ 0 & 0 \end{pmatrix} \quad \in \mathbb{R}[\iota].$$

It is obvious that $M(0,0) < M(0,1)$, so that ι is a positive element of $\mathbb{R}[\iota]$; on the other hand, for any positive real number r the inequality $M(0,1) < M(r,0)$ is also satisfied, so that we can write

$$\forall r \;\; [r \in \mathbb{R}^+ \;\; \Rightarrow \;\; 0 < \iota < r],$$

which means that ι is *a positive infinitesimal number* in $\mathbb{R}[\iota]$. Finally, from the matrix equation $M(0,1) \times M(0,1) = M(0,0)$ it follows that we must have

$$\iota^2 = 0 \ . \tag{2.1}$$

With these properties while we certainly have $\iota \in \mathbb{R}[\iota]$, it cannot be true that ι belongs to the field $\mathbb{R}$.[1]

Each element $M(a,b) \in \mathbb{R}[\iota]$ admits a decompositon

$$M(a,b) = \begin{pmatrix} a & b \\ 0 & a \end{pmatrix} = a \begin{pmatrix} 1 & 0 \\ 0 & 1 \end{pmatrix} + b \begin{pmatrix} 0 & 1 \\ 0 & 0 \end{pmatrix}$$

from which we deduce that the matrices,

$$1 \equiv \begin{pmatrix} 1 & 0 \\ 0 & 1 \end{pmatrix} \quad \text{and} \quad \iota \equiv \begin{pmatrix} 0 & 1 \\ 0 & 0 \end{pmatrix}$$

which are linearly independent (over $\mathbb{R}$), constitute a base for $\mathcal{M}(\mathbb{R}^2)$. Every matrix of this set can therefore be expressed in the form

$$M(a,b) = a + b\iota$$

which we shall call the *algebraic representation* of the elements of $\mathbb{R}[\iota]$. These elements will generally be referred to as the **ι-numbers** or, when there is no danger of confusion, simply as the **numbers** of $\mathbb{R}[\iota]$. In this algebraic form the arithmetic of the ι-numbers is similar to that of the complex numbers, bearing in mind that the equality (2.1) replaces the equality $i^2 = -1$.

The elements of the set

$$\mathcal{I}[\iota] = \{b\iota : b \in \mathbb{R}\}$$

are, in modulus, smaller than any positive real number (it being understood that for the modulus of $b\iota$ we naturally mean the number defined for $|b|\iota$). Note that 0 belongs to $\mathcal{I}[\iota]$, this being the only real number with that property. The elements of $\mathcal{I}[\iota]$, all of which are infinitesimal, have no inverses.

Exercise 2.1 *Show that $\mathcal{I}[\iota]$ is a maximal ideal of $\mathbb{R}[\iota]$ and that $\mathbb{R}[\iota]/\mathcal{I}[\iota]$ is a field isomorphic to $\mathbb{R}$.*

Each element $x \in \mathbb{R}[\iota]$ admits a unique decomposition of the following form

$$x = a + \epsilon$$

---

[1] In 1694-95 the Spanish mathematician and philosopher Bernhard Nieuwentijdt proposed an interpretation of the Leibnizian calculus which followed from the adoption of two hypotheses for the infinitesimal dx:

$$dx \neq 0,$$
$$dxdx = 0.$$

This interpretation was not accepted by Leibniz (see [13], p.146)

where $a \in \mathbb{R}$ and $\epsilon \in \mathcal{I}[\iota]$. The number $a \in \mathbb{R}$ will be called the *ι-standard part* of $x \in \mathbb{R}[\iota]$ and will be written as ${}^0x$. Given any set $A \subset \mathbb{R}$ we call the **ι-extension** of A the set $A[\iota] \subset \mathbb{R}[\iota]$ defined by

$$A[\iota] = \{x \in \mathbb{R}[\iota] : \quad {}^0x \in A\} \ .$$

2.2.1.1 ι-extension of functions of one variable

If $a \in \mathbb{R}$, $\epsilon \equiv b\iota \in \mathcal{I}[\iota]$ and $n \in \mathbb{N}$ then we can apply the Newtonian binomial formula to $(a+\epsilon)^n$, noting that $\epsilon^j = 0$ for every $j \geq 2$. Alternatively, taking into account the matricial representation of ι-numbers and using induction, it can be seen that

$$(a+\epsilon)^n = \begin{pmatrix} a & b \\ 0 & a \end{pmatrix}^n = \begin{pmatrix} a^n & na^{n-1}b \\ 0 & a^n \end{pmatrix}$$

$$= a^n + (na^{n-1})\epsilon \quad \left(\equiv \sum_{j=0}^{n} {}^nC_j a^{n-j}\epsilon^j \right) \tag{2.2}$$

Interpreting powers with negative exponent in the usual way, the formula (2.2) remains valid for any $n \in \mathbb{Z}$ (where, for $n \leq 0$, we have $a \neq 0$). Supposing now that $n \in \mathbb{N}$ and $a > 0$, an expression $(a+\epsilon)^{1/n}$ will have meaning in $\mathbb{R}[\iota]$ if there exist numbers $c \in \mathbb{R}$ and $\delta \in \mathcal{I}[\iota]$ such that the equality $a + \epsilon = (c+\delta)^n$ is satisfied. From (2.2), we see that $a + \epsilon = c^n + (nc^{n-1})\delta$, whence we obtain $c = a^{1/n}$ and $\delta = \frac{1}{n}a^{\frac{1}{n}-1}\epsilon$. Consequently, we can write

$$(a+\epsilon)^{\frac{1}{n}} = a^{\frac{1}{n}} + \left(\frac{1}{n}a^{\frac{1}{n}-1}\right)\epsilon \tag{2.3}$$

and, from (2.2) and (2.3), we can deduce the more general expression

$$(a+\epsilon)^q = a^q + \left(qa^{q-1}\right)\epsilon \tag{2.4}$$

for any exponent $q \in \mathbb{Q}$ and any real number $a > 0$. These calculations show that every function of the form $f(x) = x^q, \quad x > 0, \quad (q \in \mathbb{Q})$ can be extended to $\mathbb{R}^+[\iota] \equiv \{x + \epsilon : x \in \mathbb{R}^+, \quad \epsilon \in \mathcal{I}[\iota]\}$. Denoting this extension by ${}^\iota f$ we see that

$${}^\iota f(x+\epsilon) = x^q + (qx^{q-1})\epsilon$$

which may be written in the form

$${}^\iota f(x+\epsilon) = f(x) + f'(x)\epsilon \tag{2.5}$$

(where $f'(x)$ is the derivative of f at the point x). This result extends immediately to the whole class of functions which can be expressed in terms of a finite number of sums and products of functions of the type x^q. For simplicity we denote this class by $\mathcal{A}_0$. Now let $f : A \to \mathbb{R}$ and $g : B \to \mathbb{R}$ be two functions in $\mathcal{A}_0$ and suppose in addition that $g(B) \subset A$. Then the equations

$${}^\iota f(x+\epsilon) = f(x) + f'(x)\epsilon, \quad x \in A$$

$$^{\iota}g(x+\epsilon) = g(x) + g'(x)\epsilon, \quad x \in B$$

are easy to verify since $^{\iota}g(B[\iota]) \subset A[\iota]$. Since the composed function $f \circ g : B \to \mathbb{R}$ belongs to $\mathcal{A}_0$ it can also be extended to the ι-numbers: taking $\delta = g'(x)\epsilon \in \mathcal{I}[\iota]$ with $x \in \mathbb{R}$ we obtain successively

$$\begin{aligned} ^{\iota}[f \circ g](x+\epsilon) = \; & ^{\iota}f[^{\iota}g(x+\epsilon)] = \; ^{\iota}f[g(x)+g'(x)\epsilon] = \; ^{\iota}f[g(x)+\delta] \\ & = f[g(x)] + f'[g(x)]\delta = f[g(x)] + \{f'[g(x)]g'(x)\}\epsilon \\ & = [f \circ g](x) + [f \circ g]'(x)\epsilon. \end{aligned}$$

By induction we can generalise this result for an arbitrary finite number of compositions. This allows us to extend the formula (2.5) to the class of all functions which can be expressed in terms of finitely many sums, products and compositions of functions of the type $x^q, \quad x \in A \subset \mathbb{R}, \quad (q \in \mathbb{Q})$. This class of functions, viz. the **algebraic functions**, will be denoted by $\mathcal{A}_1$. Thus, we have the result,

Theorem 2.2 Every algebraic function f can be extended to the ι-numbers by the equation

$$^{\iota}f(x+\epsilon) = f(x) + f'(x)\epsilon \tag{2.6}$$

at all points $x \in \mathbb{R}$ at which f is defined and differentiable.

Reciprocally, as shown in the following examples, we can use this theorem to determine, by a purely algebraic process, the derivatives of algebraic functions.

Example 2.3

1. Extending the real function of a single real variable defined by

$$f(x) = \frac{1}{x}, \quad x > 0$$

to the set $\mathbb{R}^+[\iota]$, we see that

$$\begin{aligned} ^{\iota}f(x+\epsilon) &= \frac{1}{x+\epsilon} \\ &= \frac{x-\epsilon}{x^2} = \frac{1}{x} + \left(-\frac{1}{x^2}\right)\epsilon \end{aligned}$$

from which we derive the (familiar) result

$$f'(x) = -\frac{1}{x^2}, \quad x > 0.$$

2. Considering now the function defined by

$$f(x) = \sqrt{1+x^3}, \quad x > -1$$

we have

$$\begin{aligned} ^{\iota}f(x+\epsilon) &= (1+(x+\epsilon)^3)^{1/2} = ((1+x^3)+3x^2\epsilon)^{1/2} \\ &= \sqrt{1+x^3} + \frac{1}{2}(1+x^3)^{-1/2}(3x^2\epsilon) \end{aligned}$$

and so, as would be expected, we obtain

$$f'(x) = \frac{3x^2}{2\sqrt{1+x^3}}, \quad x > -1 \ .$$

Generalisation: It is possible to enlarge further the class of functions which can be extended to the ι-numbers. In particular, if D is an open set in $\mathbb{R}$, we can consider the class of **real analytic functions** on D: a function f is said to be real analytic in an open set $D \subset \mathbb{R}$ if it is indefinitely differentiable on D and if for each $x \in D$ there exists a real number $r_x > 0$ for which the equation

$$f(x+h) = f(x) + \sum_{j=1}^{\infty} \frac{f^{(j)}(x)}{j!} h^j$$

is satisfied for every $h \in \mathbb{R}$ such that $x + h \in D$ and $|h| < r_x$. Since for every $\epsilon \in \mathcal{I}[\iota]$ we have $|\epsilon| < r_x$, for any $x \in D$, f can be extended to $D[\iota]$, giving the following

$$^{\iota}f(x+\epsilon) = f(x) + \sum_{j=1}^{\infty} \frac{f^j(x)}{j!}\epsilon^j = f(x) + f'(x)\epsilon \ .$$

Every analytic function is real analytic: this latter class, however, does contain other functions, namely, those functions algebraic on some open set in $\mathbb{R}$.

The function $\log(1+x)$, real analytic in the open interval $(-1, 1)$, admits the following development around the origin

$$\log(1+x) = x - \frac{x^2}{2} + \frac{x^3}{3} - \frac{x^4}{4} + \dots$$

For any $\delta \in \Pi[\iota]$ we then get $^{\iota}\log(1+\delta) = \delta$, a result which allows us to extend power functions with any real exponent to the ι-numbers. In fact, for $f(x) = x^{\alpha}, \ \ \alpha > 0$ where $\alpha \in \mathbb{R}$, we have

$$\begin{aligned} ^{\iota}f(x+\epsilon) = & \ ^{\iota}\exp[\alpha^{\iota}\log(x+\epsilon)] = \ ^{\iota}\exp\left[\alpha.^{\iota}\log\left(x(1+\frac{\epsilon}{x})\right)\right] \\ & = \exp[\alpha \log x].^{\iota}\exp\left(\alpha\frac{\epsilon}{x}\right) \\ & = x^{\alpha}\left(1 + \frac{\alpha}{x}\epsilon\right) = x^{\alpha} + \left(\alpha x^{\alpha-1}\right)\epsilon \end{aligned}$$

which shows that the formula (2.5) still applies in this case.

Exercise 2.4 *Using the standard definitions of the functions* sin *and* cos *when the argument is a matrix* $M \in \mathbb{R}^{2\times 2}$, *show that in* $\mathbb{R}[\iota]$ *the trigonometric formulas*

$$\begin{aligned} ^{\iota}\sin(\alpha+\beta) = & \ ^{\iota}\sin\alpha.^{\iota}cos\beta +^{\iota}\cos\alpha.^{\iota}\sin\beta \\ ^{\iota}\cos(\alpha+\beta) = & \ ^{\iota}\cos\alpha.^{\iota}\cos\beta -^{\iota}\sin\alpha.^{\iota}\sin\beta \end{aligned}$$

are satisfied for any $\alpha, \beta \in \mathbb{R}[\iota]$ *and that, for arbitrary* $\epsilon \in I[\iota]$, *we also have*

$$^{\iota}\cos\epsilon = 1, \quad \text{and} \quad ^{\iota}\sin\epsilon = \epsilon \ .$$

Applying the above results, determine, by algebraic means, the derivative of the function

$$f(x) = \sin\sqrt{1+x^3}, \qquad x > -1 \ .$$

Exercise 2.5 *Given that a differentiable function $f : I \subset \mathbb{R} \to \mathbb{R}$ is invertible, and that its inverse $g \equiv f^{-1}$ is also differentiable in $f(I) \subset \mathbb{R}$, determine algebraically an expression for the derivative of the inverse function in terms of f'.*

2.2.1.2 ι-extension of functions of several variables

The above type of calculation can also be applied to functions of several variables, allowing similar determination of algebraic expressions for partial derivatives of real analytic functions. For example, a function f of two variables for which partial derivatives of the first order exist in an open subset D of $\mathbb{R}^2$, together with at least the mixed derivatives of the second order, can be extended to the set $D[\iota \times \iota]$ defined by

$$D[\iota \times \iota] = \{(x+\epsilon, y+\delta) \in \mathbb{R}[\iota] \times \mathbb{R}[\iota] : (x,y) \in D \ \text{ and } \ (\epsilon, \delta) \in I[\iota]^2\}.$$

Considering $f(x,y)$ first as a partial function of x and then as a partial function of y, for any $\epsilon, \delta \in \mathcal{I}[\iota]$, we get successively

$$\begin{aligned} {}^{\iota}f(x+\epsilon, y+\delta) = & \ {}^{\iota}f(x, y+\delta) + \left\{ {}^{\iota}\left(\frac{\partial f}{\partial x}\right)(x, y+\delta) \right\} \epsilon \\ & = f(x,y) + \left\{ \frac{\partial f}{\partial y}(x,y) \right\} \delta + \left\{ \frac{\partial f}{\partial x}(x,y) \right\} \epsilon + \left\{ \frac{\partial^2 f}{\partial x \partial y}(x,y) \right\} \epsilon\delta \\ & = f(x,y) + \left\{ \frac{\partial f}{\partial x}(x,y) \right\} \epsilon + \left\{ \frac{\partial f}{\partial y}(x,y) \right\} \delta \\ & \equiv f(x,y) + \left(\epsilon \frac{\partial}{\partial x} + \delta \frac{\partial}{\partial y} \right) f(x,y) \ . \end{aligned} \tag{2.7}$$

Conversely the equation (2.7) could be used for the calculation of $\frac{\partial f}{\partial x}(x,y)$ and $\frac{\partial f}{\partial y}(x,y)$ by simple algebraic manipulation.

Example 2.6 Consider, for example, the function defined by

$$f(x,y) = \exp(x^2 + \sqrt{y}), \qquad (x,y) \in \mathbb{R} \times \mathbb{R}^+$$

Extending this function to $\mathbb{R}[\iota] \times \mathbb{R}^+[\iota]$ we get

$$\begin{aligned} {}^{\iota}f(x+\epsilon, y+\delta) = & \ {}^{\iota}\exp\left[(x+\epsilon)^2 + \sqrt{y+\delta}\right] \\ = & \ {}^{\iota}\exp\left[x^2 + 2x\epsilon + \sqrt{y} + \frac{1}{2\sqrt{y}}\delta\right] \\ = & \ \exp(x^2+\sqrt{y}).{}^{\iota}\exp(2x\epsilon).{}^{\iota}\exp\left(\frac{1}{2\sqrt{y}}\delta\right) \end{aligned}$$

$$= f(x,y)\left\{(1+2x\epsilon)\left(1+\frac{1}{2\sqrt{y}}\delta\right)\right\}$$

$$= f(x,y) + \{2x.\exp(x^2+\sqrt{y})\}\epsilon + \left\{\frac{1}{2\sqrt{y}}\exp(x^2+\sqrt{y})\right\}\delta$$

whence we obtain immediately

$$\frac{\partial f}{\partial x}(x,y) = 2x.\exp(x^2+\sqrt{y})$$

$$\frac{\partial f}{\partial y}(x,y) = \frac{1}{2\sqrt{y}}\exp(x^2+\sqrt{y}),$$

where $(x,y) \in \mathbb{R}\times\mathbb{R}^+$.

We could also use this process to obtain formulas for derivatives of functions composed of functions extended to the ι-numbers. For example, let $f : A \to \mathbb{R}^2$ be the function defined by $f(t) = (\phi_1(t), \phi_2(t))$ where $\phi_1, \phi_2 : A \subset \mathbb{R} \to \mathbb{R}$ are functions extendable to $A[\iota]$. First, for $\epsilon \in \mathcal{I}[\iota]$,

$$\begin{aligned}{}^{\iota}f(t+\epsilon) &= ({}^{\iota}\phi_1(t+\epsilon),\ {}^{\iota}\phi_2(t+\epsilon))\\ &= (\phi_1(t)+\phi_1'(t)\epsilon, \phi_2(t)+\phi_2'(t)\epsilon)\\ &= (\phi_1(t),\phi_2(t)) + (\phi_1'(t),\phi_2'(t))\,\epsilon = f(t)+f'(t)\epsilon\ .\end{aligned}$$

Then, given a function $g : D \subset \mathbb{R}^2 \to \mathbb{R}$ which can be extended to $D[\iota\times\iota]$, and assuming that D contains the codomain of f, we have

$$\begin{aligned}{}^{\iota}[g\circ f](t+\epsilon) &= {}^{\iota}g[{}^{\iota}f(t+\epsilon)] = {}^{\iota}g[f(t)+f'(t)\epsilon]\\ &= {}^{\iota}g((\phi_1(t)+\phi_1'(t)\epsilon), (\phi_2(t)+\phi_2'(t)\epsilon))\\ &= g(\phi_1(t),\phi_2(t)) + \frac{\partial g}{\partial x}(\phi_1,\phi_2(t))\{\phi_1'(t)\epsilon\}+\\ &\qquad + \frac{\partial g}{\partial y}(\phi_1(t),\phi_2(t))\{\phi_2'(t)\epsilon\}\\ &= g\circ f(t) + \left\{\left(\frac{\partial g}{\partial x}\circ f(t), \frac{\partial g}{\partial y}\circ f(t)\right)\bullet f'(t)\right\}\end{aligned}$$

from which, as a result,

$$\begin{aligned}\frac{d}{dt}[g\circ f](t) &= \frac{\partial g}{\partial x}(\phi_1(t),\phi_2(t))\,\phi_1'(t) + \frac{\partial g}{\partial y}(\phi_1(t),\phi_2(t))\,\phi_2'(t)\\ &= \left(\frac{\partial g}{\partial x}\circ f\right)(t)\phi_1'(t) + \left(\frac{\partial g}{\partial y}\circ f\right)(t)\phi_2'(t)\\ &= [(\mathbf{grad}g)\circ f](t)\bullet f'(t)\end{aligned}$$

where $\bullet$ denotes the usual scalar product of vectors.

Exercise 2.7 *Supposing that we now have $g(D) \subset I$, determine algebraically the expressions for the partial derivatives*

$$\frac{\partial}{\partial x}[f\circ g] \quad\text{and}\quad \frac{\partial}{\partial y}[f\circ g]$$

at a general point $(x,y) \in D$.

2.2.2 ι-numbers of higher order

The construction of the ring of (first order) ι-numbers, $\mathbb{R}[\iota] \equiv \mathbb{R}_1[\iota]$, can be formally generalised to obtain infinitesimals of order higher than the first (that is, infinitesimals whose quotient with respect to a non-zero infinitesimal of $\mathbb{R}[\iota]$ is still an infinitesimal). Let n be any number in $\mathbb{N}$ and consider the subset $\mathbf{M}(\mathbb{R}^{n+1})$ of all square matrices of order $n+1$ of the form

$$M(a_0, a_1, \ldots, a_n) = \begin{pmatrix} a_0 & a_1 & a_2 & \ldots & a_{n-1} & a_n \\ 0 & a_0 & a_1 & \ldots & a_{n-2} & a_{n-1} \\ . & . & . & \ldots & . & . \\ 0 & 0 & 0 & \ldots & a_1 & a_2 \\ 0 & 0 & 0 & \ldots & a_0 & a_1 \\ 0 & 0 & 0 & \ldots & 0 & a_0 \end{pmatrix}$$

where $a_0, a_1, \ldots, a_n$ are arbitrary real numbers. Then, equipped with the usual operations for matrices, $\mathbf{M}(\mathbb{R}^{n+1})$ is a commutative algebra with unit element. Introducing a strict order relation defined for two arbitrary matrices $M(a_0, a_1, \ldots, a_n)$ and $M(b_0, b_1, \ldots, b_n)$ in $\mathbf{M}(\mathbb{R}^{n+1})$ by

$$M(a_0, a_1, \ldots, a_n) < M(b_0, b_1, \ldots, b_n)$$

if and only if $a_j < b_j$ for the first index $j \in \{0, 1, \ldots, n\}$ for which $a_j \neq b_j$, and then defining the relation $\leq$ as usual, the structure $\mathbb{R}_n[\iota] \equiv (\mathbf{M}(\mathcal{R}^{n+1}), +, \times, \leq)$ is a totally ordered algebra. The set $\mathbb{R}$, with its field structure, can be embedded in $\mathbb{R}_n[\iota]$ by means of the injective mapping $a \mapsto M(a, 0, \ldots, 0) = aI_{n+1}$, where I_{n+1} denotes the identity matrix of order $n+1$. We now write

$$\iota = \begin{pmatrix} 0 & 1 & 0 & \ldots & 0 & 0 \\ 0 & 0 & 1 & \ldots & 0 & 0 \\ . & . & . & \ldots & . & . \\ 0 & 0 & 0 & \ldots & 1 & 0 \\ 0 & 0 & 0 & \ldots & 0 & 1 \\ 0 & 0 & 0 & \ldots & 0 & 0 \end{pmatrix}$$

$$\equiv M(0, 1, 0, \ldots, 0),$$

and calculation of the successive powers of the square matrices in the usual way yields the following results:

$$\iota^2 = \begin{pmatrix} 0 & 0 & 1 & \ldots & 0 & 0 \\ 0 & 0 & 0 & \ldots & 0 & 0 \\ . & . & . & \ldots & . & . \\ 0 & 0 & 0 & \ldots & 0 & 1 \\ 0 & 0 & 0 & \ldots & 0 & 0 \\ 0 & 0 & 0 & \ldots & 0 & 0 \end{pmatrix}$$

$$\begin{matrix} . & . \\ . & . \\ . & . \end{matrix}$$

$$\iota^n = \begin{pmatrix} 0 & 0 & 0 & \dots & 0 & 1 \\ 0 & 0 & 0 & \dots & 0 & 0 \\ . & . & . & \dots & . & . \\ 0 & 0 & 0 & \dots & 0 & 0 \\ 0 & 0 & 0 & \dots & 0 & 0 \\ 0 & 0 & 0 & \dots & 0 & 0 \end{pmatrix}$$

and

$$\iota^{n+j} = \begin{pmatrix} 0 & 0 & 0 & \dots & 0 & 0 \\ 0 & 0 & 0 & \dots & 0 & 0 \\ . & . & . & \dots & . & . \\ 0 & 0 & 0 & \dots & 0 & 0 \\ 0 & 0 & 0 & \dots & 0 & 0 \\ 0 & 0 & 0 & \dots & 0 & 0 \end{pmatrix}, \qquad j = 1, 2, \dots.$$

If $M(a_0, a_1, \dots, a_n)$ is any element of $\mathbb{R}_n[\iota]$ then we have

$$M(a_0, \dots, a_n) = a_0 \begin{pmatrix} 1 & 0 & \dots & 0 & 0 \\ 0 & 1 & \dots & 0 & 0 \\ . & . & \dots & . & . \\ 0 & 0 & \dots & 1 & 0 \\ 0 & 0 & \dots & 0 & 1 \end{pmatrix} + \dots + a_n \begin{pmatrix} 0 & 0 & \dots & 0 & 1 \\ 0 & 0 & \dots & 0 & 0 \\ . & . & \dots & . & . \\ 0 & 0 & \dots & 0 & 0 \\ 0 & 0 & \dots & 0 & 0 \end{pmatrix}$$

from which , if we write $a_0 M(1, 0, \dots, 0) \equiv a_0 I_{n+1}$ simply as a_0, we can represent the general element of $\mathbb{R}n[\iota]$ in the algebraic form

$$M(a_0, a_1, \dots, a_n) = a_0 + a_1\iota + a_2\iota^2 + \dots + a_n\iota^n.$$

Each element of $\mathbb{R}_n[\iota]$ can thus be identified with a polynomial in the indeterminate ι of degree not greater than n. The algebraic operations for elements of $\mathbb{R}_n[\iota]$ can be defined as for polynomials, taking care to eliminate from the final result all terms of higher degree than n in the indeterminate ι.

The set

$$\mathcal{I}_n[\iota] = \{\epsilon \equiv a_1\iota + \dots + a_n\iota^n : a_1, \dots, a_n \in \mathbb{R}\}$$

consists of all the infinitesimals of $\mathbb{R}_n[\iota]$, that is, of all the numbers which, in modulus, are less than any positive real number. The order $o(\epsilon)$ of an infinitesimal is equal to the first index $j \in \{1, \dots, n\}$ for which we have $a_j \neq 0$, setting for convenience $o(\epsilon) = +\infty$ when $\epsilon = 0$.

2.2.2.1 ι-extension of functions of one variable

In a way similar to that employed in the case of $\mathbb{R}_1[\iota]$, we can extend real analytic functions belonging to $\mathcal{A}$ to subsets of the ring $\mathbb{R}_n[\iota]$, obtaining the following final result,

$$^{\iota}f(x + \epsilon) = f(x) + f'(x)\epsilon \dots + \frac{f^n(x)}{n!}\epsilon^n \tag{2.8}$$

where x is a real number belonging to the intersection of the domains of the functions $f, f', \dots, f^{(n)}$ and ϵ is an arbitrary infinitesimal.[2] This result thus allows us to calculate the first n derivatives of a function of $\mathcal{A}$ by purely algebraic means.

[2] It is clear that if $o(\epsilon) > 1$ then some of the powers of ϵ in (2.8) will be null.

Example 2.8 Consider the function

$$f(x) = \sqrt{1+x^2}, \qquad x \in \mathbb{R}$$

and let $\epsilon \in \mathbb{R}_2[\iota]$ be an infinitesimal of the first order. Then

$${}^{\iota}f(x+\epsilon) = f(x) + f'(x)\epsilon + \frac{1}{2!}f''(x)\epsilon^2$$

and therefore

$$\sqrt{1+(x+\epsilon)^2} = \sqrt{(1+x^2)} + f'(x)\epsilon + \frac{1}{2!}f''(x)\epsilon^2$$

Squaring both sides and using the fact that $\epsilon^j = 0$ for $j \geq 3$, we get

$$(1+x^2) + 2x\epsilon + \epsilon^2 = (1+x^2) + 2\sqrt{1+x^2}f'(x)\epsilon + [(f'(x))^2 + \sqrt{1+x^2}f''(x)]\epsilon^2$$

from which, by identifying coefficients of powers of ϵ we obtain

$$f'(x) = \frac{x}{\sqrt{1+x^2}}$$

$$f''(x) = \frac{1}{(1+x^2)\sqrt{1+x^2}}$$

for any $x \in \mathbb{R}$.

The same result can be obtained by a similar treatment to that used in the case of infinitesimals of the first order. Thus

$$\begin{aligned}
{}^{\iota}f(x+\epsilon) &= \sqrt{1+(x+\epsilon)^2} = \sqrt{(1+x^2)+\epsilon(2x+\epsilon)} = \sqrt{(1+x^2)+\delta} \\
&= \sqrt{1+x^2} + \frac{1}{2}(1+x^2)^{-1/2}\delta + \frac{1}{2!}\frac{1}{2}(1-\frac{1}{2})(1+x^2)^{-3/2}\delta^2 \\
&= \sqrt{1+x^2} + \frac{1}{2}(1+x^2)^{-1/2}(2x\epsilon+\epsilon^2) - \\
&\qquad\qquad -\frac{1}{8}(1+x^2)^{-3/2}(2x\epsilon+\epsilon^2)^2 \\
&= \sqrt{1+x^2} + \frac{x}{\sqrt{1+x^2}}\epsilon + \frac{1}{2\sqrt{(1+x^2)^3}}\epsilon^2 \quad .
\end{aligned}$$

Exercise 2.9 *Let $f : A \subset \mathbb{R} \to \mathbb{R}$ and $g : B \subset \mathbb{R} \to \mathbb{R}$ be two functions each n-times differentiable throughout their domains.*

1. Supposing that $A \cap B \neq \emptyset$, deduce algebraically the Leibniz formula for the derivative of order n of the product fg at the points of $A \cap B$ at which both the functions are n-times differentiable.

2. Supposing that $g(B) \subset A$ deduce algebraically the expressions for the derivatives

$$(f \circ g)', (f \circ g)'', \ldots, (f \circ g)^{(n)}$$

at each point $x \in B$ at which the function $f \circ g$ will be n-times differentiable.

2.2.2.2 ι-Extension of functions of several variables

A real analytic function of two variables $f : D \subset \mathbb{R}^2 \to \mathbb{R}$ can be extended to $D[\iota \times \iota]$, giving the following result,

$$ {}^{\iota}f(x+\epsilon, y+\delta) = f(x,y) + \sum_{j=1}^{n} \frac{1}{j!}\left(\epsilon\frac{\partial}{\partial x} + \delta\frac{\partial}{\partial y}\right)^{[j]} f(x,y) $$

where ϵ, δ are any two infinitesimals of first order and $(\ldots)^{[j]}$ denotes the symbolic jth power of the differential operator $\epsilon\frac{\partial}{\partial x} + \delta\frac{\partial}{\partial y}$.

The generalisation to more than two variables is immediate.

Exercise 2.10 *Let $f : D \subset \mathbb{R}^2 \to \mathbb{R}$ and $g : I \subset \mathbb{R} \to \mathbb{R}^2$ be two functions extended to the sets $D_2[\iota \times \iota] \subset \mathbb{R}_2[\iota \times \iota] \times \mathbb{R}_2[\iota \times \iota]$ and $I_2[\iota] \subset \mathbb{R}_2[\iota]$ respectively.*

1. Supposing that $g(I) \subset D$, determine the expression for $(f \circ g)''(t)$.

2. Supposing that $f(D) \subset I$, obtain the expressions for $\frac{\partial^2}{\partial x^2}[g \circ f]$, $\frac{\partial^2}{\partial x \partial y}[g \circ f]$ and $\frac{\partial^2}{\partial y^2}[g \circ f]$ at a point $(x, y) \in D$.

2.2.3 ι-numbers of infinite order

The above development can be carried a little further if we take a natural next step and consider *semi-infinite* upper triangular matrices of the form

$$ M(a_0, a_1, a_2, \ldots, a_n, \ldots) = \begin{pmatrix} a_0 & a_1 & a_2 & \cdots & a_n & \cdots \\ 0 & a_0 & a_1 & \cdots & a_{n-1} & \cdots \\ 0 & 0 & a_0 & \cdots & a_{n-2} & \cdots \\ \cdot & \cdot & \cdot & \cdots & \cdot & \cdots \\ \cdot & \cdot & \cdot & \cdots & \cdot & \cdots \\ \cdot & \cdot & \cdot & \cdots & \cdot & \cdots \end{pmatrix}. $$

where $(a_n)_{n \in \mathbb{N}_0}$ is an arbitrary sequence of real numbers. We denote by $\mathbf{M}(\mathbb{R}^{\mathbb{N}_0})$ the set of all matrices of this type. With the usual operations for semi-infinite matrices this forms a commutative algebra with unit element. Introducing a strict order relation

$$ M(a_0, a_1, \ldots, a_n, \ldots) < M(b_0, b_1, \ldots, b_n, \ldots) $$

if and only if $a_j < b_j$ for the first index $j \in \mathbb{N}_0$ for which $a_j \neq b_j$, and defining the relation $\leq$ as usual, $\mathbb{R}_\infty[\iota] \equiv (\mathbf{M}(\mathbb{R}^{\mathbb{N}_0}), +, \times, \leq)$ is a totally ordered commutative algebra. This algebra contains an ordered subalgebra $\mathbb{R}_n[\iota]$ for each $n = 0, 1, 2, 3, \ldots$. In fact the mapping

$$ \phi : \mathbb{R}_n[\iota] \to \mathbb{R}_\infty[\iota] $$
$$ M(a_0, \ldots, a_n) \mapsto M(a_0, a_1, \ldots, a_n, \ldots, 0, \ldots, 0, \ldots) $$

constitutes an embedding of the structures concerned. In this sense we may write

$$ \mathbb{R}_n[\iota] \subset \mathbb{R}_\infty[\iota] \qquad n = 0, 1, 2, \ldots $$

the inclusion being strict. It follows that we have

$$ \bigcup_{n \in \mathbb{N}_0} \mathbb{R}_n[\iota] \subset \mathbb{R}_\infty[\iota] $$

where $\mathbb{R}_0[\iota] \equiv \mathbb{R}$. By analogy with the previous cases we set

$$\iota \equiv M(0,1,0,0,\ldots,0,\ldots) = \begin{pmatrix} 0 & 1 & 0 & 0 & \ldots & 0 & \ldots \\ 0 & 0 & 1 & 0 & \ldots & 0 & \ldots \\ 0 & 0 & 0 & 1 & \ldots & 0 & \ldots \\ . & . & . & . & \ldots & . & . \\ . & . & . & . & \ldots & . & . \\ . & . & . & . & \ldots & . & . \end{pmatrix}$$

and then obtain successively

$$\iota^2 = M(0,0,1,0,\ldots,0,\ldots) = \begin{pmatrix} 0 & 0 & 1 & 0 & \ldots & 0 & \ldots \\ 0 & 0 & 0 & 1 & \ldots & 0 & \ldots \\ 0 & 0 & 0 & 0 & \ldots & 0 & \ldots \\ . & . & . & . & \ldots & . & \ldots \\ . & . & . & . & \ldots & . & \ldots \\ . & . & . & . & \ldots & . & \ldots \end{pmatrix}$$

$$\iota^3 = M(0,0,0,1,\ldots,0,\ldots) = \begin{pmatrix} 0 & 0 & 0 & 1 & \ldots & 0 & \ldots \\ 0 & 0 & 0 & 0 & \ldots & 0 & \ldots \\ 0 & 0 & 0 & 0 & \ldots & 0 & \ldots \\ . & . & . & . & \ldots & . & \ldots \\ . & . & . & . & \ldots & . & \ldots \\ . & . & . & . & \ldots & . & \ldots \end{pmatrix}$$

$$\iota^n = M(0,0,0,0,\ldots,n,\ldots) = \begin{pmatrix} 0 & 0 & 0 & 0 & \ldots & 1 & \ldots \\ 0 & 0 & 0 & 0 & \ldots & 0 & \ldots \\ 0 & 0 & 0 & 0 & \ldots & 0 & \ldots \\ . & . & . & . & \ldots & . & \ldots \\ . & . & . & . & \ldots & . & \ldots \\ . & . & . & . & \ldots & . & \ldots \end{pmatrix}$$

Identifying $a \in \mathbb{R}$ with the matrix $M(a,0,0,0,\ldots,0,\ldots)$ we have for any element of $\mathbb{R}_\infty[\iota]$:

$$M(a_0, a_1, \ldots, a_n, \ldots) = a_0 + \sum_{j=1}^{\infty} a_j \iota^j$$

where $a_0 \in \mathbb{R}$ is the real part and $\sum_{j=1}^{\infty} a_j \iota^j$ is the infinitesimal part of that number. The set

$$\mathcal{I}_\infty[\iota] = \left\{ \epsilon = \sum_{j=1}^{\infty} a_j \iota^j \in \mathbb{R}_\infty[\iota] : (a_j)_{j=1,2,\ldots} \in \mathbb{R}^{\mathbf{N}} \right\}$$

consists of all the infinitesimals of $\mathbb{R}_\infty[\iota]$.

Note that $\mathcal{I}_\infty[\iota]$ contains infinitesimals of arbitrarily large order. Also, the number $0 \in \mathcal{I}_\infty[\iota]$ has order greater than every natural number; it is an infinitesimal of infinite order.

2.2.3.1 ι-Extensions of functions of one variable

A real analytic function $f : A \subset \mathbb{R} \to \mathbb{R}$ extends naturally to $A[\iota] \subset \mathbb{R}_\infty[\iota]$ under the following formula

$$ {}^\iota f(x+\epsilon) = f(x) + \sum_{j=1}^{\infty} \frac{f^j(x)}{j!}\epsilon^j \tag{2.9}$$

for any $x \in A$ and $\epsilon \in \mathcal{I}_\infty[\iota]$.

Example 2.11 Considering the function

$$f(x) = \sin x, \qquad x \in \mathbb{R}$$

we get

$$ {}^\iota f(x+\epsilon) = \sin x . {}^\iota \cos\epsilon + {}^\iota \sin\epsilon . \cos x$$

from which, replacing ${}^\iota \cos\epsilon$ by $1-\epsilon^2/2!+\epsilon^4/4!+\ldots$ and ${}^\iota \sin\epsilon$ by $\epsilon-\epsilon^3/3!+\epsilon^5/5!-\ldots$, we obtain

$$ {}^\iota f(x+\epsilon) = \sin x + \epsilon\cos x - \frac{\epsilon^2}{2!}\sin x - \frac{\epsilon^3}{3!}\cos x + \frac{\epsilon^4}{4!}\sin x - \ldots$$

from which we can derive the results

$$f^j(x) = \sin(x + j\frac{\pi}{2}), \qquad j = 0,1,2,\ldots$$

2.3 Fields of ι-numbers

2.3.1 The field of the superreal numbers

All elements of the form $M(a_0, a_1, \ldots, a_n, \ldots) \in \mathbb{R}_\infty[\iota]$ with $a_0 \neq 0$ are invertible in $\mathbb{R}_\infty[\iota]$. In fact the matrix equation

$$\begin{pmatrix} a_0 & a_1 & \ldots \\ 0 & a_0 & \ldots \\ . & . & \ldots \\ . & . & \ldots \end{pmatrix} \begin{pmatrix} x_0 & x_1 & \ldots \\ 0 & x_0 & \ldots \\ . & . & \ldots \\ . & . & \ldots \end{pmatrix} = \begin{pmatrix} 1 & 0 & \ldots \\ 0 & 1 & \ldots \\ . & . & \ldots \\ . & . & \ldots \end{pmatrix}$$

leads to the infinite system of linear equations

$$a_0 x_0 = 1$$

$$a_0 x_1 + a_1 x_0 = 0$$

$$a_0 x_2 + a_1 x_1 + a_2 x_0 = 0$$

$$\ldots\ldots\ldots\ldots$$

which is solvable and determinate. The solution of this system then allows us to write:

$$\left(a_0 + \sum_{j=1}^{\infty} a_j \iota^j\right)^{-1} = \frac{1}{a_0} - \frac{a_1}{a_0^2}\iota - \frac{1}{a_0^3}(a_0 a_2 - a_1^2)\iota^2 - \ldots$$

This process fails completely if $M(a_0, a_1, \ldots, a_n, \ldots) \in \mathbb{R}_\infty[\iota]$ is an infinitesimal (that is, if $a_0 = 0$). Such a number is not invertible in $\mathbb{R}_\infty[\iota]$. Accordingly in order to obtain a *field* in which an embedding of $\mathbb{R}_\infty[\iota]$ would be possible it is first necessary to define inverses for the non-null infinitesimal elements. Let $\mathbf{M}_0(\mathbb{R}^{\mathbf{N}_0})$ be the subset of $\mathbf{M}(\mathbb{R}^{\mathbf{N}_0})$ formed by all the invertible elements, that is all those matrices whose elements of the first row and first column are different from zero. Then consider the Cartesian product

$$\mathbf{M}_{0,Z}(\mathbb{R}) = \mathbb{Z} \times \mathbf{M}_0(\mathbb{R}^{\mathbf{N}_0}) \quad .$$

We algebraicise this set by defining the following operations:

Algebraic operations Given two ordered pairs (p, M_0), (q, N_0) of this set we define an operation of multiplication, denoted by $\otimes$, in the following way:

$$(p, M_0) \otimes (q, N_0) = (p + q, M_0 N_0)$$

where $M_0 N_0 \in \mathbf{M}(\mathbb{R}^{\mathbf{N}_0})$ denotes the usual product of matrices and $p+q$ is the sum of p and q in $\mathbb{Z}$. With the $\otimes$-product the set $\mathbf{M}_{0,\mathbb{Z}}(\mathbb{R})$ is an abelian group whose unit element is the pair $(0, I_\infty)$ where I_∞ denotes the identity matrix of $\mathbf{M}_0(\mathbb{R}^{\mathbf{N}_0})$. Each pair $(p, M_0) \in \mathbf{M}_{0,\mathbb{Z}}(\mathbb{R})$ possesses an inverse: in fact, from the equation

$$(p, M_0) \otimes)(x, X_0) = (0, I_\infty)$$

there results the system

$$\begin{aligned} p + x = 0, &\quad \text{in } \mathbb{Z} \\ M_0 X_0 = I_\infty &\quad \text{in } \mathbf{M}_0(\mathbb{R}^{\mathbf{N}_0}) \end{aligned}$$

from which we obtain $x = -p$ and a matrix $X_0 = M_0^{-1}$ which always exists in $\mathbf{M}_0(\mathbb{R}^{\mathbf{N}_0})$. Then

$$(p, M_0)^{-1} = (-p, M_0^{-1}) \quad .$$

To each non-null matrix of $\mathbf{M}(\mathbb{R}^{\mathbf{N}_0})$ it is possible to associate a unique element of the set $\mathbf{M}_{0,\mathbb{Z}}(\mathbb{R})$ as follows: given a matrix $M \equiv M(a_0, a_1, \ldots) \neq 0$ of $\mathbf{M}_0(\mathbb{R}^{\mathbf{N}_0})$, let $p \in \mathbb{N}_0$ be the index of the first non-null element of the sequence $(a_0, a_1, \ldots)$, that is, such that

$$a_0 = a_1 = \ldots = a_{p-1} = 0, \quad a_p \neq 0 \quad .$$

This matrix can then be decomposed into blocks of the form

$$[\mathbf{O}_p | M_0]$$

where

$$\mathbf{O}_p = \begin{pmatrix} 0 & 0 & \ldots & 0 \\ 0 & 0 & \ldots & 0 \\ 0 & 0 & \ldots & 0 \end{pmatrix}, \qquad (p \text{ columns})$$

and

$$M_0 = \begin{pmatrix} a_p & a_{p+1} & a_{p+2} & \cdots \\ 0 & a_p & a_{p+1} & \cdots \\ \cdot & \cdot & \cdot & \cdots \\ \cdot & \cdot & \cdot & \cdots \end{pmatrix} \qquad (a_p \neq 0).$$

Given this notation we shall now write

$$M = {}^{p}M_0, \qquad p \geq 0$$

(where the element of the first row and first column of M_0 is always different from zero). Using these conventions we define the mapping

$$\phi : \mathbf{M}(\mathbb{R}^{\mathbf{N}_0}) \to \mathbf{M}_{\mathbb{Z}}(\mathbb{R}) \equiv \mathbf{M}_{0,\mathbb{Z}}(\mathbb{R}) \cup \{(0, \mathbf{0})\}$$

putting

$$\phi(M) = (p, M_0) \text{ if } M \neq \mathbf{0} \text{ and } \phi(\mathbf{0}) = (0, \mathbf{0}) \ .$$

It can be shown that this mapping is injective. Then since $M_0 \equiv M_0(a_0, a_1, \ldots)$ is a member of $\mathbf{M}_0(\mathbb{R}^{\mathbf{N}_0})$ with $a_0 \neq 0$, it follows that M_0 is invertible with $M_0^{-1} \equiv M_0^{-1}(a_0', a_1', \ldots)$ where $a_0' = a_0^{-1} \neq 0$. Thus,

$$\phi(M_0) = (0, M_0) = {}^{0}M_0 \quad \text{and} \quad \phi(M_0^{-1}) = (0, M_0^{-1}) = {}^{0}(M_0^{-1})$$

and therefore

$$\phi(M_0)^{-1} = \phi(M_0^{-1}).$$

If we were to have $M \equiv M(0, a_1, a_2, \ldots)$ then M would not be invertible in $\mathbf{M}_0(\mathbb{R}^{\mathbf{N}_0})$; however, provided that $M \neq \mathbf{0}$ there will exist an integer $p \geq 1$ such that

$$\phi(M) = (p, M_0)$$

and therefore

$$\phi(M)^{-1} = (-p, M_0^{-1})$$

where M_0^{-1} is a well-defined matrix in $\mathbf{M}_0(\mathbb{R}^{\mathbf{N}_0})$. That is, although $M \neq 0$ will not itself be a matrix invertible in $\mathbf{M}(\mathbb{R}^{\mathbf{N}_0})$, it must have an image under ϕ in $\mathbf{M}_{\mathbb{Z}}(\mathbb{R})$. This fact allows us to define inverses of non-null infinitesimals in a field which adequately extends the ring $\mathbb{R}_\infty[\iota]$.

Taking in particular $M = \iota$, we get

$$\phi(\iota) = (1, I_\infty)$$

and consequently

$$\phi(\iota)^{-1} = (-1, I_\infty) \ .$$

That is, although ι is not invertible in $\mathbb{R}[\iota]$, $\phi(\iota)^{-1}$ is an element well defined in $\mathbf{M}_{\mathbb{Z}}(\mathbb{R})$.

It now makes sense to define any integer power of $\phi(\iota)$. Thus, for any $p \in \mathbb{Z}$,

$$\phi(\iota)^p = (1, I_\infty)^p \equiv (p, I_\infty)$$

and in this sense we can write ι^p in place of $\phi(\iota)^p$, greatly simplifying the notation.

Any element $(p, M_0) \in \mathbf{M}_{\mathbb{Z}}(\mathbb{R})$ may be represented in the form

$$(p, M_0) = (p, I_\infty) \otimes (0, M_0)$$

$$= \iota^p \otimes (0, M_0) \equiv \iota^p \otimes {}^{0}M_0$$

while

$$(0, \mathbf{0}) = \iota^p \otimes \ {}^0\mathbf{0}$$

with $p \in \mathbb{Z}$ chosen arbitrarily. This mode of representation is important for defining in $\mathbf{M}_{\mathbb{Z}}(\mathbb{R})$ an operation of addition, denoted by $\oplus$. Thus, given $(p, M_0), (q, N_0) \in \mathbf{M}_{\mathbb{Z}}(\mathbb{R})$, we get

$$(p, M_0) = \iota^p \otimes (0, M_0) \equiv \iota^p \otimes \ {}^0M_0,$$
$$(q, N_0) = \iota^q \otimes (0, N_0) \equiv \iota^q \otimes \ {}^0N_0 \ .$$

Before forming the sum of these elements it is necessary to reduce them to a form in which they have the same iota-coefficients. First, given two integers $p, q \in \mathbb{Z}$ let $(p \wedge q) \equiv min\{p, q\}$. Then, setting

$$(p, M_0) = \iota^p \otimes (0, M_0) = \iota^{p\wedge q} \otimes \ {}^{(p-p\wedge q)}M_0$$

and

$$(q, N_0) = \iota^q \otimes (0, N_0) = \iota^{p\wedge q} \otimes \ {}^{(q-p\wedge q)}N_0,$$

we can now define the sum of these two elements by putting

$$(p, M_0) \oplus (q, N_0) = \iota^{p\wedge q} \otimes \left({}^{(p-p\wedge q)}M_0 + {}^{(q-p\wedge q)} N_0\right)$$

$$\equiv \left(p \wedge q, {}^{(p-p\wedge q)} M_0 + {}^{(q-p\wedge q)} N_0\right)$$

where ${}^{(p-p\wedge q)}M_0 + {}^{(q-p\wedge q)}N_0 \equiv [\mathbf{0}_{(p-p\wedge q)}|M_0] + [\mathbf{0}_{(q-p\wedge q)}|N_0]$ denotes the usual matrix sum. Note that if we were to have $p = q$ then this matrix might not be in $\mathbf{M}_0(\mathbb{R}^{\mathbf{N}_0})$. It would then be necessary to effect another, supplementary, transformation in order to express the final result in canonical form. [If one or other of the matrices M_0 and N_0 is null then the calculations simplify, noting that by convention

$$\iota^p \otimes \ {}^0\mathbf{0} = \iota^q \otimes \ {}^0\mathbf{0}$$

for any $p, q \in \mathcal{Z}$.]

The operations $\otimes$ and $\oplus$ in $\mathbf{M}_{\mathbb{Z}}(\mathbb{R})$ enjoy the following properties

- they are always possible and regular,
- they are associative, and
- they are commutative.

Additionally the operation $\otimes$

- possesses a unit element $(0, I_\infty)$, and is such that
- each non-null element has an inverse,

while the operation $\oplus$

- possesses a zero $(0, \mathbf{0})$, and is such that
- to each element (p, M_0) there corresponds a symmetric element $(p, -M_0)$.

The sum of (p, M_0) and $(p, -M_0)$ is the element $(p, \mathbf{0})$ which is always represented conventionally as $(0, \mathbf{0})$. Finally, we can easily verify that the two operations satisfy the following form of distributivity:

$$(p, M_0) \otimes [(q, N_0) \oplus (r, P_0)] = [(p, M_0) \otimes (q, N_0)] \oplus [(p, M_0) \otimes (r, P_0)]$$

where (p, M_0),(q, N_0) and (r, P_0) are arbitrary elements of $\mathbf{M}_{\mathbb{Z}}(\mathbb{R})$.

Order relation An arbitrary element (p, M_0) of $\mathbf{M}_{\mathbb{Z}}(\mathbb{R})$ with $M_0 \equiv M_0(a_0, a_1, \ldots)$ is said to be **positive** if and only if $a_0 > 0$. We can then introduce a strict order relation in $\mathbf{M}_{\mathbb{Z}}(\mathbb{R})$, putting

$$(p, M_0) < (q, N_0)$$

if and only if $(q, N_0) \oplus (p, -M_0)$ is positive. After this we can define the relation $\leq$ in the usual way, that is

$$(p, M_0) \leq (q, N_0)$$

if and only if $(p, M_0) < (q, N_0)$ or $(p, M_0) = (q, N_0)$.

The structure $\mathbb{R}]\iota[\equiv (\mathbf{M}_{\mathbb{Z}}(\mathbb{R}), \oplus, \otimes, \leq)$ is an ordered field and will be called the **field of superreal numbers.**

The set of the real numbers, together with its own ordered field structure, can be embedded in $\mathbb{R}]\iota[$ by the mapping

$$\theta : \mathbb{R} \to \mathbb{R}]\iota[$$

$$a \mapsto \theta(a) = (0, M_0(a, 0, 0, \ldots))$$

allowing us to write, in this sense, $\mathbb{R} \subset \mathbb{R}]\iota[$.

To simplify the notation, and when there is no danger of confusion, we make the identification

$$a \equiv (0, M_0(a, 0, 0, \ldots))$$

The field $\mathbb{R}]\iota[$ contains elements other than the real numbers and we can classify its members as follows:

the ι-numbers

$$(m, M_0(a_0, a_1, \ldots)) \equiv \iota^m \otimes (0, M_0) \equiv \iota^m \otimes \ {}^0M_0$$

are said to be

- **finite** if $m \geq 0$. If we have $m = 0$ and, in addition, $a_j = 0$ for all $j > 0$ then the corresponding number is real.

- **infinitesimal** if we have $m > 0$, or if we have zero,

- **infinite** if we have $m < 0$.

Just as the set of real numbers can be embedded in $\mathbb{R}]\iota[$ so also can the rings of ι-numbers studied earlier, so that we may write (in this sense)

$$\mathbb{R}_n[\iota] \subset \mathbb{R}]\iota[\ \text{ and } \ \mathbb{R}_\infty[\iota] \subset \mathbb{R}]\iota[.$$

Representing the operations $\oplus$ and $\otimes$ by the ordinary symbols $+$ and $\times$ (or, in the latter case, simply by juxtaposition), any element $\alpha(\iota)$ of $\mathbb{R}]\iota[$ can be represented in the form,

$$\begin{aligned}\alpha(\iota) &\equiv (p, M_0(a_0, a_1, \ldots, a_n, \ldots))\\ &= (p, M_0(a_0, 0, \ldots, 0, \ldots)) + (p+1, M_0(a_1, 0, \ldots, 0, \ldots))\\ &\qquad + \ldots + (p+n, M_0(a_n, 0, \ldots, 0, \ldots)) + \ldots\\ &= a_0\iota^p + a_1\iota^{p+1} + \ldots + a_n\iota^{p+n} + \ldots.\end{aligned}$$

To obtain an even more suggestive algebraic representation we can re-define the real coefficients of α in accordance with the following rules:

- if $p > 0$ set $b_0 = \ldots b_{p-1} = 0$, and $b_p = a_0$, $b_{p+1} = a_1, \ldots$, so as to obtain

$$\alpha(\iota) \equiv \sum_{j=p}^{\infty} b_j\iota^j,$$

- if $p = 0$ set $b_0 = a_0, b_1 = a_1, \ldots$ so as to obtain

$$\alpha(\iota) \equiv b_0 + \sum_{j=1}^{\infty} b_j\iota^j = \sum_{j=0}^{\infty} b_j\iota^j,$$

(where $\iota^0 = 1$),

- if $p = -m$ (with $m > 0$) set $b_{-m} = a_0, b_{-m+1} = a_1, \ldots$ and $b_0 = a_m, b_1 = a_{m+1}, b_2 = a_{m+2}, \ldots$ so as to obtain

$$\alpha(\iota) \equiv b_{-m}\iota^{-m} + \ldots + b_0 + \sum_{j=1}^{\infty} b_j\iota^j = \sum_{j=-m}^{\infty} b_j\iota^j.$$

An ι-number $\alpha(\iota) \equiv (p, M_0(a_0, a_1, \ldots)) \in \mathbb{R}]\iota[$ then admits a general representation of the form

$$\alpha(\iota) = \sum_{j=p}^{\infty} b_j\iota^j$$

where p may be any integer. If p is positive $\alpha(\iota)$ is infinitesimal, if p is zero $\alpha(\iota)$ is finite and finally, if p is negative then $\alpha(\iota)$ is an infinite number. We call this representation of the elements of $\mathbb{R}]\iota[$ the **algebraic representation**. By analogy with the decimal representation of a real number we could write the ι-number $\alpha(\iota) \equiv (p, M_0(a_0, a_1, \ldots)) \in \mathbb{R}]\iota[$ conventionally as

$$\alpha(\iota) \equiv b_{-m} \cdots b_0, b_1 b_2 \ldots b_n, \ldots.$$

This might be called the **iotimal representation** of α. (Note, however, that the b's are real numbers, not just simple digits.) It is possible to define formally algorithms for the operations of addition and multiplication which are for the most part identical with the usual operations with real numbers, but without carrying over in this case.

The algebraic representation of the ι-numbers shows that $\mathbb{R}]\iota[$ is a field isomorphic with the field introduced by D.Tall in [51] and for which he originally coined the name of the field of superreal numbers.

2.3.2 The fields of Levi-Civita

The field $\mathbb{R}]\iota[$ of the superreal numbers of D.Tall is not algebraically closed. Moreover, the passage from $\mathbb{R}]\iota[$ to $\mathbb{C}]\iota[$, the field of supercomplex numbers, (formally identical with $\mathbb{R}]\iota[$, but with coefficients in $\mathbb{C}$) does not resolve this problem: we cannot, for example, define the square root of the number ι. In fact, putting $\iota \equiv (1, I_\infty)$, if there did exist in $\mathbb{R}]\iota[$ an element which could legitimately be denoted by $\sqrt{\iota} \equiv (p, M_0)$ then, squaring both sides of the equation, we would get

$$(1, I_\infty) = (2p, M_0^2)$$

and this does not have a solution in $\mathbf{M}_{\mathbb{Z}}(\mathbb{R})$, nor in $\mathbf{M}_{\mathbb{Z}}(\mathbb{C})$. This example, however, does suggest that we consider the sets

$$\mathbf{M}_{\mathcal{Q}}(\mathcal{K}) \equiv \mathcal{Q} \times \mathbf{M}_0(\mathcal{K}^{\mathcal{N}_0}) \cup \{(0, \mathbf{0})\}$$

with $\mathcal{K} = \mathbb{R}$ or $\mathcal{K} = \mathbb{C}$. It remains to use these to resolve the problem of defining the power of positive superreal numbers for an arbitrary fractional exponent. Thus, taking for example

$$\sqrt{\iota} = (1/2, I_\infty)$$

it would be natural to denote this by $\iota^{1/2}$. Proceeding in a similar way we would also obtain

$$\sqrt{1+\iota^2} = \pm(\iota^{1/2} + \frac{1}{2}\iota^{3/2} - \frac{1}{8}\iota^{5/2} + \frac{1}{16}\iota^{7/2} - \frac{1}{128}\iota^{9/2} + \ldots),$$

and it would be possible to generalise this process for roots of any integer index.

Each one of these "numbers" determines (and is determined by) two sequences: a sequence $(a_j)_{j=0,1,2,\ldots}$ of real (or complex) coefficients with $a_0 \neq 0$, and a sequence of rational exponents $(q_j)_{j=0,1,2,\ldots}$. To make sense of this we would next need to consider the set of all elements of the form

$$\alpha(\iota) \equiv \sum_{j=0}^{\infty} a_j \iota^{q_j}$$

which is a series of generalised powers formed with rational exponents.

2.3.2.1 Preliminaries

The construction suggested above allows us to obtain non-Archimedean fields which have been studied by Berz in [2]. This author calls them **Levi-Civita fields** in honour of T.Levi-Civita (1873-1941) who appears to have been the first to make a rigorous study of such matters in the beginning of the 20th century. The field of Levi-Civita (real or complex) is the field generated by

$$\mathbf{M}_{\mathbb{Q}}(\mathcal{K}) \equiv \mathbb{Q} \times \mathbf{M}_0(\mathcal{K}^{\mathbf{N}_0}) \cup \{(0, \mathbf{0})\}$$

(where $\mathcal{K}$ is the field $\mathbb{R}$ or the field $\mathbb{C}$). The "numbers" created by Berz are presented in the form of certain functions on $\mathbb{Q}$ into $\mathbb{R}$ or $\mathbb{C}$. Before giving formal definition of these we need to introduce some preliminary concepts.

Definition 2.12 A subset M of rational numbers is said to be **finite to the left** if for every rational number $q \in \mathbb{Q}$ there exists in M only a finite number of elements less than q. We denote by $\mathcal{F}_e$ the family of all subsets of rational numbers which are finite to the left.

The properties of the elements of $\mathcal{F}_e$ which are established in the following two Lemmas are important in what follows:

Lemma 2.13 Let $M \in \mathcal{F}_e$. If the set M is not empty then its elements can be arranged in ascending order and M has a minimal element. If M is infinite then the strictly increasing monotone sequence which is obtained by this process tends to $+\infty$.

Proof If $M \subset \mathbb{Q}$ is finite then it is a totally ordered set and its elements can always be arranged in increasing order. Suppose then that M is infinite. For each $n \in \mathbb{N}$ the set defined by

$$M_n = \{q \in M : q \leq n\}$$

is finite (since $M \in \mathcal{F}_e$), and moreover

$$M = \bigcup_{n=1}^{\infty} M_n \quad .$$

Then we can arrange the elements of M_1 in increasing order, add the elements of $M_2 \backslash M_1$ also arranged in increasing order, and so on successively; the result is a strictly increasing monotone sequence.

If the sequence thus obtained were to be bounded then there would exist a rational number beneath which there would be infinitely many rational numbers of M. This would contradict the hypothesis that M is finite to the left. Thus the sequence of elements of M must be unbounded above and therefore it tends to $+\infty$. •

Lemma 2.14 Every subset of a set of $\mathcal{F}_e$ belongs to $\mathcal{F}_e$. If M, N are two sets of $\mathcal{F}_e$ then $M \cup N$ and $M \cap N$ both belong to $\mathcal{F}_e$. Moreover, the set $M \oplus N = \{a + b : a \in M \text{ and } b \in M\}$ belongs to $\mathcal{F}_e$ and, for each $x \in M \oplus N$ there exists only a finite number of points (a, b) belonging to $M \times N$ such that $x = a + b$.

Proof The first three assertions follow directly from Definition 2.12. To show that $M \oplus N$ also belongs to $\mathcal{F}_e$, let x_M and x_N be minimal elements of M and N respectively. Given any number $q \in \mathbb{Q}$ consider the sets defined by

$$M' = \{x \in M : x < q - x_N\} \qquad N' = \{x \in N : x < q - x_M\}$$

$$M^0 = M \backslash M' \qquad N^0 = N \backslash N' \quad .$$

Then,

$$\begin{aligned} M \oplus N &= (M' \cup M^0) \oplus (N' \cup N_0) \\ &= (M' \oplus N') \cup (M^0 \oplus N') \cup (M' \oplus N^0) \cup (M^0 \oplus N^0) \\ &= (M' \oplus N') \cup (M^0 \oplus N) \cup (M \oplus N^0) \quad . \end{aligned}$$

From the definition of the sets M^0 and N^0 it is clear that the sets $(M^0 \oplus N)$ and $(M \oplus N^0)$ do not contain elements smaller than q, and therefore that all the elements of $M \oplus N$ which are smaller than q must be in $M' \oplus N'$. By the hypothesis of the theorem this is itself a finite number.

Now let $x \in M \oplus N$ be an arbitrarily given element, consider $q = x + 1$ and construct the sets M' and N' as above. Then $x \notin M^0 \oplus N$ and $x \notin M \oplus N^0$ and therefore all the pairs $(a, b) \in M \times N$ such that $x = a + b$ belong to the set $M' \times N'$ which is finite. •

2.3.2.2 Definition of algebraic operations

Given a function $\alpha : \mathbf{Q} \to \mathcal{K}$ we call the **support** of α the set M_α defined by

$$M_\alpha = \{\alpha \in \mathbf{Q} : \alpha(q) \neq 0\} \quad .$$

Definition 2.15 We denote by $\mathcal{K}_{\mathbf{Q}}]\iota[$ the set of all functions $\alpha : \mathbf{Q} \to \mathcal{K}$ whose support is finite to the left; that is,

$$\mathcal{K}_{\mathbf{Q}}]\iota[= \{\alpha \in \mathcal{K}^{\mathbf{Q}} : M_\alpha \in \mathcal{F}_e\} \quad .$$

We define two operations (addition and multiplication) on $\mathcal{K}_{\mathbf{Q}}]\iota[$, putting for each $q \in \mathbf{Q}$,

- *addition*

$$[\alpha + \gamma](q) = \alpha(q) + \gamma(q)$$

- *multiplication*

$$[\alpha\gamma](q) = \sum_{q_\alpha, q_\gamma \in Q, q_\alpha + q_\gamma = q} \alpha(q_\alpha).\gamma(q_\gamma)$$

for any $\alpha, \gamma \in \mathcal{K}_{\mathbf{Q}}\ .]\iota[$.

Since

$$M_{\alpha+\gamma} \subset M_\alpha \cup M_\gamma \quad \text{and} \quad M_{\alpha\gamma} \subset M_\alpha \oplus M_\gamma$$

then from Lemmas 2.13 and 2.14 it follows that $\mathcal{K}_{\mathbf{Q}}]\iota[$ is closed with respect to these operations of addition and multiplication.

Exercise 2.16 *Verify that $\mathcal{K}_{\mathbf{Q}}]\iota[$ is a commutative ring with zero and unit, and that it has no divisors of zero.*

In order to prove that $\mathcal{K}_{\mathbf{Q}}]\iota[$ is a field it remains to show that every non-null element possesses an inverse. This can be demonstrated by appealing to a theorem of Berz (art.cit.) First, however, we define a relation of **proximity** in $\mathcal{K}_{\mathbf{Q}}]\iota[$:

Definition 2.17 Let $x \in \mathcal{K}_{\mathbf{Q}}]\iota[$ and let M_x be the support of x. The **order** of $x \neq 0$ s the rational number $m_x = min(M_x)$: if $x = 0$ then the order of x is defined to be $+\infty$.

Given $x, y \in \mathcal{K}_{\mathbf{Q}}]\iota[$ and any number $r \in \mathcal{Q}$ we write $x =_r y$ if and only if $m_{x-y} \geq r$, or equivalently

$$x =_r y \Leftrightarrow \forall_{q \in Q}[q \leq r \Rightarrow x(q) = y(q)] \quad .$$

Exercise 2.18 *Verify that $=_r$ is an equivalence relation and show that*

$$\forall_{r,s\in\mathbb{Q}}[r > s \Rightarrow [x =_r y \Rightarrow x =_s y]]$$

for any $x, y \in \mathcal{K}_{\mathbb{Q}}]\iota[$.

Theorem 2.19 (Berz) Let $r \in \mathbb{Q}$ be an arbitrarily chosen rational number and let $A_r \subset \mathcal{K}_{\mathbb{Q}}]\iota[$ be the set of all elements $x \in \mathcal{K}_{\mathbb{Q}}]\iota[$ such that $m_x \geq r$. Consider a function $f : A_r \to \mathcal{K}_{\mathbb{Q}}[\iota]$ such that $f(A_r) \subset A_r$, and suppose that there exists $k \in \mathbb{Q}, k > 0$, such that

$$\forall_{x,y\in A_r}\forall_{q\in\mathbb{Q}}[[x =_q y \Rightarrow f(x) =_{q+k} f(y)] \ .$$

Then there exists one and only one point $x \in A_r$ such that $x = f(x)$.

Proof: Let $\alpha_0 \in A_r$ be an arbitrarily chosen element and define recursively

$$\alpha_i = f(\alpha_{i-1}), \qquad i = 1, 2, \ldots$$

Since, by hypothesis, $f(A_r) \subset A_r$, then $(\alpha_i)_{i=0,1,2,\ldots}$ is a sequence of elements of A_r. This sequence satisfies the following condition,

$$\alpha_i(p) = \alpha_{i-1}(p) \ \text{ for all } \ p \in \mathbb{Q} \ \text{ such that } \ p < (i-1)k + r. \tag{2.10}$$

In fact, with $\alpha_0, \alpha_1 \in A_r$, we have from the definition of A_r

$$\alpha_0(p) = 0 = \alpha_1(p), \quad \text{for} \quad p > r$$

and therefore (2.10) is satisfied for $i = 1$.

Suppose now that (2.10) is satisfied for some arbitrarily fixed $i \in \mathbb{N}$. Then $\alpha_i(p) = \alpha_{i-1}(p)$ whenever $p < (i-1)k + r$; that is to say, $\alpha_i =_q \alpha_{i-1}$ for all $q < (i-1)k + r$, from which in accordance with the definition of f we have

$$f(\alpha_i) =_{q+k} f(\alpha_{i-1}) \ .$$

Consequently, since $\alpha_{i+1} = f(\alpha_i)$ and $\alpha_i = f(\alpha_{i-1})$, then we have

$$\alpha_{1+1}(p) = \alpha_i(p) \ \text{ for } \ p \leq q + k < (i-1)k + r + k = ik + r$$

which, by induction, proves that (2.10) is satisfied for every $i \in \mathbb{N}$.

Let $\alpha : \mathbb{Q} \to \mathbb{C}$ be the function defined as follows: for $q \in \mathbb{Q}$ choose $i \in \mathbb{N}$ such that $(i-1)k + r > q$ and denote the value chosen by i_q; then put

$$\alpha(q) = \alpha_{i_q}(q) \ .$$

This definition does not depend on the number i_q. If i'_q were some other possible choice, and we had, for example, $i'_q > i_q$, then we would get

$$q < (i_q - 1)k + r < (i'_q - 1)k + r$$

and therefore, from (2.10),

$$\alpha(q) = \alpha_{i'_q}(q) = \alpha_{i'_q-1}(q) = \ldots = \alpha_{i_q}(q) \ .$$

On the other hand, since for any $p < q$ the choice of i_q is also possible then

$$\alpha(p) = \alpha_{i_p}(p) = \alpha_{i_q}(p)$$

and therefore

$$\alpha =_q \alpha_{i_q} \quad .$$

Since the support of α_{i_q} only contains a finite number of rational numbers smaller than q, $(\alpha_{i_q} \in A_r)$, then the support M_α of α also belongs to $\mathcal{F}_e$ and therefore $\alpha \in \mathcal{K}_{\mathbf{Q}}]\iota[$. In addition since $\alpha(p) = 0$ for all $p < r$, then $\alpha \in A_r$.

We now show that α is a solution of the fixed point equation. For $q \in \mathbf{Q}$ again let i_q be such that $(i_q - 1)k + r > q$. Then

$$\alpha =_q \alpha_{i_q} =_q \alpha_{i_q+1}$$

where the second relation $=_q$ follows since we have $\alpha_{i_q} =_{q+r} \alpha_{i_q+1}$ (see Exercise 2.18). From the contraction property we obtain

$$f(\alpha) =_{q+r} f(\alpha_{i_q})$$

which in its turn implies

$$\alpha(q) = \alpha_{i_q+1}(q) = [f(\alpha_{i_q})](q) = [f(\alpha)](q) \quad .$$

Since $q \in \mathbf{Q}$ is arbitrary it follows that $\alpha = f(\alpha)$.

It remains to prove that α is the unique fixed point of f. The contraction property of f can be expressed in the following form,

$$m_{f(x)-f(y)} \geq m_{x-y} + k$$

for all $x, y \in A_r$. This implies that if $x, y \in A_r$ were two fixed points of f then

$$m_{x-y} = m_{f(x)-f(y)} \geq m_{x-y} + k$$

which could only happen if we had $m_{x-y} = +\infty$, or equivalently, if $x = y$. •

Applying Theorem 2.19 it is now easy to verify that $(\mathcal{K}_{\mathbf{Q}}]\iota[, +, .)$ is a field.

Theorem 2.20 Every non-null element of $\mathcal{K}_{\mathbf{Q}}]\iota[$ possesses an inverse.

Proof Let $z \in \mathcal{K}_{\mathbf{Q}}]\iota[$ be any non-null element, $M_z \in \mathcal{F}_e$ the support of z, $m_z = min(M_z)$ and $a_0 = z(m_z) \neq 0$. Then

$$z = a_0 z'$$

where $z' \in \mathcal{K}_{\mathbf{Q}}]\iota[$ is such that $m_{z'} = m_z$, $z'(m_z) = 1$ and $z'(q) = z(q)$ for all $q \in \mathbf{Q}$. For each $p \in \mathbf{Q}$ define $\iota_p \in \mathcal{K}_{\mathbf{Q}}]\iota[$, by setting

$$\iota_p(q) = \begin{cases} 1, & \text{if } q = p \\ 0, & \text{if } q \neq p \end{cases}$$

and then, denoting by $z'' \in \mathcal{K}_{\mathbf{Q}}]\iota[$ the element defined for $q \in \mathbf{Q}$ by

$$z''(q) = z'(q + m_z)$$

we obtain the following decomposition[3]

$$z' = \iota_{m_z}.z'' \quad .$$

Consequently,

$$z = a_0 \iota_{m_z} z''$$

where $m_{z''} = 0$ and $z''(0) = 1$. Since the element $a_0 \in \mathcal{K}$ is not null it admits an inverse a_0^{-1}. The element $\iota_{m_z} \in \mathcal{K}_{\mathbb{Q}}]\iota[$ also admits an inverse which is the element defined by

$$\iota_{m_z}^{-1}(q) = \begin{cases} 1, & \text{if } q = -m_z \\ 0, & \text{if } q \neq -m_z \end{cases}$$

since

$$[\iota_{m_z}\iota_{m_z}^{-1}](q) = \sum_{p'+p''=q} \iota_{m_z}(p')\iota_{m_z}^{-1}(p'') = \iota_0 \equiv \begin{cases} 1, & \text{if } q = 0 \\ 0, & \text{if } q \neq 0 \end{cases} \quad .$$

In its turn, z'' admits the following decomposition

$$z'' = 1 + y$$

where $y \in \mathcal{K}_{\mathbb{Q}}]\iota[$ is such that $m_y > 0$. If $m_y = +\infty$ then $y = 0$ and therefore $z'' = 1$ admits an inverse. If $y \neq 0$ then in order to prove that z'' possesses an inverse it is enough to show that there exists $x \in \mathcal{K}_{\mathbb{Q}}]\iota[$ such that

$$(1 + x)(1 + y) = 1$$

which is equivalent to the equation

$$x = -yx - y \quad .$$

Defining $f_y : \mathcal{K}_{\mathbb{Q}}]\iota[\to \mathcal{K}_{\mathbb{Q}}]\iota[$ by

$$f_y(x) = -yx - y$$

the problem reduces to determining a fixed point for the function f_y. First we verify that f_y will completely satisfy the conditions of Theorem 2.19. Setting,

$$A_y = \{x \in \mathcal{K}_{\mathbb{Q}}]\iota[: m_x \geq m_y\}$$

we have $f_y(A_y) \subset A_y$. Given $x_1, x_2 \in A_y$ such that $x_1 =_q x_2$, we have $yx_1 =_{q+m_y} yx_2$ and therefore

$$-yx_1 - y =_{q+m_y} -yx_2 - y$$

and this means that f_y does satisfy the conditions of Theorem 2.19 for $k = m_y$. It follows immediately that there exists $x \in A_y$ such that $x = f_y(x)$ and therefore that

$$(z'')^{-1} = 1 + x \quad .$$

---

[3] In fact, for $q \in \mathbf{Q}$ we get

$$[\iota_{m_z} z''](q) = \sum_{p'+p''=q} \iota_{m_z}(p')z''(p'') = \iota_{m_z}(m_z)z''(q - m_z) = z'(q)$$

whence $z'(q) = 0$ for all $q < 0$.

From this it follows finally that

$$z^{-1} = \frac{1}{a_0}\iota_{m_z}^{-1}(1+x) \in \mathcal{K}_{\mathbb{Q}}[\iota]$$

is the inverse of $z = a_0\iota_{m_z}(1+y) \in \mathcal{K}_{\mathbb{Q}}]\iota[$. •

Rational powers Since each element $x \in \mathcal{K}_{\mathbb{Q}}]\iota[$ can be decomposed into a product of the form

$$z = a_0\iota_{m_z}.(1+y)$$

with $m_y > 0$ (possibly $m_y = +\infty$), we can easily calculate positive integral powers of z: for given $n \in \mathbb{N}$ we have

$$z^n = a_0^n.\iota_{m_z}^n.(1+y)^n \quad .$$

Now, since $a_0 \in \mathcal{K}$ then $a_0^n \in \mathcal{K}$; on the other hand $(1+y)^n = 1+y'$ with $m_{y'} > 0$, where

$$y' = \sum_{j=1}^{n} {}^nC_j y^j \quad .$$

Finally, from the definition of ι_{m_z} it follows that, by induction over $n \in \mathbb{N}$,

$$\iota_{m_z}^n = \iota_{n.m_z}$$

and therefore that

$$z^n = a_0^n.\iota_{n.m_z}.(1+y'), \quad \text{with} \quad m_{y'} > 0 \quad .$$

This formula is still valid when $n = 0$, for in this case $a_0^0 = 1$ and $y' = 0$, thus giving

$$z^0 = \iota_0 \equiv 1 \quad .$$

The powers with negative exponents are defined in the usual way by putting

$$z^{-n} = (z^{-1})^n$$

for all $z \neq 0$ and $n \in \mathbb{N}_0$.

It remains now to consider the case of powers of $z \in \mathcal{K}_{\mathbb{Q}}]\iota[$ with fractional exponents. For this it is enough to study exponents of the form $1/n$ with $n \in \mathbb{N}$.

Let $z \in \mathcal{K}_{\mathbb{Q}}]\iota[$ such that

$$z = a_0.\iota_{m_z}.(1+y), \quad \text{with} \quad m_y > 0,$$

and, so as to avoid the trivial case, suppose that $a_0 \neq 0$ (giving, therefore $z \neq 0$). If z possesses roots of order n then let $w \in \mathcal{K}_{\mathbb{Q}}]\iota[$ be one of those roots. Then w admits a decomposition

$$w = b_0.\iota_{m_w}.(1+x)$$

with $b_0 \in \mathcal{K}$ and $x \in \mathcal{K}_{\mathbb{Q}}]\iota[$ such that $m_x > 0$; we then obtain the equality

$$w^n = b_0^n.\iota_{n.m_w}.(1+x)^n w = a_0.\iota_{m_z}.(1+y) = z$$

from which there follows the system of equations

$$b_0^n = a_0, \qquad \text{in } \mathcal{K}$$

$$n.m_w = m_z, \qquad \text{in } \mathbf{Q}$$

$$(1+x)^n = (1+y), \qquad \text{in } \mathcal{K}_{\mathbf{Q}}]\iota[\ .$$

The first equation will have a solution if and only if the nth roots of a_0 exist in $\mathcal{K}$; the second equation gives $m_w = m_z/n$; the third equation will have one and only one solution with $m_x > 0$, and this follows from Theorem 2.19.

The equation $(1+x)^n = 1+y$ can take the form

$$nx + x^2P(x) = y \tag{2.11}$$

where $P(x)$ is a polynomial with integral coefficients. Since $m_x > 0$ then $m_{P(x)} \geq 0$ and $m_{x^2P(x)} = 2m_x + m_{P(x)} > m_x > 0$. Therefore from (2.11) it follows that we must have $m_x = m_y$. Writing (2.11) in the form $x = f_y(x)$ we have that, for each fixed y, f_y will be the function defined by

$$f_y(x) = \frac{1}{n}y - \frac{1}{n}x^2P(x) \ .$$

If A_y is now the set defined by

$$A_y = \{x \in \mathcal{K}_{\mathbf{Q}}]\iota[: m_x \geq m_y\}$$

then any solution of (2.11) must belong to A_y. In addition, $f_y(A_y) \subset A_y$: in fact, if $x \in A_y$ then $m_{x^2P(x)} \geq 2m_x \geq 2m_y > m_y$ and therefore $m_{f_y(x)} = m_y$ whence we conclude that $f_y(x) \in A_y$.

Now let $x_1, x_2 \in A_y$ be such that $x_1 =_q x_2$ for some $q \in \mathbf{Q}$, that is such that

$$\forall_{p\in\mathbf{Q}}[p \leq q \Rightarrow x_1(p) = x_2(p)] \ . \tag{2.12}$$

Noting that for $p \in \mathbf{Q}$ we have

$$x_1^2 = \sum_{r+s=p} x_1(r)x_1(s)$$

we shall now consider the values of p satisfying the equality $x_1^2(p) = x_2^2(p)$.

For all the values of p for which we have $r \leq q$ and $s \leq q$ we have

$$x_1^2(p) = \sum_{r+s=p} x_1(r)x_1(s) = \sum_{r+s=p} x_2(r)x_2(s) = x_2^2(p)$$

since, from (2.12), the pairs (r, s) considered in the first summation are exactly the same as those which must be considered in the second summation.

The restriction $r \leq q \wedge s \leq q$ is automatically satisfied when we have $p \leq q$; however, there are situations in which $r > q \vee s > q$ for which the equality of the two summations is still satisfied. If, for example, we were to have $r > q$ then the equality of the two sums will persist if we let $s < m_y$; in this case the product $x_1(r)x_1(s)$ is null and equal, therefore, to $x_2(r)x_2(s)$. This situation always occurs when we let $p \leq q + m_y$: if we have $r < q$ then we must let $s < m_y$; for values of p greater than $q + m_y$ there will be situations with $r > q$ and $s > m_y$ and therefore the equality of the sums will not necessarily be satisfied. In conclusion we can write

$$x_1 =_q x_2 \Rightarrow x_1^2 =_{q+m_y} x_2^2 \quad .$$

More generally, by induction, we can verify that

$$x_1 =_q x_2 \Rightarrow x_1^j =_{q+(j-1)m_y} x_2^j \quad \Rightarrow x_1^j =_{q+m_y} x_2^j \tag{2.13}$$

for any $j \in \mathbb{N}$, $j > 1$.

From (2.13) it follows that $x_1^2 P(x_1) =_{q+m_y} x_2^2 P(x_2)$ and so we have $f_y(x_1) =_{q+m_y} f_y(x_2)$. The function f_y then satisfies the conditions of Theorem 2.19 so that we can guarantee the existence of one and only one $x \in \mathcal{K}_{\mathbb{Q}}]\iota[$ with $m_x > 0$ such that $(1+x)^n = 1 + y$. Thus we may conclude that the number $z \in \mathcal{K}_{\mathbb{Q}}]\iota[$ possesses roots of order $n \in \mathbb{N}$ in $\mathcal{K}_{\mathbb{Q}}]\iota[$ if and only if the number $a_0 \in \mathcal{K}$ possesses roots of order n in $\mathcal{K}$. (If $\mathcal{K} = \mathbb{C}$ then the number z will possess precisely n distinct roots of order n).

In particular, for any $n \in \mathbb{N}$, let $\alpha = \sqrt[n]{\iota_p}$ where p is an arbitrarily fixed rational number. Since α is an element of $\mathcal{K}_{\mathbb{Q}}]\iota[$ then

$$\alpha = b_0.\iota_{m_\alpha}.(1+x)$$

from which, raising both sides to the power n, we get

$$\alpha^n = b_0^n.\iota_{m_\alpha}^n.(1+x)^n = b_0^n.\iota_{nm_\alpha}.(1+x)^n = \iota_p$$

whence we obtain

$$b_0^n = 1, \quad n.m_\alpha = p \quad \text{and} \quad x = 0 \quad .$$

Denoting by $\iota_p^{1/n}$ the nth root of ι_p which corresponds to the root $b_0 = 1$ (from the equation $b_0^n = 1$) then, since $p/n \in \mathbb{Q}$, it must be the case that

$$\iota_p^{1/n} = \iota_{p/n} \quad .$$

This can be generalised to any rational exponent $q \in \mathbb{Q}$ so that

$$\iota_p^q = \iota_{pq} \quad .$$

Note that $\iota_0 = 1$ (which is the function equal to 1 for $q = 0$ and equal to 0 for $q \neq 0$, $q \in \mathbb{Q}$) and denote simply by ι the element ι_1, that is to say let $\iota : \mathbb{Q} \to \mathcal{K}$ be defined by

$$\iota(q) = \begin{cases} 1, & \text{if } \ q = 1 \\ 0, & \text{if } \ q \neq 1 \end{cases} \quad .$$

Then for any $p \in \mathbb{Q}$ we obtain

$$\iota_p = \iota^p$$

whence we get the following representation for an arbitrary $z \in \mathcal{K}_{\mathbb{Q}}]\iota[$,

$$z = a_0.\iota^{m_z}.(1+y)$$

with $y \in \mathcal{K}_{\mathbb{Q}}]\iota[$ such that $m_y > 0$.

2.3.2.3 The ordered field $(\mathbb{R}_{\mathbb{Q}}]\iota[, +, ., \leq)$

We denote by $\mathbb{R}^+_{\mathbb{Q}}]\iota[$ the set formed by all elements x of $\mathbb{R}_{\mathbb{Q}}]\iota[$ such that $x \neq 0$ and $x(m_x) > 0$, and describe such elements as **positive**.

Definition 2.21 Given two elements $x, y \in \mathbb{R}_{\mathbb{Q}}]\iota[$ we write $x \leq y$ if and only if $x = y$, or $y - x \in \mathbb{R}^+_{\mathbb{Q}}]\iota[$.

Exercise 2.22 *Verify that $(\mathbb{R}_{\mathbb{Q}}]\iota[, +, ., \leq)$ is a totally ordered field, that is that $\leq$ is a total ordering relation compatible with the field operations or, equivalently, that*

- $x \leq y \Rightarrow x + z \leq y + z$
- $0 \leq z \Rightarrow [x \leq y \Rightarrow xz \leq yz]$

for any $x, y, z \in (\mathbb{R}_{\mathbb{Q}}]\iota[, +, ., \leq)$.

Exercise 2.23 *Verify that we can embed the ordered field of the superreal numbers in the ordered field $(\mathbb{R}_{\mathbb{Q}}]\iota[, +, ., \leq)$ and that, in particular $\mathbb{R} \subset (\mathbb{R}_{\mathbb{Q}}]\iota[, +, ., \leq)$.*

Exercise 2.24 *Verify that the element $\iota \in (\mathbb{R}_{\mathbb{Q}}]\iota[, +, ., \leq)$ satisfies the double inequality*

$$0 < \iota \leq \frac{1}{n}$$

for every $n \in \mathbb{N}$, so that it is a positive infinitesimal number.

Series representation Given a number $z \in \mathcal{K}_{\mathbb{Q}}]\iota[$, let M_z be its support. If $M_z = \{q_0, q_1, \ldots, q_m\}$ (with $q_0 < q_1 < \ldots < q_m$) then, taking into account that $z(a_j) = a_j, j = 0, 1, 2, \ldots, m$, we must have

$$z = \sum_{j=0}^{m} a_j \iota_{q_j} = \sum_{j=0}^{m} a_j \iota^{q_j} \quad .$$

If $M_z = \{q_0, q_1, q_2, \ldots\}$ is infinite (with $q_0 < q_1 < q_2 < \ldots$) we must similarly have

$$z = \sum_{j=0}^{\infty} a_j \iota_{q_j} = \sum_{j=0}^{\infty} a_j \iota^{q_j} \quad .$$

In fact for any $q \in \mathbb{Q}$, taking the case in which M_z is an infinite set, we get

$$z(q) = \sum_{j=0}^{\infty} a_j \iota_{q_j}(q)$$

and therefore

- if for all $j = 0, 1, 2, \ldots$, we were to have $q \neq q_j$ then $z(q) = 0$,
- if for some $j = 0, 1, 2, \ldots$, we were to have $q = q_j$ then $z(q) = a_j \iota_q(q) = a_j$.

This representation can be justified analytically, associating the relation of order $\leq$ with a natural topology from which we derive an appropriate notion of convergence

Definition 2.25 Let $(z_n)_{n\in\mathbb{N}}$ be a sequence of elements of $\mathcal{K}_{\mathbb{Q}}]\iota[$. We say that this sequence is **ι-convergent** to $z \in (\mathcal{K}_{\mathbb{Q}}]\iota[$ if for each $\epsilon \in \mathbb{R}^+_{\mathbb{Q}}]\iota]$ there exists $n_0 \in \mathbb{N}$ such that

$$\forall_{n\in\mathbb{N}}[n \geq n_0 \Rightarrow |a_n - a| < \epsilon] \ .$$

The notion of ι-convergence now allows us to justify the representation of the elements of $\mathcal{K}_{\mathbb{Q}}]\iota[$, under the form of generalised power series.

Theorem 2.26 The infinite ι-series

$$\sum_{j=0}^{\infty} a_j \iota^{q_j}$$

is ι-convergent to $z \in \mathcal{K}_{\mathbb{Q}}]\iota[$. We can therefore write

$$z = \sum_{j=0}^{\infty} a_j \iota^{q_j}$$

which shows that $\mathcal{K}_{\mathbb{Q}}]\iota[$ is the field of generalised power series corresponding to all sequences of coefficients $(a_j)_{j\in\mathbb{N}_0}$ of elements of $\mathcal{K}$ and all sequences of exponents $(q_j)_{j=0,1,2,\ldots}$ such that $q_0 < q_1 < q_2 < \ldots$ and $\lim_{j\to\infty} q_j = +\infty$.

Proof If $z_n = \sum\limits_{j=0}^{n} a_j \iota^{q_j}, \quad n = 1, 2, \ldots$ then $z_n \in \mathcal{K}_{\mathbb{Q}}]\iota[$ for all $n \in \mathbb{N}$.

Without loss of generality we may suppose that M_z is an infinite set. Let $\epsilon \in \mathbb{R}^+_{\mathbb{Q}}]\iota[$ be arbitrarily chosen and let $n_\epsilon \in \mathbb{N}$ be such that $\iota^{n_\epsilon} < \epsilon$. Since $q_j \to +\infty$ there exists $j_\epsilon \in \mathbb{N}_0$ such that $\forall_j[j \geq j_\epsilon \Rightarrow q_j > n_\epsilon]$ and therefore

$$\forall_{n\in\mathcal{N}}[n \geq n_\epsilon \Rightarrow \forall_{q\in\mathcal{Q}}[q \leq n_\epsilon \Rightarrow (z - z_n)(q) = 0]] \ .$$

Consequently

$$\forall_{n\in\mathcal{N}}[n \geq n_\epsilon \Rightarrow |z - z_n| < \epsilon]$$

which shows that z_n ι-converges to z. •

The Laugwitz field L

The representation of the elements of the fields of Levi-Civita in the form of infinite ι-series immediately suggests a new enlargement of the number field: that which is generated when we allow the elements of the sequences of exponents to be real numbers. We then obtain a field, which is denoted by **L** in homage to D.Laugwitz who studied it. **L** is the set formed of all formal series of generalised powers of the form

$$\alpha = \sum_{k=0}^{\infty} a_k \iota^{\nu_k}$$

where the coefficients a_k are real (or complex) numbers and the exponents $\nu_k, k = 0, 1, 2, \ldots$, are real numbers such that

$$\nu_0 < \nu_1 < \nu_2 < \cdots < \nu_k < \ldots \qquad (\nu_k \uparrow \infty).$$

If for some order p we have $a_k = 0$ for all $k > p$ then the formal series can be written in the form

$$a_0\iota^{\nu_0} + a_1\iota^{\nu_1} + \ldots + a_p\iota^{\nu_p} \quad .$$

Definition 2.27 (Equality) Two formal series of generalised powers are said to be equal if and only if for each term with coefficient a_j which occurs in one series, but not in the other, we have $a_j = 0$.

We define the following operations in **L**:

Addition The sum of two formal series of generalised powers $\sum a_k\iota^{\nu_k}$ and $\sum b_k\iota^{\mu_k}$ is the formal series

$$\sum_{k=0}^{\infty} c_k\iota^{\lambda_k} = \left(\sum_{k=0}^{\infty} a_k\iota^{\nu_k}\right) + \left(\sum_{k=0}^{\infty} b_k\iota^{\mu_k}\right)$$

whose terms $\{c_k\}$ are determined in accord with the following rules:

- $\{\lambda_k\}$ is the set union of the sets $\{\nu_k\}$ and $\{\mu_k\}$ with the elements arranged in increasing order.
- if, for some $m = 0, 1, 2, \ldots$, the exponent λ_m occurs in both the sets $\{\nu_k\}$ and $\{\mu_k\}$, that is if we have, for example, $\lambda_m = \nu_p = \mu_q$, then $c_m = a_p + b_q$.
- if, for some $m = 0, 1, 2, \ldots$, λ_m does not appear in $\{\mu_k\}$ and if, for example we have $\lambda_m = \nu_p$ then $c_m = a_p$; if λ_m does not appear in $\{\nu_k\}$ and if, for example, we have $\lambda_m = \mu_q$, then $c_m = b_q$).

Multiplication We define the product of two formal series of generalised powers $\sum a_k\iota^{\nu_k}$ and $\sum b_k\iota^{\mu_k}$ by setting

$$\left(\sum_{k=0}^{\infty} a_k\iota^{\nu_k}\right) \cdot \left(\sum_{k=0}^{\infty} b_k\iota^{\mu_k}\right) = \left(\sum_{k=0}^{\infty} c_k\iota^{\lambda_k}\right)$$

where $\{\lambda_k\}$ is the set formed by all sums of the form $\nu_p + \mu_q$ with the elements arranged in increasing order, and

$$c_k = \sum a_p b_q,$$

$p, q \in \mathbb{N}_0$ being such that $\nu_p + \mu_q = \lambda_k$.

Exercise 2.28 *Show that* **L** *is a commutative ring in which*

$$0 \equiv 0\iota^0 + 0\iota^1 + 0\iota^2 + \ldots, \qquad 1 \equiv 1\iota^0 + 1\iota^1 + 1\iota^2 + \ldots$$

are respectively the zero and unit elements.

In order to prove that **L** is a field it is necessary to show now that every non-null element (that is, every element with at least one of the coefficients different from zero) possesses a multiplicative inverse.

Let $\tau \in \mathbf{L}$ be non-null,

$$\tau = \sum_{k=0}^{\infty} a_k \iota^{\nu_k}$$

and suppose that $a_0 \neq 0$. We can then write

$$\tau = a_0 \iota^{\nu_0} \alpha$$

where

$$\alpha = 1 + \sum_{k=1}^{\infty} b_k \iota^{\mu_k}$$

with $b_k = a_k / a_0$ and $\mu_k = \nu_k - \nu_0$ for $k = 1, 2, \ldots$. If we show that α possesses an inverse, α^{-1}, then we can also prove that τ possesses an inverse of the following form:

$$\tau^{-1} = a_0^{-1} \iota^{-\nu_0} \alpha^{-1} \quad .$$

Thus to prove that $\mathbf{L}$ is a field it is enough to show that elements of the form

$$\alpha = 1 + \sum_{k=1}^{\infty} b_k \iota^{\nu_k} \qquad \text{with} \quad 0 < \nu_1 < \nu_2 < \ldots < \nu_k < \ldots$$

(where $\nu_k \to \infty$) possess inverses in $\mathbf{L}$.

Associated with the natural topology which can be introduced into $\mathbf{L}$, we can define a notion of ι-convergence which allows us to show that the sequence

$$1 - \left(\sum_{k=1}^{\infty} b_k \iota^{\nu_k} \right) + \left(\sum_{k=1}^{\infty} b_k \iota^{\nu_k} \right)^2 - \ldots + (-1)^n \left(\sum_{k=1}^{\infty} b_k \iota^{\nu_k} \right)^n , \quad n = 0, 1, 2, \ldots$$

ι-converges to α^{-1}, thereby showing that every element of $\mathbf{L}$ (different from zero) does possess an inverse.

Note 2.29 Various articles exist in the specialised literature in which some aspects of Mathematical Analysis are studied in the context of this number system. However it is not technically very easy to work with objects of this type.

NSA, as we shall see, allows us to work with infinite and infinitesimal quantities of a much simpler form. For this reason we do not pursue the above subject to any further extent.

Chapter 3

Introduction to Nonstandard Analysis

Infinites and indivisibles transcend our finite comprehension, the first due to their greatness, the last for their smallness; imagine what could happen when they are combined among themselves.

GALILEO GALILEI "Dialogues relative to two new sciences "

3.1 NSA: an Elementary Model

It should be clear from what has gone before that simply adding new, ideal elements to the real number system will not in itself be enough to give a well-founded extension of that system which includes infinite numbers and infinitesimals. Such a procedure would not generally guarantee sound results and would generate structures (such as the "standard"systems described earlier) which are difficult to understand and of limited applicability, **Nonstandard Analysis** actually offers a good deal more: it claims to provide an extension of $\mathbb{R}$ which

1. *will be a proper extension*, and at the same time,

2. *will possess,* **in a sense**, *all the properties of* $\mathbb{R}$.

Such a proper extension of $\mathbb{R}$, would necessarily be non-Archimedian, and thus offer the possibility of a satisfactory definition of the infinite and infinitesimal elements postulated by Leibniz. On the other hand $\mathbb{R}$ itself is an Archimedian field, and so requirements 1. and 2. would appear to be contradictory. These contradictions, however, vanish in the construction used by Robinson wherein the expression *the same properties* is restricted to just a part of those valid for the real number system - namely, to those properties which can be expressed in terms of an appropriate and adequate formal language. Undoubtedly the most important (and also the least elementary) part of the original programme of Leibniz as developed by Robinson was the delimitation of this collection of properties. Several other versions of non-Archimedian fields which properly contain $\mathbb{R}$ had indeed been constructed during the twentieth century, but their usefulness did not go beyond Algebra, for it was not found possible to extend Real Analysis in a sufficiently interesting form.

3.1.1 Preliminary aspects

To begin with it is convenient to recall, very briefly, the process, devised by Georg Cantor (1845-1918), of constructing the set of real numbers from sequences of rational numbers. As will be seen later on, a very similar process can be used to construct an elementary model of NSA.

3.1.1.1 Cantorian construction of the real numbers

Let $\mathbf{Q}^{\mathbf{N}}$ be the set of all sequences of rational numbers equipped with termwise definitions of the usual arithmetic operations. Denote by $\mathcal{F}_{\mathbf{Q}}$ the subset of $\mathbf{Q}^{\mathbf{N}}$ consisting of all Cauchy sequences $(q_n)_{n\in\mathbf{N}}$. That is to say, $\mathcal{F}_{\mathbf{Q}}$ is the set of all sequences $(q_n)_{n\in\mathbf{N}}$ of rational numbers which satisfy the Cauchy condition, $\lim_{m,n\to\infty}(q_m - q_n) = 0$ (where m and n tend to infinity independently of each other): equivalently we have

$$\forall_{k\in\mathbf{N}}\exists_{p\in\mathbf{N}}\forall_{m,n\in\mathbf{N}}[m,n \geq p \Rightarrow |q_m - q_n| < 1/k] \ .$$

Such sequences will not necessarily converge in $\mathbf{Q}$: consider, for example, the sequence whose general term is defined by the familiar algorithm for calculating square roots applied to the number 2,

$$1; 1.4; 1.41; 1.414; 1.4142; \ldots \tag{3.1}$$

This sequence is, clearly, a Cauchy sequence, since for $m > n$ the terms q_m and q_n are equal up to the nth decimal place, and can only differ, at most, from that point on. Nevertheless $(q_n)_{n\in\mathbf{N}}$ does not tend to any rational whatsoever.

Exercise 3.1 *Show that there exists no rational number whose square will be equal to 2.*

It is easy to show that a sequence of rational numbers convergent in $\mathbf{Q}$ is always a Cauchy sequence: on the other hand, the above example shows that the converse is not true. This lack of symmetry, which makes the set $\mathbf{Q}$ inadequate for the treatment of many questions of mathematical analysis, is resolved if $\mathbf{Q}$ is given new elements which *behave* as limits of those Cauchy sequences of rational terms which do not converge in $\mathbf{Q}$. These new elements can be defined in terms of the particular Cauchy sequences which generate them. Thus, for the limit of the sequence (3.1) there is created a "new number ", denoted by $\sqrt{2}$, which is defined by this specific sequence.

Since it is desired that the set of such "new numbers" should constitute an extension of $\mathbf{Q}$, it becomes necessary to identify each rational number with an appropriate sequence. The correspondence, however, is not unique: there are many sequences of rational numbers which converge to the same given rational number.

Exercise 3.2 *Show that if $(q_n)_{n\in\mathbf{N}}$ and $(q'_n)_{n\in\mathbf{N}}$ are two sequences in $\mathcal{F}_{\mathbf{Q}}$ such that*

$$\lim_{n\to\infty}(q_n - q'_n) = 0 \tag{3.2}$$

then, if $(q_n)_{n\in\mathbf{N}}$ converges (in $\mathbf{Q}$) to a number $q \in \mathbf{Q}$, $(q'_n)_{n\in\mathbf{N}}$ also converges to the same limit.

Taking account of exercise 3.2, it is natural to identify all the sequences of rational numbers which satisfy (3.2). We make this identification with respect to a relation R defined as follows: given two sequences $(q_n)_{n\in\mathbf{N}}, (q'_n)_{n\in\mathbf{N}} \in \mathcal{F}_{\mathbb{Q}}$ we write

$$(q_n)_{n\in\mathbf{N}}\mathrm{R}(q'_n)_{n\in\mathbf{N}} \quad \Leftrightarrow \quad \lim_{n\to\infty}(q_n - q'_n) = 0$$

(where the term "limit" must be interpreted in the sense of $\mathbb{Q}$). This relation is actually an equivalence relation, and the quotient

$$\mathcal{F}_{\mathbb{Q}}/\mathrm{R}$$

denoted by $\mathbb{R}$, then constitutes a model for the *set of real numbers.* Each element of $\mathbb{R}$ is an equivalence class

$$[(q_n)_{n\in\mathbf{N}}] = \{(q'_n)_{n\in\mathbf{N}} \in \mathcal{F}_{\mathbb{Q}} : (q'_n)_{n\in\mathbf{N}}\mathrm{R}(q_n)_{n\in\mathbf{N}}\}$$

which can be represented arbitrarily by any one of its elements. In particular, for any $q \in \mathbb{Q}$, the class $[(q, q, q, \ldots)]$ can stand not only for the constant sequence but also for all sequences of rational numbers whose limit in $\mathbb{Q}$ is equal to q. The set $\mathbb{Q}$ can be embedded in $\mathbb{R}$ by means of the mapping

$$\phi : \mathbb{Q} \to \mathbb{R}$$

$$q \mapsto [(q_n)_{n\in\mathbf{N}}]$$

where $(q_n)_{n\in\mathbf{N}}$ is any sequence of $\mathcal{F}_{\mathbb{Q}}$ which converges to $q \in \mathbb{Q}$; (on occasion it may be convenient to take for $(q_n)_{n\in\mathbf{N}}$ the constant sequence in which $q_n = q$ for all $n \in \mathbb{N}$). Noting that for arbitrary $q, q' \in \mathbb{Q}$ we have

$$\phi(q) = \phi(q') \quad \Rightarrow \quad [(q_n)_{n\in\mathbf{N}}] = [(q'_n)_{n\in\mathbf{N}}]$$

$$\Rightarrow (q)_{n\in\mathbf{N}}\mathrm{R}(q')_{n\in\mathbf{N}} \quad \Rightarrow \quad q = q',$$

it follows that ϕ is actually injective.

The operations of addition and multiplication and the relation of order in $\mathbb{Q}$ can be extended to $\mathbb{R}$ (by means of their representative sequences), allowing us to prove that

$$(\mathbb{R}, +, ., \leq)$$

is an **ordered field** of which $(\mathbb{Q}, +, ., \leq)$ is a subordered field.[1]

Exercise 3.3 *Define the algebraic operations and ordering relation in $\mathbb{R}$, explicitly so that the form which they have, when restricted to $\mathbb{Q}$, coincides with the usual algebraic operations and ordering relation of the field $\mathbb{Q}$.*

3.1.1.2 The ring ${}^b\mathbb{R}$

After constructing the field of the real numbers there arises the question of whether a construction of the same type, now applied to $\mathbb{R}$ instead of $\mathbb{Q}$, might allow us to

[1]Strictly the field $\mathbb{Q}$ should be replaced by the field $\phi(\mathbb{Q})$ to which it is isomorphic.

obtain a further amplification of the number system. (This is surely a question of real interest from every point of view.) However, the response is in the negative, and is given by

Theorem A sequence of real numbers is convergent if and only if it is a Cauchy sequence.

which will be proved in Chapter 4 (Theorem 4.13). This result simply makes explicit the fact that $(\mathbb{R}, +, .+, \leq)$ is a **complete field**, whereas $(\mathbb{Q}, +, ., \leq)$ is, on the contrary, not complete. Thus nothing further can be obtained in this way. However, by applying the Cantorian process without the restriction to Cauchy sequences, it might nevertheless be possible to obtain new enlargements of $\mathbb{R}$ in terms of sequences of real numbers. For if, on the one hand, we relax the condition that these sequences must be Cauchy then, on the other hand, we may be able to modify the corresponding equivalence relation appropriately. In fact, the process to be described below does not allow us to obtain anything more than a partially ordered ring; however, it does constitute a good preliminary exercise for the eventual construction of the *field of hyperreal numbers.*

Let $\mathbb{R}^{\mathbf{N}}$ be the set of all sequences of real numbers

$$\mathbf{x} \equiv (x_n)_{n \in \mathbf{N}} = (x_1, x_2, x_3, \ldots) \quad .$$

This set can be provided with a natural algebraic structure by defining the operations of addition and multiplication of sequences term by term,

$$\mathbf{x} \oplus \mathbf{y} \equiv (x_n)_{n \in \mathbf{N}} \oplus (y_n)_{n \in \mathbf{N}} = (x_n + y_n)_{n \in \mathbf{N}},$$

$$\mathbf{x} \otimes \mathbf{y} \equiv (x_n)_{n \in \mathbf{N}} \otimes (y_n)_{n \in \mathbf{N}} = (x_n . y_n)_{n \in \mathbf{N}} \quad .$$

With these definitions, addition and multiplication in $\mathbb{R}^{\mathbf{N}}$ satisfy the usual laws of arithmetic - associativity. commutativity and distributivity. The sequences

$$\mathbf{0} \equiv (0, 0, \ldots, 0, \ldots), \qquad \mathbf{1} \equiv (1, 1, \ldots, 1, \ldots)$$

constitute the unit elements of addition and multiplication respectively. and, therefore, $(\mathbb{R}^{\mathbf{N}}, +, .)$ is a ring with unit element. The mapping $\phi : \mathbb{R} \to \mathbb{R}^{\mathbf{N}}$, defined by

$$\phi(x) = (x, x, x, \ldots, x, \ldots)$$

gives an embedding of $\mathbb{R}$ (as a field structure) into $\mathbb{R}^{\mathbf{N}}$.

This algebra of $\mathbb{R}^{\mathbf{N}}$ has divisors of zero, and also constitutes a set much larger than is necessary or desirable for mathematical analysis. Thus, even two very similar elements such as, for example,

$$(1, 0, 0, \ldots, 0, \ldots) \quad \text{and} \quad (0, 1, 0, \ldots,),$$

have to be considered as distinct objects in the context of $\mathbb{R}^{\mathbf{N}}$. As is usual in such cases, we can address this problem by identifying sequences which are "sufficiently similar" by means of a suitable equivalence relation. Consider first in the set $\mathbb{R}^{\mathbf{N}}$ a relation, denoted by $[\equiv]$, defined as follows:

Two real sequences $(a_n)_{n\in\mathbf{N}}$ *and* $(a'_n)_{n\in\mathbf{N}}$ *are said to be* $[\equiv]$*-equivalent if and only if the set*

$$\{n \in \mathbb{N} : a_n = a'_n\}$$

is a cofinite subset of $\mathbb{N}$ *(that is, a subset whose complement in* $\mathbb{N}$ *is finite): we then write* $(a_n)_{n\in\mathbf{N}}[\equiv](a'_n)_{n\in\mathbf{N}}$.

Exercise 3.4 *Verify that the relation* $[\equiv]$ *is, in fact, an equivalence relation (that is, is reflexive, symmetric and transitive).*

The equivalence relation $[\equiv]$ divides the set $\mathbb{R}^{\mathbf{N}}$ into equivalence classes. We denote by ${}^\flat\mathbb{R}$ the set of those classes

$${}^\flat\mathbb{R} \equiv \mathbb{R}^{\mathbf{N}}/[\equiv]$$

and by $[(a_n)_{n\in\mathbf{N}}]$ the equivalence class representative of the sequence $(a_n)_{n\in\mathbf{N}}$,

$$[(a_n)_{n\in\mathbf{N}}] \equiv \{(a'_n)_{n\in\mathbf{N}} \in \mathbb{R}^{\mathbf{N}} : (a'_n)_{n\in\mathbf{N}}[\equiv](a_n)_{n\in\mathbf{N}}\} \quad .$$

The set ${}^\flat\mathbb{R}$ can be given an algebraic structure by means of those operations of addition and multiplication defined termwise for sequences in $\mathbb{R}^{\mathbf{N}}$. That is to say, we can write,

$$[(a_n)_{n\in\mathbf{N}}] + [(b_n)_{n\in\mathbf{N}}] \equiv [(a_n + b_n)_{n\in\mathbf{N}}]$$

and

$$[(a_n)_{n\in\mathbf{N}}].[(b_n)_{n\in\mathbf{N}}] \equiv [(a_n.b_n)_{n\in\mathbf{N}}]$$

it being easy to prove that the final result of these operations does not depend on the particular representatives chosen from each equivalence class. The system can even be (*partially*) ordered by a relation $\leq$ defined in the following way

$$[(a_n)_{n\in\mathbf{N}}] \leq [(b_n)_{n\in\mathbf{N}}]$$

if and only if $\{n \in \mathbb{N} : \quad a_n \leq b_n\}$ is cofinite in $\mathbb{N}$. It is similarly easy to show that this partial[2] ordering relation is compatible with the equivalence relation $[\equiv]$. The resulting structure

$$({}^\flat\mathbb{R}, +, ., \leq),$$

is a commutative, partially ordered ring with unit element.

The mapping $\phi : \mathbb{R} \to {}^\flat\mathbb{R}$ defined for each $x \in \mathbb{R}$ by

$$\phi(x) = [(x, x, x, \ldots, x, \ldots)]$$

gives an embedding of $\mathbb{R}$ into ${}^\flat\mathbb{R}$ allowing us to write, in this sense, $\mathbb{R} \subset {}^\flat\mathbb{R}$.

Exercise 3.5 *The ring* ${}^\flat\mathbb{R}$ *contains other elements in addition to the real numbers. Verify, through examples, that in* ${}^\flat\mathbb{R}$ *there exist elements which, in modulus, are less than every positive real number, and elements which, in modulus, are greater than every positive real number, that is to say, infinitely small elements and infinitely large elements respectively.*

[2] The elements $[(0, 1, 0, 1, \ldots)]$ and $[(1, 0, 1, 0, \ldots)]$, for example, are not comparable.

${}^b\mathbb{R}$ *is not a field*, since it possesses divisors of zero. For example, neither of the sequences

$$(1, 0, 1, 0, 1, \ldots) \quad \text{and} \quad (0, 1, 0, 1, 0, \ldots)$$

belong to the equivalence class of zero, which shows that the elements which they represent must *both* be different from zero. However, the product of these sequences is the null sequence and, therefore, the product of the corresponding equivalence classes will be the null element. This is a result of the fact that the equivalence relation introduced in $\mathbb{R}^{\mathbb{N}}$ is not strong enough to eliminate all the divisors of zero. An equivalence relation strong enough to do this would open the way to the construction of an ordered field of which $(\mathbb{R}, +, ., \leq)$ is a proper subordered field.

Note 3.6 It is possible to consider some aspects of mathematical analysis in the context of the set of numbers ${}^b\mathbb{R}$ by using reasoning of an infinitesimal type. But the fact that ${}^b\mathbb{R}$ is not totally ordered (and also not even a field) creates problems. We need to modify this type of construction until we do obtain a totally ordered field which contains a subordered field isomorphic with $\mathbb{R}$. This will be carried out in the following section.

3.1.2 The hyperreal numbers

The definition of the relation $[\equiv]$ in the preceding section is based, implicitly, on a classification of the subsets of $\mathbb{N}$ into two large classes, namely the subsets which are cofinite and those which are not. Every cofinite set is infinite, and so we can say, informally, that cofinite sets are *large sets*: two sequences of real numbers are $[\equiv]$-equivalent if there is a large set of indices corresponding to equal terms. On the other hand, a finite set of indices may be reasonably described as a *small set*: two sequences which coincide only for a finite number of indices will not belong to the same class.

We may now ask what happens in the case of two sequences which coincide for an infinite number of indices but not for a cofinite set of them. The relation $[\equiv]$ cannot cope with this kind of situation. There certainly do exist infinite sets which are not cofinite, such as, for example, the set of all even numbers. Such sets, under the criterion being presently used, do not fit into the classification of sets of indices as large and small. This fact is the cause of the serious difficulties arising in the otherwise desirable structure ${}^b\mathbb{R}$. Now, in order to obtain a suitable equivalence relation which would resolve these difficulties, we need to preserve the idea that finite sets of indices are *small* while their complements in $\mathbb{N}$ are *large*, and that two sequences are equivalent if they coincide in a large subset of $\mathbb{N}$. It is not true that every subset of $\mathbb{N}$ is either finite or cofinite and so it is now necessary to extend the idea of "largeness" in such a way as to allow the classification of *all* the subsets of $\mathbb{N}$ into *"small sets"* and *"large sets"*. Intuitively it seems desirable that such a classification should obey the following basic rules:

1. Every subset is either large or else small (but not both).

2. Every finite subset is small (or, equivalently, every cofinite subset is large).

3. The union of any two small subsets is small (or, equivalently, the intersection of any two large sets is large).

3. Every subset which contains a large subset is necessarily itself a large subset.[3]

Two sequences in $\mathbb{R}^{\mathbb{N}}$ will then be identified whenever their terms coincide for a "large set" of indices.

In order to formalise these ideas mathematically we introduce some basic notions concerning *filters* and *ultrafilters* over $\mathbb{N}$.

Definition 3.7 A **filter** on $\mathbb{N}$ is any non-empty family $\mathcal{F}$ of subsets of $\mathbb{N}$ which satisfies the following properties:
1) $\emptyset \notin \mathcal{F}$
(2) If $A \in \mathcal{F}$ and $B \in \mathcal{F}$ then $A \cap B \in \mathcal{F}$
(3) If $A \in \mathcal{F}$ and $A \subset B$ then $B \in \mathcal{F}$.

It follows immediately from (3) that the set $\mathbb{N}$ belongs to $\mathcal{F}$ and also that if A and B are two subsets of $\mathbb{N}$ such that $A \cap B \in \mathcal{F}$ then, necessarily, $A \in \mathcal{F}$ and $B \in \mathcal{F}$.

Exercise 3.8 *Verify that the families of subsets of* $\mathbb{N}$ *defined below are filters.*

1. The family $\mathcal{F}^i = \{A \subset \mathbb{N} : i \in A\}$, *where* i *is a given element of* $\mathbb{N}$, *is a filter over* $\mathbb{N}$.

2. $\mathcal{H}$ *being a non-empty family of subsets of* $\mathbb{N}$, *the family*

$$\mathcal{F}^{\mathcal{H}} = \{A \subset \mathbb{N} : A \supset \bigcap_{i=1}^{n} B_i \text{ for some } n \in \mathbb{N} \text{ and } B_i \in \mathcal{H}, 1 \leq i \leq n\}$$

is a filter which contains $\mathcal{H}$, *and is called the filter generated by* $\mathcal{H}$. *If* $\mathcal{H} = \emptyset$ *then for the sake of generality, we set* $\mathcal{F}^{\mathcal{H}} = \mathbb{N}$. *In the case where we have* $\mathcal{H} = \{\{i\}\}$ *we obtain the filter* $\mathcal{F}^i$ *described previously.*

3. Let $\{\mathcal{F}_x : x \in X\}$ *be a family of filters over* $\mathbb{N}$ *totally ordered by the relation* $\subset$, *that is, such that for all* $x, y \in X$ *we have either* $\mathcal{F}_x \subset \mathcal{F}_y$ *or* $\mathcal{F}_y \subset \mathcal{F}_x$. *Then*

$$\bigcup_{x \in X} \mathcal{F}_x = \{A \subset \mathbb{N} : \exists_x [x \in X \wedge A \in \mathcal{F}_x]\}$$

is a filter over $\mathbb{N}$.

4. $\mathcal{F}_0 = \{A \subset \mathbb{N} : card(A^c) \in \mathbb{N}\}$ *is the family of cofinite subsets of* $\mathbb{N}$, *and forms a filter over* $\mathbb{N}$.

The family $\mathcal{F}_0$ described in the last line of Exercise 3.8, is a particularly important example of a filter over $\mathbb{N}$, and is usually referred to as the **Fréchet filter**.

If, in the definition of a filter we substitute for (1) as follows

(1′) $\mathcal{F}$ contains no finite sets,

then $\mathcal{F}$ is said to be a **free filter**. Thus, the Fréchet filter is free while the filter $\mathcal{F}^i$ described in the previous exercise is not free.

---

[3] It must be admitted that the classification of the subsets of $\mathbb{N}$ into these two categories, rather than that of finite and cofinite, is a little artificial, but this is unavoidable.

Using the notion of Fréchet filter, we can redefine the relation $[\equiv]$ in $\mathbb{R}^{\mathbf{N}}$ in the following, more elegant form:

$$(a_n)_{n\in\mathbf{N}}[\equiv](a'_n)_{n\in\mathbf{N}} \Leftrightarrow \{n \in \mathbb{N} : \quad a_n = a'_n\} \in \mathcal{F}_0 \quad .$$

As already remarked above, this relation does not lead to the construction of a field. In order to obtain an equivalence relation sufficiently strong to do this we now need to introduce the following further definition.

Definition 3.9 A filter over $\mathbb{N}$ is said to be an **ultrafilter** if, for every subset A of $\mathbb{N}$, we have

(4) $$A \in \mathcal{U} \quad \text{or} \quad A^c \in \mathcal{U} \quad .$$

In the particular case of the Fréchet filter, for example, neither the set of all odd numbers

$$\{1, 3, 5, 7, \ldots\}$$

nor its complement, the set of all even numbers, belongs to $\mathcal{F}_0$ and therefore $\mathcal{F}_0$ is not an ultrafilter. However, if $\mathcal{U}$ is an ultrafilter over $\mathbb{N}$ then *one of these sets and only one must belong to* $\mathcal{U}$.

Exercise 3.10 *Let $U_1, U_2, \ldots, U_n$ be disjoint subsets of $\mathbb{N}$. Show that, if the union $U_1 \cup U_2 \cup \ldots \cup U_n$ belongs to the ultrafilter $\mathcal{U}$, then there exists one and only one index $k = 1, 2, \ldots, n$ such that $U_k \in \mathcal{U}$.*

An ultrafilter is a filter: if it is a free filter, then the ultrafilter will also be a **free ultrafilter**. In order to lead to the required results when constructing the hyperreal numbers we shall only have to consider free ultrafilters. Now it is not immediately obvious that there do exist free ultrafilters over an infinite set. The question of their existence is actually a matter requiring serious consideration at a fundamental level. In the event, however, the following theorem, derived by Tarski in 1930, offers a sound basis for the construction of a system of hyperreal numbers:

Tarski's Theorem Every free filter can be extended to a free ultrafilter.

A proof of Tarski's theorem will be given later, in a more general context.

If $\mathcal{U}_{\mathbf{N}}$ is a free ultrafilter on $\mathbb{N}$ then, necessarily, $\mathcal{F}_0 \subset \mathcal{U}_{\mathbf{N}}$: in fact, if A belongs to $\mathcal{F}_0$ then A^c is a finite set which therefore cannot belong to $\mathcal{U}_{\mathbf{N}}$: it follows immediately that $(A^c)^c \equiv A \in \mathcal{U}_{\mathbf{N}}$.

3.1.2.1 Definition of the hyperreal numbers

In what follows, $\mathcal{U} \equiv \mathcal{U}_{\mathbf{N}}$ denotes an ultrafilter over $\mathbb{N}$ which contains $\mathcal{F}_0$, and which we suppose fixed once and for all. Using this ultrafilter $\mathcal{U}$ in place of the Fréchet filter $\mathcal{F}_0$ we can now introduce an equivalence relation on $\mathbb{R}^{\mathbf{N}}$ which is much stronger than $[\equiv]$, though of the same type.

Definition 3.11 Two sequences $(a_n)_{n\in\mathbf{N}}$, $(a'_n)_{n\in\mathbf{N}}$ belonging to $\mathbb{R}^{\mathbf{N}}$ will be said to be $\mathcal{U}$-equivalent if and only if the set $\{n \in \mathbb{N} : \quad a_n = a'_n\}$ belongs to $\mathcal{U}$: we then write

$$(a_n)_{n\in\mathbf{N}} \sim_{\mathcal{U}} (a'_n)_{n\in\mathbf{N}}$$

(or sometimes simply, $(a_n)_{n\in\mathbf{N}} \sim (a'_n)_{n\in\mathbf{N}}$).

The relation introduced by Definition 3.11 is, as can easily be verified, an equivalence relation allowing us to divide the set $\mathbb{R}^{\mathbf{N}}$ into equivalence classes. The quotient set is denoted by

$$^*\mathbb{R} = \mathbb{R}^{\mathbf{N}} / \sim_{\mathcal{U}}$$

and the equivalence class which contains $(a_n)_n \in \mathbb{N}$ is denoted by $[(a_n)_{n\in\mathbf{N}}]$ (or sometimes, more simply, just by $[(a_n)]$). $^*\mathbb{R}$ is known as the **set of hyperreal numbers**.

That the algebraic structure of $^*\mathbb{R}$ can be given by a slight extension of the arithmetic operations and the order relation already defined in $\mathbb{R}$, will be seen from the theorem which follows.

Theorem 3.12 Let $(a_n)_{n\in\mathbf{N}}$, $(a'_n)_{n\in\mathbf{N}}$, $(b_n)_{n\in\mathbf{N}}$, $(b'_n)_{n\in\mathbf{N}}$, be four sequences of real numbers such that $(a_n)_{n\in\mathbf{N}} \sim (a'_n)_{n\in\mathbf{N}}$ and $(b_n)_{n\in\mathbf{N}} \sim (b'_n)_{n\in\mathbf{N}}$. Then

(a) $(a_n + b_n)_{n\in\mathbf{N}} \sim (a'_n + b'_n)_{n\in\mathbf{N}}$,

(b) $(a_n.b_n)_{n\in\mathbf{N}} \sim (a'_n.b'_n)_{n\in\mathbf{N}}$,

(c) $a_n \leq b_n, ^{\mathcal{U}}\forall_{n\in\mathbf{N}} \quad \Rightarrow a'_n \leq b'_n, \quad ^{\mathcal{U}}\forall_{n\in\mathbf{N}}$

where $^{\mathcal{U}}\forall_{n\in\mathbf{N}}$, (for $\mathcal{U}$-nearly all $n \in \mathbb{N}$), means that the property in question is satisfied over a subset of indices belonging to the ultrafilter $\mathcal{U}$.

Proof Consider the following sets

$$\begin{aligned}
A &= \{n \in \mathbb{N} \ : a'_n = a_n\} \\
B &= \{n \in \mathbb{N} \ : b'_n = b_n\} \\
C &= \{n \in \mathbb{N} \ : a_n + b_n = a'_n + b'_n\} \\
D &= \{n \in \mathbb{N} \ : a'_n.a_n = b_n.b'_n\} \\
E &= \{n \in \mathbb{N} \ : a_n \leq b_n\} \\
F &= \{n \in \mathbb{N} \ : a'_n \leq b'_n\}
\end{aligned}$$

Since $A \cap B \subset C$, $A \in \mathcal{U}$ and $B \in \mathcal{U}$, it follows that $A \cap B \in \mathcal{U}$ and, therefore, that $C \in \mathcal{U}$, which proves (a). Result (b) is obtained similarly, using the fact that $A \cap B \subset D$. Finally, since $(A \cap B) \cap E \subset F$ then, from $A \cap B \in \mathcal{U}$ and $E \in \mathcal{U}$ it follows that $(A \cap B) \cap E \in \mathcal{U}$ and so that $F \in \mathcal{U}$, which implies (c). •

This theorem completely justifies the definition which follows.

Definition 3.13 Given two numbers $\alpha \equiv [(a_n)_{n\in\mathbf{N}}]$ and $\beta \equiv [(b_n)_{n\in\mathbf{N}}]$ of $^*\mathbb{R}$, we define the **sum**, $\alpha + \beta$, and **product**, $\alpha.\beta$, by

$$\alpha + \beta \equiv [(a_n)_{n\in\mathbf{N}}] + [(b_n)_{n\in\mathbf{N}}] = [(a_n + b_n)_{n\in\mathbf{N}}]$$

$$\alpha.\beta \equiv [(a_n)_{n\in\mathbf{N}}].[(b_n)_{n\in\mathbf{N}}] = [(a_n.b_n)_{n\in\mathbf{N}}] \ .$$

In addition we introduce an ordering relation by writing

$$\alpha \leq \beta \Leftrightarrow \{n \in \mathbb{N} : \quad a_n \leq b_n\} \in \mathcal{U} \ .$$

With these operations of addition and multiplication and with the relation of order defined as above, it is easy to show that

$$({}^*\mathbb{R}, +, ., \leq)$$

is a **totally ordered field** with zero $\mathbf{0} \equiv [(0,0,0,\ldots)]$ and unit $\mathbf{1} \equiv [(1,1,1,\ldots)]$.

Note 3.14 This construction for ${}^*\mathbb{R}$ has been made on the basis of an ultrafilter $\mathcal{U}$ on $\mathbb{N}$, whose existence is guaranteed by the theorem of Tarski, but of which we have not given any explicit description. Nor is it possible to do this since it would be equivalent to formulating an infinity of special definitions of membership: that is, one for each of the sets of the ultrafilter $\mathcal{U}$, and this has an uncountably infinite cardinal. Moreover, there is not just one possible ultrafilter over $\mathbb{N}$ to be taken into account; to each particular such ultrafilter there will correspond a specific set of hyperreal numbers. We know, for example, either that $[(0,1,0,1,0,\ldots)] \equiv \mathbf{1}$ and $[(1,0,1,0,1,\ldots)] \equiv \mathbf{0}$ or else that $[(0,1,0,1,0,\ldots)] \equiv \mathbf{0}$ and $[(1,0,1,0,1,\ldots)] \equiv \mathbf{1}$, but we do not know which alternative obtains. That depends on whether we have $\{1,3,5,7,\ldots\} \in \mathcal{U}$ or, alternatively, $\{2,4,6,8,\ldots\} \in \mathcal{U}$. This fact, although it may be a little disturbing during the first encounter with NSA, surprisingly comes to seem relatively unimportant in the practical context of doing mathematical analysis.

3.1.2.2 Properties of the hyperreals

The mapping $* : \mathbb{R} \rightarrow {}^*\mathbb{R}$ defined by

$$a \mapsto {}^*a = [(a,a,a,\ldots)]$$

constitutes, as may be verified without difficulty, an injective homomorphism of fields (monomorphism) which preserves the order relation; that is to say,

$$\forall_{a,b}[a, b \in \mathbb{R} \Rightarrow {}^*(a+b) = {}^*a + {}^*b, \text{ and}$$

$$ {}^*(a.b) = {}^*a.{}^*b, \quad \text{and}$$

$$a \leq b \Rightarrow {}^*a \leq {}^*b]$$

(where we adopt the same notation for the operations and order relation in $\mathbb{R}$ as in ${}^*\mathbb{R}$). This mapping therefore defines an *embedding* of the ordered field $\mathbb{R}$ into the ordered field ${}^*\mathbb{R}$. Identifying $\mathbb{R}$ with the set $\{{}^*a \in {}^*\mathbb{R} : a \in \mathbb{R}\}$ allows us to write $\mathbb{R} \subset {}^*\mathbb{R}$, considering the real numbers here as elements of ${}^*\mathbb{R}$. These elements are generally called the **standard elements** of ${}^*\mathbb{R}$ and, for the sake of simplicity of notation, are usually written without the star. However, ${}^*\mathbb{R}$ contains many more elements than these standard ones which we identify with the real numbers. For example, consider the hyperreal number

$$\epsilon = [(1/n)_{n \in \mathbf{N}}] = [(1, 1/2, 1/3, \ldots)]$$

which, as will be shown below, is a positive infinitesimal. That is, ϵ is a number which is larger than zero but less than any positive real number whatsoever. Consequently, it does not belong to the set $\mathbb{R}$ and is, therefore, a **nonstandard number**.

That ϵ really is a positive infinitesimal can be shown easily and quite briefly. Denoting the sequence which represents ϵ by $(e_n)_{n\in\mathbb{N}}$, we clearly have

$$\{n \in \mathbb{N} : e_n = 0\} = \emptyset \notin \mathcal{U} \text{ and } \{n \in \mathbb{N} : e_n > 0\} = \mathbb{N} \in \mathcal{U},$$

so that, $\epsilon > 0$. On the other hand, let $r \in \mathbb{R}$ be an entirely arbitrary positive (real) number. Since e_n tends to zero in $\mathbb{R}$), then $\{n \in \mathbb{N} : he_n < r\}$ is cofinite and so

$$\{n \in \mathbb{N} : e_n < r\} \in \mathcal{U} \quad .$$

Therefore, given that $r \in \mathbb{R}^+$ is arbitrary, we can write

$$\forall_r [r \in \mathbb{R}^+ \Rightarrow 0 < \epsilon < r] \tag{3.3}$$

which completes the proof.

Using the formula (3.3), and setting successively $r = 1/n$, with $n = 1, 2, \ldots$, we can deduce that with $\epsilon \in {}^*\mathbb{R}$, $\epsilon > 0$, then for every $n \in \mathbb{N}$ we have $n\epsilon < 1$. This formally confirms that that ${}^*\mathbb{R}$ is a non-archimedian field.

Exercise 3.15 *Verify that the numbers defined by*

$$\omega \equiv [(n)_{n\in\mathbf{N}}] = [(1,2,3,\ldots)] = \epsilon^{-1}, \quad \omega^2 \equiv [(n^2)_{n\in\mathbf{N}}] = [(1,4,9,\ldots)],$$

$$\sqrt{\omega} \equiv [(\sqrt{n})_{n\in\mathbf{N}}], \quad 2^\omega \equiv [(2^n)_{n\in\mathbf{N}}]$$

are all infinite positive hyperreals, each majorising every positive real number and so not belonging to $\mathbb{R}$.

Exercise 3.16 *Let* $x = [(x_n)_{n\in\mathbf{N}}]$ *be any hyperreal number whatsoever. We define the* ****-modulus*** *of* $x \in {}^*\mathbb{R}$ *by the following formula:*

$$^*|x| = [(|x_n|)_{n\in\mathbf{N}}]$$

which of course is also a hyperreal number.

1. *Verify that the *-modulus of a hyperreal number satisfies the usual properties of the modulus function, that is to say, that we have*
   (a) $^*|x| \geq 0 \wedge [^*|x| = 0 \Leftrightarrow x = 0]$
   (b) $^*|xy| = {}^*|x|.{}^*|y|$
   (c) $^*|x+y| \leq {}^*|x| + {}^*|y|$
   whatever the numbers $x, y \in {}^*\mathbb{R}$.

2. *Verify that* $^*|.|$ *is an extension to* ${}^*\mathbb{R}$ *of the modulus function of a real number.*

Note: In order to simplify notation we will just write $^*|.|$ as $|.|$, using the same symbol to denote modulus, whether we deal with a real number or with a hyperreal number.

As has already been established, there are (many) more elements in ${}^*\mathbb{R}$ than in $\mathbb{R}$. Given the inclusion of $\mathbb{R}$ in ${}^*\mathbb{R}$, we can classify the hyperreal numbers (relative to the real numbers) in the following way:

Definition 3.17 The hyperreal number x is said to be

finite if and only if $|x| < r|$ for *some* real number $r > 0$.
infinite if and only if $|x| > r$ for *every* real number $r > 0$.
infinitesimal if and only if $|x| < r$ for *every* real number $r > 0$.

We describe as **appreciable** any number which is finite but not infinitesimal.

For example, the three hyperreal numbers

$$0 \equiv [(0, 0, 0, \ldots)], \quad \epsilon \equiv [(1, 1/2, 1/3, \ldots)]. \quad \sqrt{\epsilon} \equiv [(1, 1/\sqrt{2}, 1/\sqrt{3}, \ldots)]$$

are all infinitesimals; they differ one from the other and satisfy the ordering $0 < \epsilon < \sqrt{\epsilon}$, as a result of the fact that the corresponding sequences which represent them converge to zero at different rates.[4]
Note that 0 is the unique infinitesimal which belongs to $\mathbb{R}$.

By contrast, elements of the type

$$\omega \equiv [(n)_{n \in \mathbf{N}}], \quad \pi^{\omega} \equiv [(\pi^{n})_{n \in \mathbf{N}}] \quad \text{and} \quad -\sqrt{\omega} \equiv [(-\sqrt{n})_{n \in \mathbf{N}}]$$

are infinite numbers, the first two positive and the third negative. Finally, a number like $2 + \epsilon$ is an example of an appreciable hyperreal number.

Exercise 3.18 *Show that a number $x \in {}^*\mathbb{R}$ is appreciable if and only if it satisfies the double inequality*

$$\frac{1}{n} < |x| < n$$

for some $n \in \mathbb{N}$.

We denote by ${}^*\mathbb{R}_b$ the set of all finite hyperreal numbers and by **mon**(0) the set of all infinitesimal numbers. The set of all infinite numbers, denoted by ${}^*\mathbb{R}_\infty$, can be further divided, when necessary, into the set of the positive infinite numbers and that of the negatives. Thus, we have

$${}^*\mathbb{R} = {}^*\mathbb{R}_b \cup {}^*\mathbb{R}_\infty = {}^*\mathbb{R}_b \cup {}^*\mathbb{R}_\infty^+ \cup {}^*\mathbb{R}_\infty^- \ .$$

Every real number is a finite hyperreal number, so that we have the inclusion $\mathbb{R} \subset {}^*\mathbb{R}_b$.
In the theorem which follows it is shown that the algebra of ${}^*\mathbb{R}$ does satisfy the elementary arithmetic rules which we would wish it to have.

Theorem 3.19 The set of the finite hyperreal numbers and that of the infinitesimals, with the usual operations, are both subrings of ${}^*\mathbb{R}$. That is to say, sums, differences and products of finite numbers are finite and sums, differences and products of infinitesimal numbers are infinitesimal. The infinitesimals form an ideal of ${}^*\mathbb{R}_b$: that is to say, the difference of two infinitesimals, and the product of an infinitesimal and a finite number, are both infinitesimal.

---

[4] The smaller a positive infinitesimal is, the more rapidly will any sequence which represents it tend to zero.

Proof Let $\alpha, \gamma \in {}^*\mathbb{R}_b$. Then there exists a real number $a > 0$ such that $|\alpha| < a/2$ and $|\gamma| < a/2$ and therefore

$$|\alpha \pm \gamma| \leq |\alpha| + |\gamma| < a, \qquad |\alpha\gamma| = |\alpha|.|\gamma| < b$$

where $b = a^2/4$ is a positive real number: these inequalities mean that $\alpha \pm \gamma$ and $\alpha\gamma$ really do belong to ${}^*\mathbb{R}_b$. Suppose now that $\epsilon, \delta \in mon(0)$; then, for every positive real number $r \in \mathbb{R}^+$, we have $|\epsilon| < r/2$ and $|\delta| < r/2$ and therefore

$$|\epsilon \pm \delta| \leq |\epsilon| + |\delta| < r \quad .$$

On the other hand, since $r \in \mathbb{R}^+$ implies that $\sqrt{r} \in \mathbb{R}^+$, we also have $|\epsilon| < \sqrt{r}$, and $|\delta| < \sqrt{r}$ and therefore

$$|\epsilon\delta| = |\epsilon|.|\delta| < r \quad .$$

It follows at once that $\epsilon \pm \delta, \epsilon\delta \in mon(0)$. Now let $\alpha \in {}^*\mathbb{R}_b$. Then there exists $a \in \mathbb{R}^+$ such that $|\alpha| < a$ and, for all $r \in \mathbb{R}^+$, $|\epsilon| < r/a$. Consequently,

$$|\alpha\epsilon| < r,$$

or $\alpha\epsilon \in mon(0)$, which completes the proof of the theorem. •

Exercise 3.20 *Show that the inverse of a non-null infinitesimal is an infinite number and that the inverse of an infinite number is infinitesimal. Show also that the inverse of an appreciable number is appreciable.*

Exercise 3.21 *Show that the ideal $mon(0)$ is maximal in ${}^*\mathbb{R}_b$ and consequently that the quotient ring*

$${}^*\mathbb{R}_b/mon(0)$$

is a field. This field, in which each element is an equivalence class which consists of finite hyperreals x, y such that $x - y \in mon(0)$, is isomorphic to $\mathbb{R}$.

The elements of ${}^*\mathbb{R}_b$ have a particularly simple structure which allows them to be written in the special form indicated in the following theorem.

Theorem 3.22 Every finite hyperreal number $x \in {}^*\mathbb{R}_b$ can be decomposed, in a unique way, into a sum of the type

$$x = a + \delta$$

where $a \in \mathbb{R}$ and δ is infinitesimal.

Proof: We prove, to begin with, that such a decomposition exists for any $x \in {}^*\mathbb{R}_b$ whatsoever. The subset of real numbers defined by

$$\{b \in \mathbb{R} : b < x\}$$

is not empty and is bounded above, and hence it possesses a supremum:

$$a = \sup\{b \in \mathbb{R} : b < x\} \quad .$$

It remains now to prove that $x - a$ is infinitesimal. Suppose, if possible, that $x - a$ is not infinitesimal. Then there exists $r \in \mathbb{R}$ such that $0 < r \leq |x - a|$ and, in consequence

if $x - a$ were positive then we would have $a + r \leq x$, which contradicts the definition of a;

if $x - a$ were negative then $x \leq a - r$ which, similarly, also contradicts the definition of a.

Thus $x - a \in mon(0)$ which completes the first part of the proof.

As for the uniqueness, suppose that the decomposition in question is not unique, so that we have $x = a + \delta = a' + \delta'$, with $a, a' \in \mathbb{R}$ and $\delta,\ \delta' \in mon(0)$. This would give the equality $a - a' = \delta - \delta'$. As $a - a'$ belongs to $\mathbb{R}$ and $\delta - \delta'$ is infinitesimal, so $a - a' \in mon(0)$, or $a = a'$ (since 0 is the only real infinitesimal). Hence we must also have $\delta = \delta'$. •

Definition 3.23 Two numbers $x, y \in {}^*\mathbb{R}$ are said to be **infinitely close** if and only if $x - y$ is infinitesimal: we then write $x \approx y$.

Definition 3.24 Let $x \in {}^*\mathbb{R}_b$ (that is, let x be a finite hyperreal number). The unique number $a \in \mathbb{R}$ such that $x \approx a$ is called the **standard part** (or **shadow**) of x and is denoted by $\mathbf{st}x$. Further, if $a \in \mathbb{R}$, the set of hyperreal numbers defined by

$$mon(a) = \{x \in {}^*\mathbb{R} : \mathbf{st}x = a\}$$

is called the **monad** of a.

(The term "monad"in the above definition is chosen in homage to Leibniz who introduced this term into philosophy.)

Using the definition of monad we can write

$${}^*\mathbb{R}_b = \bigcup_{a \in \mathbb{R}} mon(a)$$

an equality which suggests another possible interpretation of theorem 3.22.

The mapping $\mathbf{st}$: ${}^*\mathbb{R}_b \to \mathbb{R}$ which to each $x \in {}^*\mathbb{R}_b$ associates $\mathbf{st}(x) \in \mathbb{R}$ will exhibit the following behaviour with respect to the usual arithmetic operations and ordering relation:

Theorem 3.25 Let $x, y \in {}^*\mathbb{R}$. Then,

(a) $\mathbf{st}(x + y) = \mathbf{st}(x) + \mathbf{st}(y)$ and $\mathbf{st}(x - y) = \mathbf{st}(x) - \mathbf{st}(y)$

(b) $\mathbf{st}(x.y) = \mathbf{st}(x).\mathbf{st}(y)$ and $y \notin mon(0) \Rightarrow \mathbf{st}(x/y) = \mathbf{st}(x)/\mathbf{st}(y)$

(c) $x \leq y \Rightarrow \mathbf{st}(x) \leq \mathbf{st}(y)$ and $[x < y \wedge y - x \notin mon(0)] \Rightarrow \mathbf{st}(x) < \mathbf{st}(y)$.

Proof If $x, y \in {}^*\mathbb{R}_b$ then there exist $a, b \in \mathbb{R}$ and $\epsilon, \delta \approx 0$ such that $x = a + \epsilon,\quad y = b + \delta$. Consequently

$$x + y = (a + b) + (\epsilon + \delta)$$

and therefore

$$\mathbf{st}(x+y) = \mathbf{st}(x) + \mathbf{st}(y)$$

which proves (a). •

Exercise 3.26 *Complete the proof of theorem 3.25.*

The theorem which follows shows that there is a natural relation between the limiting behaviour of a convergent sequence of real numbers and the hyperreal number which it represents.

Theorem 3.27 Let $(a_n)_{n\in\mathbb{N}}$ be a sequence of real numbers which converges to the number $\alpha \in \mathbb{R}$. Then $[(a_n)_{n\in\mathbb{N}}] \approx \alpha$.

Proof From the usual definition of limit it is known that the set

$$\{n \in \mathbb{N} : \alpha - r < a_n < \alpha + r\}$$

is cofinite in $\mathbb{N}$ whatever the number $r \in \mathbb{R}^+$, and therefore that it belongs to $\mathcal{U}$. Then, $[(a_n)_{n\in\mathbb{N}}] - \alpha < r$ for any real number $r > 0$ and so $[(a_n)_{n\in\mathbb{N}}] \approx \alpha$. •

The converse of this theorem is not true. From the relation $[(a_n)_{n\in\mathbb{N}}] \approx \alpha$ the most that we can conclude is that $(a_n)_{n\in\mathbb{N}}$ possesses a subsequence $(a_{r_n})_{n\in\mathbb{N}}$ converging to α such that $\{r_n : n \in \mathbb{N}\} \in \mathcal{U}$.

Definition 3.28 Two numbers $x, y \in {}^*\mathbb{R}$ are said to be **finitely close** if and only if $x - y$ belongs to ${}^*\mathbb{R}_b$. The set of all hyperreal numbers finitely close to a given number $x \in {}^*\mathbb{R}$ is called the **galaxy** of x and denoted by $\mathbf{Gal}(x)$.

It can be verified that $\mathbf{Gal}(x) = \mathbf{Gal}(y)$ if and only if $x - y \in {}^*\mathbb{R}_b$. Thus, for any $x \in {}^*\mathbb{R}_b$, we must have $\mathbf{Gal}(x) = \mathbf{Gal}(0)$. This result can also be written as ${}^*\mathbb{R}_b = \mathbf{Gal}(0)$.

3.2 The Elementary Leibniz Principle

3.2.1 Canonical extensions of sets and functions

Given a system of hyperreal numbers constructed on the basis of some specific free ultrafilter, we can now consider, in the usual way, sets of hyperreal numbers and functions of a hyperreal variable. In order to explore the relations existing between NSA and elementary Real Analysis it is interesting to consider, in particular, those sets and functions which are, in some sense, the *"extensions"* to ${}^*\mathbb{R}$ of ordinary sets of real numbers and of real functions of a real variable. We will now treat these in some detail.

Nonstandard extension of a set Let A be a (standard) subset of $\mathbb{R}$.

Definition 3.29 The nonstandard extension of A is that subset ${}^*A$ of ${}^*\mathbb{R}$ which consists of all hyperreal numbers $x \equiv [(x_n)_{n\in\mathbb{N}}]$ such that $x_n \in A$ for $\mathcal{U}$-nearly all $n \in \mathbb{N}$ (that is, for a set of indices $n \in \mathbb{N}$ which belongs to the ultrafilter $\mathcal{U}$). Thus, we have

$$\forall_x[x \in {}^*\mathbb{R} : \quad [x \equiv [(x_n)_{n\in\mathbb{N}}] \in {}^*A \Leftrightarrow \{n \in \mathbb{N} : x_n \in A\} \in \mathcal{U}]].$$

The sets A and $^*A$ have exactly the same real elements. In fact, for any $a \in A$, the element $[(a, a, a, \dots)]$, which is identified with a, belongs to $^*A$ and, therefore,

$$A \subset {}^*A \ .$$

On the other hand it is easy to verify from the definition of nonstandard extension of a set that there cannot exist in $^*A$ any standard element which does not belong to A. However, in the general case, $^*A$ possesses many more elements than A. In fact the following result holds:

Theorem 3.30 For any set A whatever of real numbers we have $A \subset {}^*A$, with equality holding if and only if the set A is finite.

Proof The inclusion of A in $^*A$ has already been demonstrated above. It remains to prove the second part of the theorem. Suppose that A is a finite set, that is, that we have $A = \{a_1, a_2, \dots, a_p\}$ for some natural number p. Let $[(x_n)_{n\in\mathbb{N}}]$ be arbitrary. Taking account of the fact that

$$\{n \in \mathbb{N} : x_n \in A\} = \bigcup_{j=1}^{p} \{n \in \mathbb{N} : x_n = a_j\}$$

and noting that *one and only one* of the sets on the right-hand side belongs to $\mathcal{U}$ (see Exercise 3.10), it follows that there exists a number $r = 1, 2, \dots, p$ such that

$$\{n \in \mathbb{N} : \quad x_n = a_r\} \in \mathcal{U} \ .$$

Hence, $[(x_n)_{n\in\mathbb{N}}] = {}^*a_r \equiv a \in A$.

Suppose now that A is infinite and define the element $x \equiv [(x_n)_{n\in\mathbb{N}}]$ in such a way that $x_n \in A$ for all $n \in \mathbb{N}$ and $x_n \neq x_m$ whenever $n \neq m$. Then $x \in {}^*A$ but since, for any $a \in A$, the set $\{n \in \mathbb{N} : x_n = a\}$ is finite (empty or a singleton), we must have $x \neq a$. •

Example 3.31 The Hypernatural Numbers The set $\mathbb{N}$ of the natural numbers is a subset of $\mathbb{R}$ which has a nonstandard extension whose elements are called **hypernatural numbers.** $^*\mathbb{N}$ is the subset of all those hyperreal numbers of the form $\nu \equiv [(m_n)_{n\in\mathbb{N}}]$ where m_n belongs to $\mathbb{N}$ for $\mathcal{U}$-nearly all $n \in \mathbb{N}$.

$\mathbb{N}$ is a subset of $^*\mathbb{N}$ since to every $m \in \mathbb{N}$ we can make correspond the number $[(m)_{n\in\mathbb{N}}]$ which belongs to $^*\mathbb{N}$. On the other hand $\mathbb{N}$ coincides with the set of all the *finite* elements of $^*\mathbb{N}$. In fact, if $\nu = [(m_n)_{n\in\mathbb{N}}]$ is a finite element of $^*\mathbb{N}$ then there exists some $p \in \mathbb{N}$ such that $m_n \leq p$ for $\mathcal{U}$-nearly all $n \in \mathbb{N}$. For $k = 1, 2, \dots, p$ let $A_k = \{n \in \mathbb{N} : m_n = k\}$. Since $A_k \cap A_j = \emptyset$ if $k \neq j$ and $\bigcup_{k=1}^{p} A_k = \{n \in \mathbb{N} : m_n \leq p\}$ belongs to $\mathcal{U}$, it follows that one and only one of the sets A_k belongs to $\mathcal{U}$. That is, there exists one and only one natural number $r \leq p$ such that $A_r \in \mathcal{U}$. Consequently, $\nu \equiv [(m_n)_{n\in\mathbb{N}}]$ is a *transformation of the number $r \in \mathcal{N}$.

The set $^*\mathbb{N}$ possesses many of the properties of $\mathbb{N}$, such as, for example, those listed below:

1. $^*\mathbb{N}$ is a discrete set;

2. $^*\mathbb{N}$ is closed under addition and multiplication;

3. if $\nu \in {}^*\mathbb{N}$ then ν has an immediate successor, $\nu + 1$, which also belongs to ${}^*\mathbb{N}$;

4. if $\nu \in {}^*\mathbb{N}$ is such that $\nu > 1$, then it possesses an immediate predecessor, $\nu - 1$, which also belongs to ${}^*\mathbb{N}$;

The set ${}^*\mathbb{N}\backslash\mathbb{N}$ consists of all the infinite natural numbers, and is generally denoted by ${}^*\mathbb{N}_\infty$.

Example 3.32 As another example of a nonstandard extension, consider an interval $[a, b]$ of $\mathbb{R}$. Its nonstandard extension ${}^*[a, b]$ is an interval of ${}^*\mathbb{R}$ which contains all the hyperreal numbers $x \equiv [(x_n)_{n\in\mathbb{N}}]$ such that $a \leq x \leq b$ for $\mathcal{U}$-nearly all $n \in \mathbb{N}$. For instance the hyperreal interval ${}^*[0, 1]$ contains all the hyperreal numbers such that $0 \leq x \leq 1$ including, in particular, all the non-negative infinitesimals. We define it formally as

$$ {}^*[0, 1] = \{x \in {}^*\mathbb{R} : 0 \leq x \leq 1\} \ . $$

Finally note that ${}^*\mathbb{R}$ is, of course, itself the nonstandard extension of the standard set $\mathbb{R}$.

The nonstandard extensions of sets of real numbers possess some important properties, some of which are left as exercises in the following:

Exercise 3.33 *Let A and B be any two subsets of $\mathbb{R}$. Show that*

1. $A \subset B \Leftrightarrow {}^*A \subset {}^*B$ and $A = B \Leftrightarrow {}^*A = {}^*B$
2. ${}^*(A \cup B) = {}^*A \cup {}^*B$ and ${}^*(A \cap B) = {}^*A \cap {}^*B$
3. ${}^*A \cap \mathbb{R} = A$ and ${}^*\emptyset = \emptyset$.

Study the relation between the sets

$$ {}^*\left(\bigcup_{n=1}^{\infty} A_n\right) \quad \text{and} \quad \bigcup_{n=1}^{\infty} {}^*A_n $$

when $(A_n)_{n\in\mathbb{N}}$ denotes an arbitrary sequence of subsets of $\mathbb{R}$. Examine similarly the case of the intersection of a countable family of sets.

Nonstandard extension of a function

We can now define the nonstandard extension of an arbitrary standard function.

Definition 3.34 Let f be a function defined on a subset A of $\mathbb{R}$. The nonstandard extension of f is a function ${}^*f$ defined on ${}^*A \subset {}^*\mathbb{R}$ by

$$ {}^*f(x) = [(f(x_n))_{n\in\mathbb{N}}] $$

for each $x \equiv [(x_n)_{n\in\mathbb{N}}]$. Note that this extension is well defined: that is, it does not depend on the particular representation considered for each $x \in {}^*A$. In fact suppose that we had

$$ x \equiv [(x_n)_{n\in\mathbb{N}}] = [(x'_n)_{n\in\mathbb{N}}] $$

then the inclusion

$$\{n \in \mathbb{N} : x_n = x'_n\} \subset \{n \in \mathbb{N} : f(x_n) = f(x'_n)\}$$

would be satisfied and, therefore, (since, by hypothesis, $\{n \in \mathbb{N} : x_n = x'_n\} \in \mathcal{U}$)

$$[(f(x_n))_{n\in\mathbf{N}}] = [(f(x'_n))_{n\in\mathbf{N}}] \quad .$$

On the other hand $^*f$ really is an extension of f, since for $r \in \mathbb{R}$ we have

$$^*f(r) = [(f(r))_{n\in\mathbf{N}}] = {}^*(f(r)) = f(r) \quad .$$

For example, the nonstandard extension of the real function defined by

$$f(x) = \frac{1}{x}, \quad x \in \mathbb{R}^+$$

is a hyperreal function defined on $^*\mathbb{R}^+$ by

$$^*f(x) = \frac{1}{x}, \quad x \in {}^*\mathbb{R}^+$$

which takes infinite hyperreal values when x is a positive infinitesimal number and is infinitesimal when x is a positive infinite number.

Another example is the function $f(x) = \sin x$ defined on the entire real line, whose nonstandard extension is a function $^*f(x) = {}^*\sin x$ which is defined on the whole hyperreal line. For $x = [(x_n)_{n\in\mathbf{N}}]$ we have

$$^*\sin x = [(\sin x_n)_{n\in\mathbf{N}}] \in {}^*[-1,1] \quad .$$

As is well known, the following statement is valid in $\mathbb{R}$,

$$\forall_{a,b}[a, b \in \mathbb{R} \Rightarrow \sin(a+b) = \sin a . \cos b + \sin b . \cos a] \quad . \tag{3.4}$$

Then if $[(x_n)_{n\in\mathbf{N}}]$ and $[(y_n)_{n\in\mathbf{N}}]$ are any two hyperreal numbers, we obtain successively

$$\begin{aligned} ^*\sin(x+y) \equiv [(\sin(x_n+y_n)_{n\in\mathbf{N}}] &= [(\sin x_n . \cos y_n + \sin y_n \cos x_n)_{n\in\mathbf{N}}] \\ &= [(\sin x_n)_{n\in\mathbf{N}}].[(\cos y_n)_{n\in\mathbf{N}}] + [(\sin y_n)_{n\in\mathbf{N}}].[(\cos x_n)_{n\in\mathbf{N}}] \\ &= {}^*\sin x . {}^*\cos y + {}^*\sin y . {}^*\cos x \end{aligned}$$

which we can therefore write as

$$\forall_{x,y}[x, y \in {}^*\mathbb{R} \quad \Rightarrow {}^*\sin(x+y) = {}^*\sin x . {}^*\cos y + {}^*\sin y . {}^*\cos x] \quad . \tag{3.5}$$

Note that (3.5) could have been obtained from (3.4), simply by replacing $\mathbb{R}$ by $^*\mathbb{R}$ and the functions sin and cos by their extensions $^*\sin$ and $*\cos$ respectively.

For simplicity of notation it is customary to use for such very familiar functions as $|.|, \sin, \cos, \arctan, \log$ etc. the same designation (without stars) whether the domain is standard or nonstandard. Thus, for example, in the case of $f(x) = \sin(x), \quad x \in \mathbb{R}$, we would normally still write $^*f(x) = \sin(x)$ when $x \in {}^*\mathbb{R}$. This simplification should not cause any problems, since the context will indicate in each case which is the universe we work in.

Exercise 3.35 *Let A, B be two subsets of $\mathbb{R}$ and let $f : A \to B$ be an injective mapping. Show that $^*f : {}^*A \to {}^*B$ is also injective. What can be asserted in the case of a surjective mapping?*

Exercise 3.36 $\mathbf{1}_A$, *the characteristic function of the set $A \subset \mathbb{R}$, is defined by*

$$\mathbf{1}_A(x) = \begin{cases} 1 & \text{for } x \in A \\ 0 & \text{for } x \notin A \end{cases}$$

Show that $^*(\mathbf{1}_A) = \mathbf{1}_{^*A}$.

Exercise 3.37 *Let $f : \mathbb{R} \to \mathbb{R}$ be a given function. Show that $^*f(^*A) = {}^*(f(A))$ and $(^*f)^{-1}(^*B) = {}^*(f^{-1}(B))$ for any subsets A, B of $\mathbb{R}$.*

Nonstandard extension of a relation Let $\mathcal{R}$ be a relation defined on a subset A of $\mathbb{R}$. We can extend this relation to the set $^*A$, obtaining a relation $^*\mathcal{R}$ of the following form: if $[(x_n^1)], [(x_n^2)], \ldots, [(x_n^k)]$ are k elements of $^*A$ then

$$^*\mathcal{R}\left([(x_n^1)], [(x_n^2)], \ldots, [(x_n^k)]\right) \Leftrightarrow \{n \in \mathbb{N} : \mathcal{R}(x_n^1, \ldots, x_n^k)\} \in \mathcal{U} \quad . \tag{3.6}$$

If, in particular, $r_1, r_2, \ldots, r_k$ are k real numbers, then we have

$$^*\mathcal{R}([(r_1)], [(r_2)], \ldots, [(r_k)]) \Leftrightarrow \{n \in \mathbb{N} : \mathcal{R}(r_1, r_2, \ldots, r_k)\} \in \mathcal{U}$$

$$\Leftrightarrow \mathrm{R}(r_1, r_2, \ldots, r_k)$$

which shows that $^*\mathcal{R}$ does actually form an extension of $\mathcal{R}$.

For $k = 1$ we obtain the unitary relation corresponding to the definition of the set A, that is to say, $\mathcal{R}(x) \Leftrightarrow x \in A$. In this case $^*\mathcal{R}$ coincides with the nonstandard extension of the set A. The nonstandard extensions of binary relations generalise the cases already treated of the relations of equality or of the order defined in $\mathbb{R}$.

Exercise 3.38 *Let A, B be two subsets of real numbers. Using the definition of the nonstandard extension of a binary relation, show that*

$$^*(A \times B) = {}^*A \times {}^*B \quad .$$

Generalise this to the case of n subsets of $\mathbb{R}$. In particular, show that we have

$$^*(\mathbb{R}^n) = (^*\mathbb{R})^n$$

so that we can write $^*\mathbb{R}^n$ without any danger of ambiguity.

3.2.2 The Leibniz Principle relative to $\mathbb{R}$

As has been already remarked at the beginning of this chapter, one of the objectives of NSA is to achieve an extension of mathematical analysis in a form which specifically includes infinitely large numbers and infinitely small ones. Such an extension, however, cannot be carried out in an arbitrary manner: we claim that there exists an ordered field ${}^*\mathbb{R}$ which possesses, *in a certain sense*, the "same properties" as $\mathbb{R}$. But, as clearly indicated in the preceding sections, $\mathbb{R}$ and ${}^*\mathbb{R}$ are distinct sets, and, by that very fact, they cannot possess exactly the "same properties". The solution found by Robinson to this apparent difficulty is based on the significance of the expression *in a certain sense* emphasised above. Thus when we say that ${}^*\mathbb{R}$ must possess the same properties as $\mathbb{R}$, we cannot literally be referring to *all* the possible properties, but only to a certain subclass of those properties. In the next chapter, this subclass will be given a rigorous characterisation. However, we have already made some study of a class of properties which certainly *do* transfer from $\mathbb{R}$ to ${}^*\mathbb{R}$ (and *vice-versa*). We were able to do this by restricting that study to those propositions which involve only numbers, subsets of numbers, relations between numbers and numerical functions of numerical variables.

In the previous section we showed how to transfer from $\mathbb{R}$ to ${}^*\mathbb{R}$ the sine summation formula. In a similar way we have also already considered properties relating to the ordered field structure of $\mathbb{R}$ and ${}^*\mathbb{R}$. The statement

$$(\mathbb{R}, +, ., <) \text{ IS AN ORDERED FIELD}$$

can be formalised in the following way:

1- $\forall_{a,b}[a, b \in \mathbb{R} \Rightarrow \exists_c[c \in \mathbb{R} \wedge c = a + b]]$
2- $\forall_{a,b,c}[a, b, c \in \mathbb{R} \Rightarrow (a + b) + c = a + (b + c)]$
3- $\exists_0[0 \in \mathbb{R} \wedge [\forall_a[a \in \mathbb{R} \Rightarrow a + 0 = a]]]$
4- $\forall_a[a \in \mathbb{R} \Rightarrow [\exists_{-a}[(-a) \in \mathbb{R} \wedge a + (-a) = 0]]]$
5- $\forall_{a,b}[a, b \in \mathbb{R} \Rightarrow [a + b = b + a]]$

6- $\forall_{a,b}[a, b \in \mathbb{R} \Rightarrow [\exists_c[c \in \mathbb{R} \wedge c = ab]]]$
7- $\forall_{a,b,c}[a, b, c \in \mathbb{R} \Rightarrow (ab)c = a(bc)]$
8- $\exists_1[1 \in \mathbb{R} \wedge [\forall_a[a \in \mathbb{R} \Rightarrow a1 = a]]$
9- $\forall_{a,b}[a, b \in \mathbb{R} \Rightarrow ab = ba]$
10- $\forall_a[[a \in \mathbb{R} \wedge a \neq 0] \Rightarrow [\exists_{a^{-1}}[a^{-1} \in \mathbb{R} \wedge aa^{-1} = 1]]]$

11- $\forall_{a,b,c}[a, b, c \in \mathbb{R} \Rightarrow a(b + c) = ab + ac]$

12- $\forall_{a,b,c}[a, b, c \in \mathbb{R} \Rightarrow [[a < b \wedge b < c] \Rightarrow a < c]]]$
13- $\forall_{a,b}[a, b \in \mathbb{R} \Rightarrow [a \neq b \Rightarrow [a < b \vee b < a]]]$
14- $\forall_{a,b,c}[a, b, c \in \mathbb{R} \Rightarrow [a < b \Rightarrow a + c < b + c]]$
15- $\forall_{a,b}[a, b \in \mathbb{R}] \Rightarrow [[0 < a \wedge 0 < b \Rightarrow 0 < ab]]]$

If in all these propositions we replace the constant $\mathbb{R}$ (a constant of the formal language appropriate for making statements in standard mathematical analysis) by the constant ${}^*\mathbb{R}$ we obtain the following properties:

1^*- $\forall_{a,b}[a, b \in {}^*\mathbb{R} \Rightarrow \exists_c[c \in \mathbb{R} \wedge c = a + b]]$
2^*- $\forall_{a,b,c}[a, b, c \in {}^*\mathbb{R} \Rightarrow (a + b) + c = a + (b + c)]$
3^*- $\exists_0[0 \in {}^*\mathbb{R} \wedge [\forall_a[a \in {}^*\mathbb{R} \Rightarrow a + 0 = a]]]$
4^*- $\forall_a[a \in {}^*\mathbb{R} \Rightarrow [\exists_{-a}[(-a) \in {}^*\mathbb{R} \wedge a + (-a) = 0]]]$

5*- $\forall_{a,b}[a, b \in {}^*\mathbb{R} \Rightarrow [a + b = b + a]]$

6*- $\forall_{a,b}[a, b \in {}^*\mathbb{R} \Rightarrow [\exists_c[c \in {}^*\mathbb{R} \wedge c = ab]]]$
7*- $\forall_{a,b,c}[a, b, c \in {}^*\mathbb{R} \Rightarrow (ab)c = a(bc)]$
8*- $\exists_1[1 \in {}^*\mathbb{R} \wedge [\forall_a[a \in {}^*\mathbb{R} \Rightarrow a1 = a]]$
9*- $\forall_{a,b}[a, b \in {}^*\mathbb{R} \Rightarrow ab = ba]$
10*- $\forall_a[[a \in {}^*\mathbb{R} \wedge a \neq 0] \Rightarrow [\exists_{a^{-1}}[a^{-1} \in {}^*\mathbb{R} \wedge aa^{-1} = 1]]]$

11*- $\forall_{a,b,c}[a, b, c \in {}^*\mathbb{R} \Rightarrow a(b + c) = ab + ac]$

12*- $\forall_{a,b,c}[a, b, c \in {}^*\mathbb{R} \Rightarrow [[a < b \wedge b < c] \Rightarrow a < c]]]$
13*- $\forall_{a,b}[a, b \in {}^*\mathbb{R} \Rightarrow [a \neq b \Rightarrow [a < b \vee b < a]]]$
14*- $\forall_{a,b,c}[a, b, c \in {}^*\mathbb{R} \Rightarrow [a < b \Rightarrow a + c < b + c]]$
15*- $\forall_{a,b}[a, b \in {}^*\mathbb{R} \Rightarrow [[0 < a \wedge 0 < b] \Rightarrow 0 < ab]]]$

which are already known to be true since $({}^*\mathbb{R}, +, ., \leq)$ is also an ordered field (which extends $\mathbb{R}$).

In studying the Leibniz Principle restricted (or relativised) to $\mathbb{R}$, all the formulas considered must satisfy the three following conditions:

1. the operations
$$x + y, \quad xy, \quad \ldots$$
and the relations
$$x = y, \quad x \leq y, \quad \ldots$$
make no reference whatever to the notions of infinitesimal, infinity or finiteness;

2. all the variables occur as constituents of a formula of the type $x \in \mathbb{R}, y \in \mathbb{R}, \ldots$;

3. all variables and all constants are real.

This is just what obtains in the case of the Propositions 1-15 above. We pass from the first to the second set of formulas, by simply replacing the constant $\mathbb{R}$ with the constant ${}^*\mathbb{R}$, thereby obtaining a true formula of ${}^*\mathbb{R}$ from a given true formula of $\mathbb{R}$. (It is clear that we also have other constants going over from $\mathbb{R}$ to ${}^*\mathbb{R}$ such as $+, ., <$, etc. However, for these we use the same notation in each context, for the sake of simplicity.)

Other examples which can be given of this type of *transfer* are the formulas (3.4) and (3.5) shown above: we pass from the first formula to the second, simply replacing the constants "$\mathbb{R}$", "sin", and "cos" by the constants "${}^*\mathbb{R}$", "*sin", and "*cos", respectively. The first formula is valid in $\mathbb{R}$, while the second will be satisfied in ${}^*\mathbb{R}$. Again, take the case of the property which guarantees that in $\mathbb{R}$ there cannot exist any smallest positive real number

$$\neg \exists_x[x \in \mathbb{R}^+ \wedge \forall_y[y \in \mathbb{R}^+ \Rightarrow x \leq y]].$$

Replacing $\mathbb{R}^+$ by ${}^*\mathbb{R}^+$ in this expression , we obtain

$$\neg \exists_x[x \in {}^*\mathbb{R}^+ \wedge \forall_y[y \in {}^*\mathbb{R}^+ \Rightarrow x \leq y]]$$

which means that in ${}^*\mathbb{R}$ there also cannot exist a smallest positive hyperreal number, and this again forms a valid proposition.

But $\mathbb{R}$ and ${}^*\mathbb{R}$ also have profound differences (which, as a matter of fact, one would expect to be the case): in $\mathbb{R}$ we have neither infinitesimals nor infinite numbers, while ${}^*\mathbb{R}$ does contain examples of both such elements. Now the existence of (non-null) infinitesimals, might, for example, be formulated in the following way

$$\exists_x[x \in {}^*\mathbb{R}\backslash\{0\} \wedge \forall_y[y \in \mathbb{R}^+ \Rightarrow |x| < y]].$$

This is a *mixed formula* involving both (the constant) $\mathbb{R}$ and (the constant) ${}^*\mathbb{R}$. We must treat this formula entirely differently from the previous ones: it is a formula of ${}^*\mathbb{R}$, which has *not* been obtained by starting from a formula of $\mathbb{R}$ and simply replacing this set by ${}^*\mathbb{R}$.

Another example of similar interest is what might be derived from the known *completeness* of $\mathbb{R}$: would it be possible to obtain by *transfer* a proof of the completeness of ${}^*\mathbb{R}$? Clearly this ought not to be the case! In fact, the formulation of the completeness of $\mathbb{R}$ has the following logical structure

$$\forall_A[A \subset \mathbb{R}, A \neq \emptyset \text{ and } A \text{ bounded above } \Rightarrow (\ldots)].$$

If in this formula we simply replaced $\mathbb{R}$ by ${}^*\mathbb{R}$ then we would obtain the following proposition:

"all nonempty subsets of ${}^*\mathbb{R}$ which are bounded above ... (have a supremum)".

Well, that is easily seen to be false: $\mathbb{R}$ is itself a nonempty subset of ${}^*\mathbb{R}$ which is bounded above (by any positive infinite number) and yet which has no supremum. The formula in question is not susceptible to *transfer*: the quantifier variable *ranges not over* $\mathbb{R}$ but rather over *the set of all subsets of* $\mathbb{R}$. The Leibniz Principle does not include such types of formula, or rather, the formula in question does not belong to the class of formulas susceptible to *transfer*.

In order to distinguish, at an elementary level, those formulas which can be the objects of *transfer* from those which cannot, we introduce at this stage the notion of a **proposition relative to** $\mathbb{R}$.

Definition 3.39 A proposition p (that is, a formula with free variables) is said to be relative to $\mathbb{R}$ if and only if

1. it can be formulated using only

   variables or constants which represent elements of $\mathbb{R}$,

   subsets of $\mathbb{R}$, or real functions of a real variable,

   the four algebraic operations

   the relations of equality and order

   the logical symbols $\neg, \wedge, \vee, \Rightarrow, \Leftrightarrow, \forall, \exists$,

and

2. all bound variables appear in one or other of the forms

$$\forall_x[x \in A \Rightarrow \ldots] \quad \text{or} \quad \exists_x[x \in A \wedge \ldots]$$

where A is a subset of $\mathbb{R}$ (possibly all of $\mathbb{R}$).[5]

Some examples of propositions relative to $\mathbb{R}$ are the following

$$\forall_x[x \in \mathbb{R} \Rightarrow [x \in [a,b] \Leftrightarrow (a \leq x \wedge x \leq b)]]$$

$$\forall_x[x \in \mathbb{R} \Rightarrow [\sin(a+b) = \sin a.\cos b + \sin b.\cos a]]$$

$$\forall_n[n \in \mathbb{N} \Rightarrow \exists_k[k \in \mathbb{N} \wedge [n = 2k \vee n = 2k-1]]] \quad .$$

Definition 3.40 Given a proposition p relative to $\mathbb{R}$ we define the *proposition $^*p$ as being the proposition in $^*\mathbb{R}$ which is obtained from p by replacing

1. each subset A of $\mathbb{R}$ by $^*A$ and, in particular, $\mathbb{R}$ by $^*\mathbb{R}$,
2. each (real) function (of a real variable) f by $^*f$.

The *-transforms of the propositions given above will be

$$\forall_x[x \in {}^*\mathbb{R} \Rightarrow [x \in {}^*[a,b] \Leftrightarrow (a \leq x \wedge x \leq b)]]$$

$$\forall_x[x \in {}^*\mathbb{R} \Rightarrow [\sin(a+b) = \sin a.\cos b + \sin b.\cos a]]$$

$$\forall_n[n \in {}^*\mathbb{N} \Rightarrow \exists_k[k \in {}^*\mathbb{N} \wedge [n = 2k \vee n = 2k-1]]],$$

(where, as already remarked, we omit the use of the stars in the case of real numerical constants and of the familiar functions $\sin, \cos, \ldots$). So as to indicate that a proposition p, relative to $\mathbb{R}$, expresses a true proposition of real analysis, we write $\mathbb{R} \models \mathrm{p}$. For *-transform propositions, relative to $^*\mathbb{R}$, we use a similar notation, namely writing $^*\mathbb{R} \models {}^*\mathrm{p}$ if and only if $^*\mathrm{p}$ is a true proposition of nonstandard analysis. We then have the following restricted form of the **Leibniz Principle**:[6]

Theorem 3.41 A proposition p relative to $\mathbb{R}$, is true in $\mathbb{R}$ if and only if the transformed proposition $^*p$ is true in $^*\mathbb{R}$; that is, $[\mathbb{R} \models \mathrm{p}] \Leftrightarrow [{}^*\mathbb{R} \models {}^*\mathrm{p}]$.

This theorem, we dedicate to Leibniz since it was he who first suggested the creation of a number system containing infinite numbers and infinitesimals which nevertheless

[5] Formulas of this type are very often simply written as

$$\forall_{x \in A}[\cdots] \text{ or } \exists_{x \in A}[\cdots]$$

[6] Note that such propositions as, for example,

$$\exists_x[x \in {}^*\mathbb{R} \wedge \forall_y[y \in \mathbb{R}^+ \Rightarrow |x| < y]]$$

$$\exists_x[x \in {}^*\mathbb{R} \wedge \forall_y[y \in \mathbb{R}^+ \Rightarrow |x| > y]]$$

(defined for infinitesimals and for infinite numbers) involve $\mathbb{R}$ as well as $^*\mathbb{R}$. They cannot, therefore, be used in the context of the Leibniz Principle.

continues to enjoy the same properties as the original system. It is also referred to in the literature as the **Transfer Principle**. Its importance resides in the fact that it makes it possible to infer results about certain nonstandard objects from known facts about standard objects (which generates the origin of the monomorphism *). Reciprocally, it allows us to infer results about standard objects from known facts about their nonstandard extensions. Thus the Leibniz Principle allows us to develop a very powerful technique which consists in "*transferring*" to the Universe of (standard) Analysis results obtained, almost always in a much simpler and intuitive form, from an appropriate nonstandard context. In this way, applying the transfer principle, it is possible progressively to develop the techniques of NSA without necessarily having to take account of any particular model. This aspect will be widely exemplified in the following chapter, where, from time to time, there will be offered two types of proof; one which makes direct appeal to the transfer principle, and the other which proceeds in terms of whatever (previously fixed) model of NSA, happens to be used.

The Leibniz Principle itself will be proved later, in chapter 6, in a more general context, involving objects of mathematical analysis other than numbers, sets or functions.

Exercise 3.42 *Establish to which of the propositions which follow it is possible to apply the Leibniz Principle and, in the affirmative cases, state the results obtained by this application.*

1. $\forall_x[x \in \mathbb{R} \Rightarrow \exists_n[n \in \mathbb{N} \wedge x < n]]$,

2. $\exists_x[x > 0, x\text{infinite} \wedge \forall_y[y \in \mathbb{R} \Rightarrow y < x]]$,

3. $\forall_x[x \in \mathbb{R} \Rightarrow x < 0 \vee x \geq 0]$,

4. $\forall_x[x \in \mathbb{R} \quad \Rightarrow \exists_y[y \in {}^*\mathbb{R} \wedge y \approx x \wedge y \neq x]]$.

Exercise 3.43 *Apply the transfer principle to show that*

1. $\sin\epsilon \approx 0$ *and* $\cos\epsilon \approx 1$ *for any infinitesimal* ϵ,
2. $A \subset \mathbb{R}$ *is a finite set if and only if* ${}^*A = A$,
3. *If* f *is a function of a real variable such that for some* $M > 0$ *we have* $|f(x)| \leq M$ *for all* $x \in \text{dom} f$ *then* $|{}^*f(x)| \leq M$ *for all* $x \in \text{dom}\,{}^*f$.

Chapter 4

Applications to Elementary Real Analysis

In this chapter, we formulate some basic notions of analysis and elementary calculus in nonstandard terms. As will be seen, these alternative definitions are generally easier to apply and have greater intuitive appeal than the corresponding standard ones. Examples of the application of nonstandard methods will be given throughout. The choice of subject is more or less arbitrary, but is intended to illustrate the use of nonstandard techniques in real analysis at an elementary level.

4.1 Convergence

4.1.1 Sequences of real numbers

Given an arbitrary sequence $(a_n)_{n\in\mathbb{N}}$ of real numbers we can define a mapping $a : \mathbb{N} \to \mathbb{R}$ by setting $a(n) = a_n$. This makes it clear that a sequence of real numbers is really nothing more than a real-valued function defined on the set $\mathbb{N} \subset \mathbb{R}$, and as such it admits a nonstandard extension. The result, which we shall call a **standard hypersequence**, will be a mapping ${}^*a : {}^*\mathbb{N} \to {}^*\mathbb{R}$ defined for each hypernatural number $\nu \equiv [(m_n)_{n\in\mathbb{N}}] \in {}^*\mathbb{N}$, by

$$ {}^*a(\nu) = a_\nu \equiv [(a_{m_n})_{n\in\mathbf{N}}] \in {}^*\mathbb{R} \quad . $$

If ν is finite then there exists a natural number $j \in \mathbb{N} \subset {}^*\mathbb{N}$ such that $m_n = j$ for $\mathcal{U}$-nearly all $n \in \mathbb{N}$ (that is, for every n belonging to a subset of $\mathbb{N}$ which belongs to $\mathcal{U}$) and therefore

$$ [(a_{m_n})_{n\in\mathbf{N}}] = [(a_j)_{n\in\mathbf{N}}] = {}^*(a_j) \equiv a_j \quad . $$

Thus, for finite indices, the standard hypersequence $(a_\nu)_{\nu\in{}^*\mathbf{N}}$ coincides with the original sequence, as we would expect.

Definition 4.1 A real sequence $(a_n)_{n\in\mathbf{N}}$ is said to be **S-convergent** to a real number $\alpha \in \mathbb{R}$ if and only if the relation $a_\nu \approx \alpha$ is satisfied for every hypernatural number ν that is, if and only if

$$ \forall_\nu[\nu \in {}^*\mathbb{N}_\infty \Rightarrow a_\nu \approx \alpha] \quad . $$

The sequence $(a_n)_{n\in\mathbb{N}}$ is said to be **S-divergent** if it is not S-convergent.

Theorem 4.2 A real sequence $(a_n)_{n\in\mathbb{N}}$ converges to a real number $\alpha \in \mathbb{R}$ if and only if it is S-convergent to the same number.

Proof We prove the theorem by appealing directly to the Leibniz Principle. Suppose that

$$\lim_{n\to\infty} a_n = \alpha$$

and let $\nu \in {}^*\mathbb{N}_\infty$ be arbitrarily fixed. From the standard definition of the limit of a sequence we know that, given any real number $r > 0$, there exists $n_r \in \mathbb{N}$ such that

$$\forall_n[n \in \mathbb{N} \Rightarrow n \geq n_r \Rightarrow |a_n - \alpha| < r]]$$

from which, by applying the transfer principle, we get

$$\forall_n[n \in {}^*\mathbb{N} \Rightarrow n \geq n_r \Rightarrow |a_n - \alpha| < r]] \quad .$$

Since ν is an infinite hypernatural number and n_r is always a finite number, it will certainly be true that $\nu > n_r$, and therefore that $|a_\nu - \alpha| < r$. Further, since the (real) number $r > 0$ can be chosen arbitrarily, it follows that $|a_\nu - \alpha|$ is less than every positive real number. Hence $a_\nu \approx \alpha$.

To prove the converse, suppose that $a_\nu \approx \alpha$ for every number $\nu \in {}^*\mathbb{N}$. We must now show that given any real number $r > 0$, the terms of the sequence $(a_n)_{n\in\mathbb{N}}$ except for at most a finite number, belong to the (real) interval $(\alpha - r, \alpha + r)$. Given any $\sigma \in {}^*\mathbb{N}_\infty$ the hypothesis allows us to write

$$\exists_\sigma[\sigma \in {}^*\mathbb{N} \wedge \forall_\nu[\nu \in {}^*\mathbb{N} \Rightarrow [\nu \geq \sigma \Rightarrow |a_\nu - \alpha| < r]]]$$

from which, applying the transfer principle, we obtain

$$\exists_s[s \in \mathbb{N} \wedge \forall_n[n \in \mathbb{N} \Rightarrow [n \geq s \Rightarrow |a_n - \alpha| < r]]]$$

which means that α is the limit of the a_n as $n \to \infty$. •

This result can also be proved in terms of the model of NSA constructed on the basis of a free ultrafilter $\mathcal{U}$ over $\mathbb{N}$. Although such a proof has no advantage over the preceding one it is given here as an example of the method.

Alternative proof Suppose that

$$\lim_{n\to\infty} a_n = \alpha$$

and that $\nu \equiv [(m_n)_{n\in\mathbb{N}}]$ is a hypernatural number. Given $r \in \mathbb{R}^+$ there exists $n_0 \in \mathbb{N}$ such that

$$\{n \in \mathbb{N} : m_n \geq n_0\} \subset \{n \in \mathbb{N} : |a_{m_n} - \alpha| < r\} \quad .$$

Consequently, if $\nu \in {}^*\mathbb{N}_\infty$ then $\{n \in \mathbb{N} : m_n \geq n_0\} \in \mathcal{U}$, and so from the above inclusion it follows that we have $\{n \in \mathbb{N} : |a_{m_n} - \alpha| < r\} \in \mathcal{U}$. Given that $r \in \mathbb{R}$ is arbitrary, this means that $|a_\nu - \alpha| \approx 0$: that is to say, $\nu \in {}^*\mathbb{N}_\infty \Rightarrow a_\nu \approx \alpha$.

Conversely, suppose that

$$\lim_{n\to\infty} a_n \neq \alpha \quad .$$

Then

$$\exists_r[r \in \mathbb{R}^+ \wedge \forall_n[n \in \mathbb{N} \Rightarrow \exists_m[m \in \mathbb{N} \wedge m \geq n \wedge |a_{m_n} - \alpha| \geq r]]].$$

For each $n \in \mathbb{N}$ there will be a natural number $m_n \in \mathbb{N}$ which satisfies this condition. Defining $\nu \equiv [(m_n)_{n\in\mathbb{N}}]$ we have $\nu \in {}^*\mathbb{N}_\infty$, although $a_\nu \not\approx \alpha$. Hence, the condition

$$\forall_\nu[\nu \in {}^*\mathbb{N}_\infty \Rightarrow a_\nu \approx \alpha]$$

implies that the sequence $(a_n)_{n\in\mathbb{N}}$ converges to α. •

This theorem shows that the convergence of $(a_n)_{n\in\mathbb{N}}$ to α is equivalent to saying that $mon(\alpha)$ contains all the infinite tail of the standard hypersequence $(a_n)_{n\in{}^*\mathbb{N}}$.

Note 4.3 It is very difficult for those who have not met it before to accept the equality

$$0.999999999999\ldots = 1$$

although, in $\mathbb{R}$, this is certainly true. In fact, most students, if asked, would probably say that $0.999999999999\ldots$ is a number somewhat less than 1. Now, the real number 1 is the limit (in $\mathbb{R}$) of the sequence of rational numbers

$$(0.9, 0.99, 0.999, 0.9999, 0.99999, 0.999999, \ldots)$$

which, in its turn, defines in nonstandard terms the hyperreal number

$$\alpha \equiv [(0.9, 0.99, 0.999, 0.9999, 0.99999, 0.999999, \ldots)] \quad .$$

It can then be seen that (in ${}^*\mathbb{R}$) we have $\alpha \approx 1$, and that therefore, taking standard parts, we obtain $\mathbf{st}\alpha = 1$, although, in ${}^*\mathbb{R}$, we actually have $\alpha < 1$. This gives some justification for the "natural" intuitive (but erroneous) perception of the series $0.999999999999\ldots$ as a real number which is less than 1.

Exercise 4.4 *Prove that, in* ${}^*\mathbb{R}$, *the inequality* $0.999999999999\ldots < 1$ *is really satisfied, while in* $\mathbb{R}$ *it is transformed into the equality* $0.999999999999\ldots = 1$.

Classical results of the theory of real sequences can now be derived using the definition 4.1 of S-convergence (which, for the sake of simplicity, we will now simply call convergence).

Theorem 4.5 A sequence of real numbers converges to at most one real number.

Proof Suppose that a sequence $[(a_n)_{n\in\mathbb{N}}]$ converges to two limits $a, b \in \mathbb{R}$. Then, for $\nu \in {}^*\mathbb{N}_\infty$ we have $a_\nu \approx a$ and $a_\nu \approx b$, so that $a \approx b$. Therefore, since the only infinitesimal in $\mathbb{R}$ is zero, we must have $a = b$. •

Theorem 4.6 A real sequence $[(a_n)_{n\in\mathbb{N}}]$ is **bounded** if and only if $a_\nu \in {}^*\mathbb{R}_b$, for any $\nu \in {}^*\mathbb{N}_\infty$, that is, if and only if it satisfies the inclusion $\{a_\nu : \nu \in {}^*\mathbb{N}_\infty\} \subset {}^*\mathbb{R}_b$.

Proof Let $(a_n)_{n\in\mathbb{N}}$ be a bounded sequence. Then there exists a real number $M > 0$ such that

$$\forall_n[n \in \mathbb{N} \Rightarrow |a_n| \leq M] \quad .$$

By transfer to ${}^*\mathbb{R}$ this becomes

$$\forall_n[n \in {}^*\mathbb{N} \Rightarrow |a_n| \leq M]$$

and, therefore, $a_\nu \in {}^*\mathbb{R}_b$, for all $\nu \in {}^*\mathbb{N}_\infty$.

Conversely, suppose that for all $\nu \in {}^*\mathbb{N}_\infty$ we have $a_\nu \in {}^*\mathbb{R}_b$. We can then write the following

$$\exists_M[M \in {}^*\mathbb{R}^+ \wedge \forall_\nu[\nu \in {}^*\mathbb{N} \Rightarrow |a_\nu| \leq M]]$$

(where M denotes a positive hyperreal number, possibly infinite). Transferring this time in the contrary sense, we obtain

$$\exists_M[M \in \mathbb{R}^+ \wedge \forall_\nu[n \in \mathbb{N} \Rightarrow |a_n| \leq M]]$$

which shows that $(a_n)_{n\in\mathbf{N}}$ is a bounded sequence. •

Theorem 4.7 If a real sequence $(a_n)_{n\in\mathbf{N}}$ is convergent then it is bounded.

Proof Suppose that $(a_n)_{n\in\mathbf{N}}$ converges to $\alpha \in \mathbb{R}$. Since $a_\nu \approx \alpha \in \mathbb{R}$ for every infinite hypernatural number $\nu \in {}^*\mathbb{N}_\infty$, then $a_\nu \in {}^*\mathbb{R}_b$ for every $\nu \in {}^*\mathbb{N}_\infty$. •

Monotone sequences Using nonstandard definitions and the results obtained above, we now prove an important theorem of Mathematical Analysis relating to bounded monotone real sequences.

Theorem 4.8 A sequence of real numbers which is monotone and bounded is convergent.

Proof Suppose that $(a_n)_{n\in\mathbf{N}}$ is a sequence of real numbers which is monotone increasing and bounded above. (The case of a monotone decreasing sequence bounded below can be treated similarly.)

From theorem 4.6 we know that $a_\nu \in {}^*\mathbb{R}_b$ for every $\nu \in {}^*\mathbb{N}_\infty$. For some fixed $\nu \in {}^*\mathbb{N}_\infty$, let $\alpha = \mathbf{st}a_\nu$. We need first to show that α is an upper bound of $(a_n)_{n\in\mathbf{N}}$. By hypothesis we have

$$\forall_{m,n}[m, n \in \mathbb{N} \Rightarrow [n \leq m \Rightarrow a_n \leq a_m]]$$

from which, applying the transfer principle, we get

$$\forall_{m,n}[m, n \in {}^*\mathbb{N} \Rightarrow [n \leq m \Rightarrow a_n \leq a_m]] \ .$$

In particular, for any $n \in \mathbb{N}$ we have $n < \nu$ and, therefore, $a_n \leq a_\nu \approx \alpha$. All the a_n and α itself are real numbers, and the inequality $a_n \leq \alpha$ holds for all $n \in \mathbb{N}$, which proves that α is a bound of the given sequence.

Suppose now that γ is another upper bound of the set of terms of the sequence $(a_n)_{n\in\mathbf{N}}$. Then $a_n \leq \gamma$ for every $n \in \mathbb{N}$ and, by transfer, $a_n \leq \gamma$ for every $n \in {}^*\mathbb{N}$. Consequently, $\alpha \approx a_\nu \leq \gamma$ and, therefore, since α and γ are both real numbers, we must have $\alpha \leq \gamma$. It follows at once that α is the least of the upper bounds of the sequence $(a_n)_{n\in\mathbf{N}}$ and, therefore that $\alpha = \sup_{n\in\mathbf{N}} a_n$.

If instead of the particular fixed number $\nu \in {}^*\mathbb{N}_\infty$ we were to have considered some other number $\nu' \in {}^*\mathbb{N}_\infty$ then we would have obtained as the supremum of the set of the terms of the sequence $(a_n)_{n\in\mathbf{N}}$ a real number $\alpha' = \mathbf{st}a_{\nu'}$. However, since the supremum of a set (whenever it does exist) is unique, we must have $\alpha' = \alpha$ and so we can write

$$\forall_\nu[\nu \in {}^*\mathbb{N}_\infty \Rightarrow a_\nu \approx \alpha] \ .$$

In accordance with definition 4.1, this means that $\lim_{n\to\infty} a_n = \alpha$. •

Example 4.9 a being a real number lying strictly between 0 and 1, the sequence of real numbers defined by

$$a_n = a^n, \qquad n = 1, 2, \ldots$$

is strictly decreasing and bounded below (by any real negative number). Hence, in accordance with the preceding theorem, $(a_n)_{n\in\mathbb{N}}$ must converge to some (real) number α. Further, for any $\nu \in {}^*\mathbb{N}_\infty$,

$$a^\nu \approx \alpha \text{ and } a^{\nu+1} \approx \alpha$$

and therefore

$$\alpha \approx a^{\nu+1} = a.a^\nu \approx a.\alpha \ .$$

From this it follows that $\alpha \approx a\alpha$ and therefore, since both sides are real, we must have $\alpha = a\alpha$. Finally, since $a \neq 1$, this implies that $\alpha = 0$ and so we have the result

$$\lim_{n\to\infty} a^n = 0 \ .$$

Exercise 4.10 *Prove the usual rules of the calculus of sequences of real numbers. That is, supposing that $(a_n)_{n\in\mathbb{N}}$ and $(b_n)_{n\in\mathbb{N}}$ are two sequences which converge to the real numbers a and b respectively, then*

$$\lim_{n\to\infty} (a_n \pm b_n) = a \pm b,$$

$$\lim_{n\to\infty} (a_n \times b_n) = a \times b,$$

$$b \neq 0 \Rightarrow \lim_{n\to\infty} (a_n / b_n) = a/b \ .$$

Cauchy criterion for convergence In order to apply the definition 4.1 to a convergent real sequence it is necessary to know beforehand some real number α which is a candidate for the limit of the sequence in question. This suggests that in order to show that a given sequence is *not* convergent, it would theoretically be necessary to try all the possible numbers α as the limit. However, as is well known, it is actually enough to appeal to the convergence criterion of Cauchy to determine whether a sequence is convergent or not.

Recall that a sequence $(a_n)_{n\in\mathbb{N}}$ is said to be **Cauchy** if and only if to each real number $r > 0$ it is possible to associate an integer $n_r \in \mathbb{N}$ such that

$$\forall_{m,n}[m, n \in \mathbb{N} \Rightarrow [m, n \geq n_r \Rightarrow |a_m - a_n| < r]] \ . \tag{4.1}$$

In the theorem which follows we give an equivalent nonstandard characterisation of a Cauchy sequence. This characterisation corresponds to the intuitive notion that a sequence is Cauchy if and only if the terms of infinite order of the corresponding standard hypersequence all become (infinitely) close to one another.

Theorem 4.11 A sequence $(a_n)_{n\in\mathbb{N}}$ is a Cauchy sequence if and only if it satisfies the condition

$$a_\nu \approx a_\sigma$$

for any hypernatural numbers $\nu, \sigma \in {}^*\mathbb{N}_\infty$.

Proof Transferring (4.1) to ${}^*\mathbb{R}$ we have

$$\forall_{m,n}[m,n \in {}^*\mathbb{N} \Rightarrow [m,n \geq n_r \Rightarrow |a_m - a_n| < r]] \ .$$

In particular let $m = \sigma$ and $n = \nu$, where $\sigma, \nu \in {}^*\mathbb{N}_\infty \subset {}^*\mathbb{N}$. Then, for any real number $r > 0$ we have that $\sigma, \nu > n_r$, since n_r is always finite, and therefore that $|a_\sigma - a_\nu| < r$. Hence $a_\sigma \approx a_\nu$.

Conversely, suppose that the relation $a_\sigma \approx a_\nu$ is satisfied for any infinite hypernatural numbers σ, ν. Given any real number $r > 0$ whatsoever, we can then write

$$\exists_\eta[\eta \in {}^*\mathbb{N} \wedge \forall_{\sigma,\nu}[\sigma, \nu \in {}^*\mathbb{N} \Rightarrow [\sigma, \nu > \eta \Rightarrow |a_\sigma - a_\nu| < r]]] \ .$$

By hypothesis, this is satisfied for any arbitrarily chosen $\eta \in {}^*\mathbb{N}_\infty$. Transferring to $\mathbb{R}$ we obtain

$$\exists_p[p \in \mathbb{N} \wedge \forall_{m,n}[m,n \in \mathbb{N} \Rightarrow [m,n > p \Rightarrow |a_m - a_n| < r]]]$$

which, since $r \in \mathbb{R}^+$ is arbitrary, shows that the sequence $(a_n)_{n\in\mathbb{N}}$ is Cauchy. •

Exercise 4.12 *Show that a Cauchy sequence must be bounded. (In example 5.9 below a proof of this result will be given, using only nonstandard arguments.)*

We now prove the following well-known result.

Theorem 4.13 A real sequence $(a_n)_{n\in\mathbb{N}}$ is convergent if and only if it is Cauchy.

Proof Let $(a_n)_{n\in\mathbb{N}}$ be a real sequence converging to a (real) number α. Then, given any $\sigma, \nu \in {}^*\mathbb{N}_\infty$, we have $a_\sigma \approx \alpha$ and $a_\nu \approx \alpha$ and, therefore, $a_\sigma \approx a_\nu$. That is, $(a_n)_{n\in\mathbb{N}}$ is a Cauchy sequence.

Conversely, suppose that $(a_n)_{n\in\mathbb{N}}$ is a Cauchy sequence. Then it is bounded and, therefore, for $\nu \in {}^*\mathbb{N}_\infty$ we have $a_\nu \in {}^*\mathbb{R}_b$. Consequently, setting $\alpha = \mathbf{st}a_\nu$, it follows from the hypothesis that $a_\sigma \approx \alpha$ for every $\sigma \in {}^*\mathbb{N}_\infty$, and therefore that $(a_n)_{n\in\mathbb{N}}$ converges to α. •

Exercise 4.14 *Show that the fact that "every finite hyperreal number has a standard part" implies the completeness of* $\mathbb{R}$.

Upper and lower limits of a bounded sequence. Let $(a_n)_{n\in\mathbb{N}}$ be a bounded sequence of real numbers. A **limit point** of the sequence $(a_n)_{n\in\mathbb{N}}$ is a real number s for which the set

$$\{n \in \mathbb{N} : a_n \in (s - r, s + r)\}$$

has an infinite cardinal for every real number $r > 0$.

Theorem 4.15 The number $s \in \mathbb{R}$ is a limit point of the sequence $(a_n)_{n\in\mathbb{N}}$ if and only if there exists $\nu \in {}^*\mathbb{N}_\infty$ such that $a_\nu \approx s$.

Proof Suppose that s is a limit-point of the sequence $(a_n)_{n\in\mathbb{N}}$. This statement can be expressed formally as follows:

$$\forall_r[r \in \mathbb{R}^+ \Rightarrow [\forall_m[m \in \mathbb{N} \Rightarrow \exists_n[n \in \mathbb{N} \wedge n > m \wedge |a_n - s| < r]]] \ .$$

By transfer we have

$$\forall_\epsilon[\epsilon \in {}^*\mathbb{R}^+ \Rightarrow [\forall_\nu[\nu \in {}^*\mathbb{N} \Rightarrow \exists_\sigma[\sigma \in {}^*\mathbb{N} \wedge \sigma > \nu \wedge |a_\sigma - s| < \epsilon]]]$$

which guarantees that if ϵ is a positive infinitesimal and ν an infinite hypernatural there exists $\sigma > \nu$ (so that σ is infinite) such that $|a_\sigma - s| < \epsilon \approx 0$. That is to say, $a_\sigma \approx s$.

Conversely, suppose that $a_\sigma \approx s$ for some $\sigma \in {}^*\mathbb{N}_\infty$. Let $r > 0$ be an arbitrarily chosen real number and let $m \in \mathbb{N}$. Then $\sigma > m$ and $|a_\sigma - s| < r$, and therefore

$$\exists_n[n \in {}^*\mathbb{N} \wedge n > m \wedge |a_n - s| < r]$$

is a true statement (satisfied, for example, for $n = \sigma$). Transferring this sentence to $\mathbb{R}$ we have

$$\exists_n[n \in \mathbb{N} \wedge n > m \wedge |a_n - s| < r]$$

which shows that s is a limit point of $(a_n)_{n\in\mathbb{N}}$. •

This characterisation of limit point shows that the standard parts of finite terms of infinite order of a standard hypersequence are limit points of the corresponding real sequence. If a sequence is bounded, then all the terms will be finite and, therefore, taking the standard part of every term of infinite order we will obtain the set of all the limit points of the sequence. This allows a particularly simple way to derive the following famous result of mathematical analysis:

Theorem 4.16 (Bolzano-Weierstrass) A bounded sequence of real numbers possesses at least one limit point.

In fact, if $(a_n)_{n\in\mathbb{N}}$ is a bounded sequence, then $a_\nu \in {}^*\mathbb{R}_b$ for each $\nu \in {}^*\mathbb{N}_\infty$. Consequently, $s = \mathbf{st}a_\nu$ with $\nu \in {}^*\mathbb{N}_\infty$ is a limit point of the given sequence. •

Let $(a_n)_{n\in\mathbb{N}}$ be a bounded real sequence. This sequence has at least one limit point, perhaps more than one. The set

$$A \equiv \{\mathbf{st}a_\nu : \nu \in {}^*\mathbb{N}_\infty\}$$

contains all the limit points of $(a_n)_{n\in\mathbb{N}}$. Since the real sequence is bounded, the set A is itself bounded and therefore possesses a supremum and an infimum. We then define the **upper limit**, $\limsup a_n$, and the **lower limit**, $\liminf a_n$ of the sequence $(a_n)_{n\in\mathbb{N}}$ by writing

$$\limsup_{n\to\infty} a_n = \sup A$$

$$\liminf_{n\to\infty} a_n = \inf A \quad .$$

Exercise 4.17 *Prove that* $\sup A$ *and* $\inf A$ *are elements of* A *or, equivalently, that the real numbers* $\sup A$ *and* $\inf A$ *are themselves limit points of the sequence* $(a_n)_{n\in\mathbb{N}}$

.

Using the result established in this exercise we can characterise in nonstandard terms the upper limit of a bounded sequence $(a_n)_{n\in\mathbb{N}}$ as follows:

Theorem 4.18 A real number s is equal to the upper limit of a bounded sequence $(a_n)_{n\in\mathbf{N}}$ if and only if

1. $a_\nu \leq s$ for all $\nu \in {}^*\mathbb{N}_\infty$, and
2. $a_\nu \approx s$ for at least one infinite hypernatural number ν.

Proof The first condition is satisfied if and only if $\mathbf{st}a_\nu \leq s$ for every $\nu \in {}^*\mathbb{N}_\infty$, which is equivalent to saying that s is an upper bound of the set A. The second condition says that s is itself a limit point of the sequence and, therefore, $s = \max A \equiv \sup A$. It follows at once that $s = \limsup a_n$. •

Exercise 4.19 *State and prove a theorem similar to the preceding for the case of the lower limit of a bounded sequence* $(a_n)_{n\in\mathbf{N}}$.

As a final consequence of the definitions and results concerning upper and lower limits of a bounded sequence we prove the following:

Theorem 4.20 A bounded sequence $(a_n)_{n\in\mathbf{N}}$ converges to $\alpha \in \mathbb{R}$ if and only if $\limsup a_n = \alpha = \liminf a_n$.

Proof If $\limsup a_n = \alpha = \liminf a_n$ then $A = \{\alpha\}$ and, therefore, $a_\nu \approx \alpha$ for every $\nu \in {}^*\mathbb{N}_\infty$ whence we conclude that $(a_n)_{n\in\mathbf{N}}$ converges to α. The converse is immediate. •

4.1.2 Series of real numbers

The nonstandard definition of convergence can also be used to construct a nonstandard account of the theory of infinite series,

$$\sum_{n=1}^{\infty} a_n \quad . \tag{4.2}$$

In standard terms an infinite series is not the sum of infinitely many terms but rather the limit of a sequence of sums which are all finite (the sequence of the partial sums of the series). However, using the nonstandard definition of limit of a sequence, it is possible to define sums which are literally *infinite*, and whose values have a special relation to the sum of the series (always provided that the latter exists).

Given a series (4.2) we consider the sequence $(S_n)_{n\in\mathbf{N}}$ of the partial sums of the series defined by

$$S_n = \sum_{j=1}^{n} a_j \quad .$$

This sequence can be extended to a standard hypersequence $(S_\nu)_{\nu\in{}^*\mathbf{N}}$ where for each $\nu \equiv [(m_n)_{n\in\mathcal{N}}]$ the term S_ν is defined by

$$S_\nu = \sum_{j=1}^{\nu} a_j = \left[\left(\sum_{j=1}^{m_n} a_j\right)_{n\in\mathbf{N}}\right] \quad .$$

Exercise 4.21 *Show that the hyperfinite sum*

$$\sum_{j=1}^{\nu} a_j$$

enjoys the usual properties attributed to sums of finitely many terms, that is

$$\text{(a)}\quad \sum_{j=1}^{\nu} a_j + \sum_{j=1}^{\nu} b_j = \sum_{j=1}^{\nu}(a_j + b_j)$$

$$\text{(b)}\quad \sum_{j=1}^{\nu} 1 = \nu$$

$$\text{(c)}\quad \sum_{j=1}^{\nu} a_j + a_{\nu+1} = \sum_{j=1}^{\nu+1} a_j \qquad \text{or, more generally,}$$

$$\sum_{j=1}^{\nu+\sigma} a_j = \sum_{j=1}^{\nu} a_j + \sum_{j=\nu+1}^{\nu+\sigma} a_j$$

where ν, σ are two arbitrary hypernatural numbers (finite or infinite).

In the general case a hyperfinite sum of the type

$$\sum_{j=1}^{\nu} a_j$$

depends (more or less significantly) on the number ν of terms. Consider for example,

$$\sum_{j=1}^{\nu}(-1)^{k+1} = \left[\left(\sum_{j=1}^{m_n}(-1)^{k+1}\right)_{n\in\mathbf{N}}\right].$$

If ν is an even number, for example $\nu = 2\omega \equiv [(2n)_{n\in\mathbf{N}}]$, then

$$\sum_{j=1}^{2\omega}(-1)^{k+1} = \left[\left(\sum_{j=1}^{2n}(-1)^{j+1}\right)_{n\in\mathbf{N}}\right] = 0$$

while when $\nu = 2\omega + 1$ we have

$$\sum_{j=1}^{2\omega+1}(-1)^{k+1} = \left[\left(\sum_{j=1}^{2n+1}(-1)^{j+1}\right)_{n\in\mathbf{N}}\right] = 1 \ .$$

Returning to the general case of the series (4.2) it may happen that we have

$$\sum_{j=1}^{\nu} a_j \approx \sum_{j=1}^{\sigma} a_j$$

or any hypernatural numbers $\nu, \sigma \in {}^*\mathbb{N}_\infty$. We would then have $S_\nu \approx S_\sigma$ for ny $\nu, \sigma \in {}^*\mathbb{N}_\infty$ whatever, which is equivalent to saying that $(S_n)_{n\in\mathbb{N}}$ is a Cauchy equence and, therefore that the sum

$$\sum_{j=1}^{\nu} a_j$$

s a finite number for every $\nu \in {}^*\mathbb{N}$. In this case, we can associate with these infinite hyperfinite) sums a well defined real number, namely their (common) standard part. Thus we would get

$$s = \mathbf{st}\left(\sum_{j=1}^{\nu} a_j\right) \quad , \quad \nu \in {}^*\mathbb{N}_\infty \quad .$$

Take for example, the telescoping series

$$\sum_{k=1}^{\infty} \frac{1}{k(k+1)}$$

nd, for any $\nu \in {}^*\mathbb{N}_\infty$ whatever, consider the sum

$$\begin{aligned}\sum_{k=1}^{\nu} \frac{1}{k(k+1)} &= \sum_{k=1}^{\nu}\left(\frac{1}{k} - \frac{1}{k+1}\right)\\ &= \left(\frac{1}{1} - \frac{1}{2}\right) + \left(\frac{1}{2} - \frac{1}{3}\right) + \ldots \left(\frac{1}{\nu} - \frac{1}{\nu+1}\right)\\ &= 1 - \frac{1}{\nu+1} \quad .\end{aligned}$$

ince $(\nu+1)^{-1} \approx 0$ for any $\nu \in {}^*\mathbb{N}_\infty$ then

$$\sum_{k=1}^{\nu} \frac{1}{k(k+1)} \approx 1$$

hich means that

$$\lim_{n\to\infty} \sum_{k=1}^{n} \frac{1}{k(k+1)} = \mathbf{st}\left(\sum_{k=1}^{\nu} \frac{1}{k(k+1)}\right) = 1,$$

r, using the familiar, standard notation in this case,

$$\sum_{k=1}^{\infty} \frac{1}{k(k+1)} = 1 \quad .$$

n general, applying the results obtained for sequences, we can write

$$\sum_{n=1}^{\infty} a_n = s \text{ (in } \mathbb{R}\text{) if and only if } \sum_{n=1}^{\nu} a_n \approx s \text{ for every } \nu \in {}^*\mathbb{N}_\infty,$$

or, taking account of the fact that a sequence is convergent if and only if it is a Cauchy sequence,

the series $\sum_{n=1}^{\infty} a_n$ converges in $\mathbb{R}$ if and only if

$$\sum_{n=\sigma}^{\nu} a_n \approx 0 \text{ for any } \sigma, \nu \in {}^*\mathbb{N}_\infty \quad (\sigma \leq \nu).$$

This corresponds to the (standard) Cauchy criterion for the convergence of a real numerical series. Taking, in particular, $\sigma = \nu$, with $\nu \in {}^*\mathbb{N}_\infty$, we obtain the well-known necessary condition for the convergence of an infinite series:

$$\text{if } \sum_{n=1}^{\infty} a_n \text{ converges then } a_\nu = \sum_{n=\nu}^{\nu} a_n \approx 0.$$

Since $\nu \in {}^*\mathbb{N}_\infty$ is arbitrary we can conclude that the convergence of the series implies the convergence to zero of the sequence of its terms.

From these results we can infer that for a convergent series we have

$$s = \sum_{n=1}^{\infty} a_n = \mathbf{st}\left(\sum_{n=1}^{\nu} a_n\right)$$

for any $\nu \in {}^*\mathbb{N}_\infty$. That is, if the series is convergent, then we can calculate its sum by taking the standard part of the sum of an infinite (hyperfinite) number of terms.

Example 4.22 Let r be a real number lying strictly between 0 and 1. The sum of a geometric series of ratio r can then be obtained easily in the following way

$$\sum_{n=1}^{\infty} r^n = \mathbf{st}\left(\sum_{n=0}^{\nu-1} r^n\right) = \mathbf{st}\left(\frac{1-r^\nu}{1-r}\right) = \frac{1}{1-r},$$

(where, as already seen above, we have $r^\nu \approx 0$ for $\nu \in {}^*\mathbb{N}_\infty$).

Exercise 4.23

1. Show that

$$\sum_{n=1}^{\infty} \frac{1}{(2k-1)(2k+1)} = \frac{1}{2} .$$

2. Let $(a_n)_{n\in\mathbb{N}}$ be a sequence of positive real numbers which is monotone decreasing. Show that if the series

$$\sum_{n=1}^{\infty} a_n$$

is convergent then

$$\nu a_\nu \approx 0$$

for every $\nu \in {}^*\mathbb{N}_\infty$.

[HINT: note that we have $pa_{k+p} < a_{k+1} + a_{k+2} + \ldots + a_{k+p}$ *for every* $p \in \mathbb{N}$*.]*

3. Show that the series

$$\sum_{n=1}^{\infty} \frac{n}{n^2+1}$$

cannot be convergent.

4.2 Topology and Continuity

4.2.1 The usual topology of the real line

The concept of the infinite closeness of finite elements of ${}^*\mathbb{R}_b \subset {}^*\mathbb{R}$ can be used to obtain simple characterisations of the usual topological notions in $\mathbb{R}$. Intuitively, a set is open in $\mathbb{R}$ if it contains every point which is "close" to any one of its members. If we were to interpret this vague idea of "closeness" in terms of the infinitely close concept in the nonstandard extension of the subset of $\mathbb{R}$ in question then, as will be seen in what follows, we would get the usual topology of the real line.

Definition 4.24 A subset A of $\mathbb{R}$ is said to be **S-open** if and only if

$$\forall_a [a \in A \Rightarrow mon(a) \subset {}^*A]$$

(or, equivalently, if and only if ${}^*A$, contains all the points of A together with all their monads).

This definition immediately gives the following:

Theorem 4.25 A subset A of $\mathbb{R}$ is S-open if and only if it is open in the usual sense of the topology on $\mathbb{R}$.

Proof Suppose first that A is open. Then for each point $a \in A$ there exists $r_a > 0$ such that $(a - r_a, a + r_a) \subset A$. Now let $x \in {}^*\mathbb{R}$ be such that $x \approx a$. Then we must have $x \in (a - r_a, a + r_a) \subset {}^*A$ for any real number $r_a > 0$ and, therefore, since x is an arbitrary point of $mon(A)$ it follows that $mon(A) \subset {}^*A$.

Conversely, suppose that for each of the $a \in A$ we have $mon(a) \subset {}^*A$. Then for arbitrarily chosen $a \in A$, we can write

$$\exists_\epsilon [\epsilon \in {}^*\mathbb{R}^+ \wedge \forall_x [x \in {}^*\mathbb{R} \Rightarrow [|x - a| < \epsilon \Rightarrow x \in {}^*A]]] \ .$$

This proposition is true: it is enough to take for ϵ a positive infinitesimal value. Transferring to $\mathbb{R}$ we have

$$\exists_r [r \in \mathbb{R}^+ \wedge \forall_x [x \in \mathbb{R} \Rightarrow [|x - a| < r \Rightarrow x \in A]]]$$

which means that A contains together with a the open ball of radius $r > 0$ which has this point as its centre. Since $a \in A$ is arbitrary it follows that A is open.

Alternative Proof The second part of this proof can also be given using the model of NSA constructed on the basis of an ultrafilter $\mathcal{U}$ over $\mathbb{N}$. Suppose that we have

$$\forall_a [a \in A \Rightarrow mon(a) \subset {}^*A]$$

but that A is not open; then there exists $b \in A$ such that for each real number $r > 0$ we have $(b-r, b+r) \cap A^c \neq \emptyset$. In particular

$$\forall_n [n \in \mathbb{N} \Rightarrow (b - 1/n, b + 1/n) \cap A^c \neq \emptyset] \ .$$

Then, for each $n \in \mathbb{N}$, let x_n be a point of $(b - 1/n, b + 1/n) \cap A^c$. The point $x = [(x_n)_{n \in \mathcal{N}}]$ belongs to $mon(b)$, but $x \notin {}^*A$ which is contrary to hypothesis. This therefore proves that A is open. •

In what follows we shall use either of the (equivalent) terms "open"and "S-open", according to whichever is the more convenient.

Theorem 4.26 From the nonstandard definition of open set (Def.4.24) we get the following results:
(1) $\emptyset, \mathbb{R}$ are open;
(2) the intersection of a finite family of open sets is open;
(3) the union of any family of open sets is open.

Proof (1) The null set $\emptyset$ trivially satisfies the criterion for openness, and $\mathbb{R}$ is open since for any $x \in \mathbb{R}$ we certainly have $mon(x) \subset {}^*\mathbb{R}$.

(2) Let $\{A_k\}_{1 \le k \le n}$ be a finite family of open sets of $\mathbb{R}$. If $\bigcap\{A_k : 1 \le k \le n\} = \emptyset$ then from (1) this intersection of open sets is open. In the contrary case, let $x \in \bigcap\{A_k : 1 \le k \le n\}$. Then $x \in A_k$ for all $k = 1, 2, \ldots, n$; since for each $k = 1, 2, \ldots, n$ the set A_k is open then $mon(x) \subset {}^*A_k$ for all $k = 1, 2, \ldots, n$ and, therefore, (see exercise 3.33)

$$mon(x) \subset \bigcap\{{}^*A_k : k = 1, 2, \ldots, n\} = {}^*\left(\bigcap\{A_k : k = 1, 2, \ldots, n\}\right)$$

which proves that $\bigcap\{A_k : k = 1, 2, \ldots, n\}$ is open.

(3) Let $\{A_\alpha\}_{\alpha \in I}$ be a family of open sets and let $x \in \bigcup_{\alpha \in I} A_\alpha$. Then $x \in A_{\alpha_0}$ for some $\alpha_0 \in I$. Since A_{α_0} is open then $mon(x) \subset {}^*A_{\alpha_0}$ and, therefore, (see exercise 3.33),

$$mon(x) \subset \bigcup\{{}^*A_\alpha : \alpha \in I\} \subset {}^*\left(\bigcup\{A_\alpha : \alpha \in I\}\right)$$

which proves that $\bigcup_{\alpha \in I} A_\alpha$ is open. •

From this theorem it follows that the family of all the open sets of $\mathbb{R}$ constitutes a topology on $\mathbb{R}$. Other topological notions and familiar results in $\mathbb{R}$ can similarly be derived from the concept of the monad of a standard point in the universe of NSA. In particular,

Theorem 4.27 A subset $B \subset \mathbb{R}$ is **closed** if and only if for every finite $x \in {}^*B$ we have $x \in B$.

Proof Suppose that B is closed and let $x \in {}^*B$ be finite. Then there exists $a \in \mathbb{R}$ such that $a =$**st**(x). If x does not belong to B then it must belong to $B^c = \mathbb{R} \backslash B$; in this case, however, we must have $mon(a) \not\subset {}^*B^c$ and therefore B^c will not be open which contradicts the hypothesis that B is closed. It follows at once that $a \in B$.

Suppose now that for every finite element x of ${}^*B$ we have $\mathbf{st}x \in B$, but that B is not closed. Then B^c is not open, and so there will exist $b \in {}^*B$ such that $mon(b) \not\subset {}^*B^c$ and, therefore, such that $mon(b) \cap {}^*B \neq \emptyset$. Let $y \in mon(b) \cap {}^*B$. Then y is finite, $y \in {}^*B$ and so $b = \mathbf{st}y \in B$, which contradicts the fact that $b \in B^c$. Hence B is closed. •

For each $a \in \mathbb{R}$ we denote by $\mathcal{O}_a$ the family of all the open sets $A \subset \mathbb{R}$ such that $a \in A$. Now consider the following:

Theorem 4.28 A subset $V \subset \mathbb{R}$ is a **neighbourhood** of a point $a \in \mathbb{R}$ if and only if $mon(a) \subset {}^*V$.

Proof If V is a neighbourhood of $a \in \mathbb{R}$, then $V \supset A$ for some $A \in \mathcal{O}_a$ and, therefore, $mon(a) \subset {}^*A \subset {}^*V$. Thus we can write,

$$\exists_\epsilon[\epsilon \in {}^*\mathbb{R}^+ \wedge \forall_x[x \in {}^*\mathbb{R} \Rightarrow [|x-a| < \epsilon \Rightarrow x \in {}^*V]]]$$

which is satisfied for every positive infinitesimal ϵ. Transferring to $\mathbb{R}$

$$\exists_r[r \in \mathbb{R}^+ \wedge \forall_x[x \in \mathbb{R} \Rightarrow [|x-a| < r \Rightarrow x \in V]]]$$

so that V contains an open set containing a, and so is a neighbourhood of a. •

Conversely, given the usual topology on $\mathbb{R}$, we can derive the definition of the monad of a real number:

Theorem 4.29 Let a be any element of $\mathbb{R}$. Then $mon(a) = \bigcap\{{}^*A : A \in \mathcal{O}_a\}$.

Proof Since for each open set $A \in \mathcal{O}_a$ we have $mon(a) \subset {}^*A$ it follows that $mon(a) \subset \bigcap\{{}^*A : A \in \mathcal{O}_a\}$. To prove the converse inclusion let $x \in {}^*\mathbb{R}$ be such that $x \not\approx a$. Then there exists a real number $r > 0$ such that $|x-a| > e$. The set $A = (a-r, a+r)$ belongs to $\mathcal{O}_a$ and is such that $x \notin {}^*A$.
Consequently, $x \notin \bigcap\{{}^*A : A \in \mathcal{O}_a\}$ and, therefore, $\bigcap\{{}^*A : A \in \mathcal{O}_a\} \subset mon(a)$. •

Other topological notions can be obtained similarly, as shown below.

Theorem 4.30 A point $a \in \mathbb{R}$ is an **accumulation point** of a set $A \subset \mathbb{R}$ if and only if $mon(a) \cap ({}^*A \backslash \{a\}) \neq \emptyset$.

Proof Let $a \in \mathbb{R}$ be an accumulation point of the set $A \subset \mathbb{R}$. Then choosing arbitrarily a real number $r > 0$, we have

$$\exists_x[x \in A\backslash\{a\} \wedge |x-a| < r] \ .$$

Applying the transfer principle we obtain the statement

$$\exists_x[x \in {}^*A\backslash\{a\} \wedge |x-a| < r]$$

which is satisfied for any (real) number $r > 0$ whatsoever. Consequently,

$$\exists_x[x \in {}^*A\backslash\{a\} \wedge x \approx a]$$

and, therefore, $mon(a) \cap ({}^*A \backslash \{a\}) \neq \emptyset$.

The converse is easily obtained by applying the transfer principle.[1] •

Theorem 4.31 The **closure** or **adherence** of a set $A \subset \mathbb{R}$ is the set $\bar{A}$ which consists of all the points $x \in \mathbb{R}$ such that $mon(x) \cap {}^*A \neq \emptyset$.

Proof If $x \in \bar{A}$ then $x \in A$ or x is a point of accumulation of A. If $x \in A$ then $x \in {}^*A$ and $x \in mon(x)$ whence $mon(x) \cap {}^*A \neq \emptyset$. If x is an accumulation point of A then, using the previous proposition, we can write $mon(x) \cap {}^*A \neq \emptyset$. •

The mapping **st**: ${}^*\mathbb{R}_b \to \mathbb{R}$ allows us to define another mapping, also denoted by **st**, of the set of subsets of ${}^*\mathbb{R}_b$ into the set of subsets of $\mathbb{R}$. This is of the following form:

$$\text{if } \mathcal{A} \subset {}^*\mathbb{R}_b \text{ then } \mathbf{st}(\mathcal{A}) = \{\mathbf{st}(x) : x \in \mathcal{A}\} \quad .$$

Theorem 4.31 can then be expressed as follows:

$$\bar{A} = \mathbf{st}({}^*A \cap {}^*\mathbb{R}_b)$$

which, in its turn, allows us to define a procedure for constructing the closure of any subset A of $\mathbb{R}$: to obtain $\bar{A}$ we determine all the finite elements of ${}^*A$ which generate elements of A under the operation **st**. Thus if, for example, $A = (0,1)$ we would get $\mathbf{st}({}^*A) = [0,1] = \bar{A}$, while for $A = (0,+\infty)$ we would see that $\mathbf{st}({}^*A \cap {}^*\mathbb{R}_b) = [0,+\infty) = \bar{A}$.

Theorem 4.32 A subset $A \subset \mathbb{R}$ is bounded if and only if ${}^*A \subset {}^*\mathbb{R}_b$.

Proof Suppose that $A \subset \mathbb{R}$ is bounded. Then there exists a real number $M > 0$ such that

$$\forall_a [a \in A \Rightarrow |a| \leq M] \quad .$$

By transfer

$$\forall_x [x \in {}^*A \Rightarrow |x| \leq M]$$

which means that

$$\forall_x [x \in {}^*A \Rightarrow x \in {}^*\mathbb{R}_b]$$

and, therefore, ${}^*A \subset {}^*\mathbb{R}_b$. Conversely, suppose that ${}^*A \subset {}^*\mathbb{R}_b$. Then the statement

$$\exists_M [M \in {}^*\mathbb{R}^+ \wedge \forall x [x \in {}^*A \Rightarrow |x| \leq M]]$$

[1] In this footnote we again give an alternative proof of theorem 4.30 based on the construction of ${}^*\mathbb{R}$ by means of a given free ultrafilter $\mathcal{U}$ over $\mathbb{N}$.

Alternative proof From the definition of accumulation point it follows that for each natural number $n \in \mathbb{N}$ there exists $x_n \in (a - 1/n, a + 1/n)$ such that $x_n \neq a$ and $x_n \in A$. Then $x \equiv [(x_n)_{n \in \mathbb{N}}] \in {}^*A$ is such that $x \in mon(a) \backslash \{a\}$. Conversely, suppose that there exists $x \in {}^*A$ such that $x \in mon(a) \backslash \{a\}$. Given any $r > 0$ we have $\{n \in \mathbb{N} : |x_n - a| < r\} \in \mathcal{U}$. On the other hand, $\{n \in \mathbb{N} : x_n \neq a \wedge x_n \in A\} \in \mathcal{U}$ and, therefore, from the definition of the ultrafilter $\mathcal{U}$ it follows that $\{n \in \mathbb{N} : |x_n - a| < r\} \cap \{n \in \mathbb{N} : x_n \neq a \wedge x_n \in A\} \in \mathcal{U}$, which means that $\{n \in \mathbb{N} : |x_n - a| < r\} \cap \{n \in \mathbb{N} : x_n \neq a \wedge x_n \in A\} \neq \emptyset$. Now let $j \in \{n \in \mathbb{N} : |x_n - a| < r\} \cap \{n \in \mathbb{N} : x_n \neq a \wedge x_n \in A\}$. Then, $x_j \in A$ is a point such that $x_j \neq a$ and $x_j \in (a - r, a + r)$. From the fact that r is arbitrary it follows that a is an accumulation point of A.

s true (it is enough to take a sufficiently large value for M, possibly a positive infinite number). Transferring this statement to $\mathbb{R}$ we get

$$\exists_M[M \in \mathbb{R}^+ \wedge \forall x[x \in A \Rightarrow |x| \leq M]]$$

and, therefore, A is a bounded set. •

Exercise 4.33 *Show that every bounded infinite set of real numbers has at least one point of accumulation. (Theorem of Bolzano-Weierstrass.)*

Compactness Let K be a subset of the real numbers. Any family $\{A_i\}_{i\in I}$ of subsets of $\mathbb{R}$ is said to be a **covering** of K if we have

$$K \subset \bigcup_{i\in I} A_i \quad .$$

If in particular all the sets A_i, $i \in I$, are open then the covering $\{A_i\}_{i\in I}$ is said to be an **open covering** of K. This covering is said to be finite if the set of indices I is finite. Let J be a subset of I such that

$$K \subset \bigcup_{j\in J} A_j \quad .$$

Then the family $\{A_j\}_{j\in J}$ (which is also a covering of K) is said to be a subcovering of the covering $\{A_i\}_{i\in I}$ of K.

Definition 4.34 A subset K is said to be **compact** if and only if from every open covering of K we can extract a finite subcovering.

This definition of compactness is very general, and is applicable to any topological space whatsoever. But it is undeniably a somewhat difficult concept to understand, when seen for the first time, and the equivalent characterisation in nonstandard terms achieved by Robinson is easier to accept and to apply. Before announcing the theorem which establishes this nonstandard characterisation, we need to introduce the following definition:

Definition 4.35 Let $A \subset \mathbb{R}$. A point $x \in {}^*A$ is said to be **near-standard** (abbreviated to **n-standard**) if there exists a point $a \in A$ such that $x \approx a$. The set of all n-standard points of ${}^*A$ will be denoted by $\mathbf{ns}({}^*A)$.

Consider, for example, the interval $(0, 1] \subset \mathbb{R}$, and let ϵ be a positive infinitesimal. Then $\epsilon \in {}^*(0, 1]$ but $\epsilon \notin \mathbf{ns}\,{}^*(0, 1]$. Moreover, the interval $(0, 1] \subset \mathbb{R}$ is not compact: the family $\{(1/n, 1+1/n)\}_{n=1,2,...}$, which is an open covering of $(0, 1]$ does not contain any finite subcovering of that interval.

Theorem 4.36 (Robinson) A subset $K \subset \mathbb{R}$ is **compact** if and only if every point $x \in {}^*K$ is n-standard, (that is, if and only if $\mathbf{ns}({}^*K) = {}^*K$.)

Proof Suppose that K is compact but that the point $x \in {}^*K$ is not n-standard. Then x does not belong to the monad of any point of K, which means that for each $a \in K$ there exists an open neighbourhood V_a such that $x \notin {}^*V_a$. The family $\{V_a\}_{a\in K}$

constitutes an open covering of K; since K is, by hypothesis, compact, this covering admits a finite subcovering. For some $k \in \mathcal{N}$ we would then have

$$K \subset V_{a_1} \cup V_{a_2} \cup \ldots \cup V_{a_k}$$

with $a_1, a_2, \ldots a_k \in K$. Accordingly it should follow that

$$^*K \subset {^*V_{a_1}} \cup {^*V_{a_2}} \cup \ldots \cup {^*V_{a_k}}$$

which, however, contradicts the fact that x belongs to $^*K$.

Suppose now that K is not compact. From a known lemma[2] there exists a covering of K formed of open intervals with rational endpoints, $\{(a_k, b_k)\}_{k \in \mathbb{N}}$, which does not contain any finite subcovering. Consequently, for each $n \in \mathbb{N}$,

$$K \not\subset (a_1, b_1) \cup (a_2, b_2) \cup \ldots \cup (a_n, b_n)$$

and this can be expressed as the following true proposition in $\mathbb{R}$,

$$\forall_n [n \in \mathbb{N} \Rightarrow \exists_x [x \in K \wedge \forall_k [k \in \mathbb{N} \Rightarrow [k \leq n \Rightarrow x \notin (a_k, b_k)]]]] \ .$$

By transfer we get

$$\forall_n [n \in {^*\mathbb{N}} \Rightarrow \exists_x [x \in {^*K} \wedge \forall_k [k \in {^*\mathbb{N}} \Rightarrow [k \leq n \Rightarrow x \notin {^*(a_k, b_k)}]]]]$$

which is also a true proposition, in $^*\mathbb{R}$.

Let $\nu \in \mathbb{N}$ be an infinite hypernatural. Then there exists $x \in {^*K}$ such that $x \notin {^*(a_k, b_k)}$ for every $k \leq \nu$. In particular, $x \notin {^*(a_k, b_k)}$ for every $k \in \mathbb{N}$ which, as will be seen, implies that $x \notin \mathbf{ns}({^*K})$. In fact, seeing that for every $r \in K$ there exists some $k \in \mathbb{N}$ such that $a_k < r < b_k$, then if there were to exist $r \in K$ with $x \approx r$ there would exist $k \in \mathbb{N}$ such that $a_k < x < b_k$. And this contradicts the hypothesis. Consequently, $x \notin \mathbf{ns}({^*K})$. •

Returning to the previous example, we see immediately that $(0, 1]$ cannot be compact, because $\epsilon \approx 0$, $\epsilon > 0$ is such that $\epsilon \in {^*(0, 1]}$, but $\epsilon \notin \mathbf{ns}({^*(0, 1]})$. In contrast, every bounded and closed interval $[a, b]$ of $\mathbb{R}$ is compact.

Theorem 4.37 (Heine-Borel) A subset $K \subset \mathbb{R}$ is compact if and only if it is closed and bounded.

Proof If K is not closed then $K^c \equiv \mathbb{R} \backslash K$ is not open and, therefore, there exists $a \in K^c$ such that $mon(a) \not\subset {^*K^c}$ or, equivalently, such that $mon(a) \cap {^*K} \neq \emptyset$. Let $y \in mon(a) \cap {^*K}$. Since $\mathbf{st}(y) = a$ it follows that $y \notin \mathbf{ns}({^*K})$ and, therefore, that K is not compact. Hence K compact implies K closed.

If K is not bounded above, for example, then for every $n \in \mathbb{N}$ we can find a possible $x \in K$ such that $x > n$: that is

$$\forall_n [n \in \mathbb{N} \Rightarrow \exists_x [x \in K \wedge x > n]]$$

is a true proposition. Applying the transfer principle we obtain

$$\forall_n [n \in {^*\mathbb{N}} \Rightarrow \exists_x [x \in {^*K} \wedge x > n]]$$

[2] Every open covering of a set $A \subset \mathbb{R}$ contains a finite subcovering if and only if from every covering of A by open intervals with rational endpoints can be extracted a finite subcovering.

which is also a true proposition. If $\nu \in {}^*\mathbb{N}_\infty$ then there exists $x \in {}^*K$ such that $x > \nu$; it follows at once that $x \notin \mathbf{ns}({}^*K)$ and, therefore, that K is not compact. Thus, K compact implies K bounded (above). A similar argument applies for K bounded below (see Exercise 4.38).

Suppose now that K is a bounded, closed set. Since K is bounded there exists $M > 0$ such that

$$\forall_y [y \in \mathbb{R} \Rightarrow [y \in K \Rightarrow |y| \leq M]]$$

from which for all $x \in {}^*K$ we have $|x| \leq M$. Then $\mathbf{st}(x) \in \bar{K} = K$ (since K is closed). Consequently, $x \in \mathbf{ns}({}^*K)$ and it follows that K is compact. •

Exercise 4.38

1. Show that
(a) if $K \subset \mathbb{R}$ is finite then K is compact,
(b) if $B \subset \mathbb{R}$ is not bounded below then it cannot be compact.

2. Using the Robinson criterion, show that every closed subset of a compact set of $\mathbb{R}$ is compact.

4.2.2 Continuous functions

The concepts of limit and continuity can similarly be characterised in nonstandard terms. In particular, we have

Definition 4.39 A function $f : A \subset \mathbb{R} \to \mathbb{R}$ is said to be **S-continuous** at a point $a \in A$ if and only if it satisfies the condition

$$\forall_x [x \in {}^*A \Rightarrow [x \approx a \Rightarrow {}^*f(x) \approx f(a)]]$$

or, equivalently, if and only if ${}^*f(mon(a) \cap {}^*A) \subset mon(f(a))$.

As expected, the concepts of S-continuity and continuity (in the sense of Weierstrass) are equivalent.

Theorem 4.40 A function $f : A \subset \mathbb{R} \to \mathbb{R}$ is continuous at a point $a \in A$ if and only if it is S-continuous at the same point.

Proof Suppose that f is continuous at a point $a \in A$ and let $r > 0$ be an arbitrary real number. The (standard) definition of continuity implies the existence of a number $d \in \mathbb{R}$ such that

$$\forall_x [x \in A \Rightarrow [|x - a| < d \Rightarrow |f(x) - f(a)| < r]] \quad .$$

Applying the transfer principle we obtain

$$\forall_x [x \in {}^*A \Rightarrow [|x - a| < d \Rightarrow |{}^*f(x) - f(a)| < r]]$$

which is a true sentence in ${}^*\mathbb{R}$.

Now let $x \in {}^*A$ be such that $x \approx a$. Then it certainly satisfies the inequality $|x - a| < d$ and, therefore, the inequality $|{}^*f(x) - f(a)| < r$ must also be satisfied.

Since $r \in \mathbb{R}^+$ is arbitrary, by hypothesis, then ${}^*f(x) \approx f(a)$. That is, from the continuity of f it follows that

$$\forall_x[x \in {}^*A \Rightarrow [x \approx a \Rightarrow {}^*f(x) \approx f(a)]] \tag{4.3}$$

as we claimed to show.

Conversely, suppose that (4.3) is satisfied and let $r \in \mathbb{R}^+$ be arbitrarily chosen. From the hypothesis it follows that

$$\exists_\delta[\delta \in {}^*\mathbb{R}^+ \wedge \forall_x[x \in {}^*A \Rightarrow [|x-a| < \delta \Rightarrow |{}^*f(x) - f(a)| < r]]]$$

(it being enough to consider $\delta \approx 0$) and, therefore, by transfer,

$$\exists_d[d \in \mathbb{R}^+ \wedge \forall_x[x \in A \Rightarrow [|x-a| < d \Rightarrow |f(x) - f(a)| < r]]]$$

which proves the continuity of f at the point $a \in A$.[3] •

A function which is not S-continuous at a point is said to S-discontinuous at that point. The function

$$\operatorname{sgn}(x) = \begin{cases} -1, & \text{if } x < 0, \\ 0, & \text{if } x = 0, \\ +1, & \text{if } x > 0, \end{cases}$$

is S-discontinuous at the origin, since if ϵ is any positive infinitesimal then we have $\operatorname{sgn}(\epsilon) = 1$, $\operatorname{sgn}(\epsilon) = -1$, and $\operatorname{sgn}(0) = 0$.

The function defined on $\mathbb{R}\backslash\{0\}$ by

$$f(x) = \sin\left(\frac{1}{x}\right)$$

cannot be extended to a function continuous at the origin. In fact, if ν is any odd infinite hypernatural number, then, taking $\epsilon = 2/\nu\pi \approx 0$, we have ${}^*f(-\epsilon) = -{}^*f(\epsilon)$ and $|{}^*f(\epsilon)| = 1$.

Exercise 4.41 *Let m be a natural number. Using nonstandard arguments show that*

1. *x^m is a function continuous on $\mathbb{R}$,*
2. *x^{-m} is a function continuous on $\mathbb{R}\backslash\{0\}$,*
3. *$\sin x$ is a function continuous on $\mathbb{R}$.*

Exercise 4.42 *Show that the Dirichlet function, defined by*

$$D(x) = \begin{cases} 1, & \text{if } x \text{ is rational} \\ 0, & \text{if } x \text{ is irrational} \end{cases}$$

[3]**Alternative proof** Suppose that f is continuous at a point $a \in A$ and let $x \equiv [(x_n)_{n\in\mathbb{N}}]$ be a hyperreal number belonging to ${}^*A$ which is infinitely close to a. Then given $r \in \mathbb{R}^+$ there exists $s \in \mathbb{R}^+$ such that given $r \in \mathcal{R}^+$ there exists $s \in \mathcal{R}^+$ such that $\{n \in \mathbb{N} : x_n \in A \wedge |x_n - a| < s\} \subset \{n \in \mathbb{N} : |f(x) - f(a)| < s\}$. Since, by hypothesis, for each $s \in \mathbb{R}$ we have $\{n \in \mathbb{N} : x_n \in A \wedge |x_n - a| < s\} \in \mathcal{U}$, then from the definition (3.7) it follows that the set $\{n \in \mathbb{N} : |f(x_n) - f(a)| < r\}$ belongs to the ultrafilter $\mathcal{U}$ for any $r \in \mathbb{R}^+$ which, in its turn, means that the condition is necessary.

On the other hand, if f were not continuous at the point $a \in \mathbb{R}$ then there would exist a real number $r > 0$ and a sequence $(v_n)_{n\in\mathbb{N}}$ of elements of A converging to a of such a form that $|f(x_n) - f(a)| \geq r$ for every $n \in \mathbb{N}$. Consequently, $x \equiv [(x_n)_{n\in\mathbb{N}}] \in {}^*A$ is infinitely close to a, but ${}^*f(a)$ is not infinitely close to $f(a)$ whence we derive the sufficiency of the nonstandard criterion for continuity.

s discontinuous everywhere.

Exercise 4.43 *Let f, g be two functions defined and continuous on $\mathbb{R}$ and let λ denote any real number whatsoever. Show that*

$$f + g, \quad \lambda f, \quad fg, \quad f \circ g$$

are continuous functions and, further, that f/g is continuous at all points of $\mathbb{R}$ except where we have $g(x) = 0$.

Theorem 4.44 A function $f : \mathbb{R} \to \mathbb{R}$ is continuous if and only if the inverse image of every set open in $\mathbb{R}$ is itself open in $\mathbb{R}$.[4]

Proof Suppose f continuous, let A be an open set in $\mathbb{R}$ and let $a \in f^{-1}(A)$. Then $f(a) \in A$. If $y \in {}^*\mathbb{R}$ is such that $y \approx a$ then from continuity we have ${}^*f(y) \approx f(a)$ and since A is open it follows that ${}^*f(y) \in {}^*A$, or equivalently, that $y \in ({}^*f)^{-1}({}^*A)$. From the result of exercise 3.37 it follows that $y \in {}^*(f^{-1}(A))$. Since $y \in mon(a)$ is arbitrary, we must have $mon(a) \subset {}^*(f^{-1}(A))$, which means that $f^{-1}(A)$ is open.

Now suppose that the inverse image of every set open in $\mathbb{R}$ is an open set in $\mathbb{R}$ and let $a \in \mathbb{R}$ and $y \in {}^*\mathbb{R}$ such that $y \approx a$. If the relation $f(a) \approx {}^*f(y)$ were not satisfied then for some $r \in \mathbb{R}^+$ we would have $|f(a) - {}^*f(y)| \geq r$. This in turn implies that ${}^*f(y)$ does not belong to the set ${}^*(f(a) - r, f(a) + r)$, or, equivalently, that $y \notin {}^*f^{-1}({}^*(f(a) - r, f(a) + r))$. Since $a \in {}^*f^{-1}({}^*(f(a) - r, f(a) + r))$, this contradicts the hypothesis that $y \approx a$. •

Uniform continuity As is well known, a function $f : A \subset \mathbb{R} \to \mathbb{R}$ is uniformly continuous on A if and only if, given a real number $r > 0$ we can always find another real number $s > 0$ for which the inequality $|f(x) - f(y)| < r$ is satisfied for *all* the numbers $x, y \in A$ which are such that $|x - a| < s$. The nonstandard characterisation of this concept of uniform continuity is given in the following theorem:

Theorem 4.45 A function f defined on a subset A of $\mathbb{R}$ is uniformly continuous there if and only if

$$\forall_{x,y}[x, y \in {}^*A \Rightarrow [x \approx y \Rightarrow {}^*f(x) \approx {}^*f(y)]]$$

(where both x and y may be nonstandard).

Proof Suppose that a function $f : A \to {}^*\mathbb{R}$ is uniformly continuous and let $r \in \mathbb{R}^+$ be chosen arbitrarily. Then there exists $d \in \mathbb{R}^+$ such that

$$\forall_{x,y}[x, y \in A \Rightarrow [|x - y| < d \Rightarrow |f(x) - f(y)| < r]]$$

from which, by transfer we obtain

$$\forall_{x,y}[x, y \in {}^*A \Rightarrow [|x - y| < d \Rightarrow |{}^*f(x) - {}^*f(y)| < r]] \ .$$

Let $x, y \in {}^*A$ such that $x \approx y$. Then we always have $|x - y| < d$ and, therefore, $|{}^*f(x) - {}^*f(y)| < r$. Since $r \in \mathbb{R}^+$ is arbitrary it follows that ${}^*f(x) \approx {}^*f(y)$, which proves that f is uniformly continuous on A:

$$\forall_{x,y}[x, y \in {}^*A \Rightarrow [x \approx y \Rightarrow {}^*f(x) \approx {}^*f(y)]] \ . \tag{4.4}$$

[4]This result can be generalised immediately to the case where the domain of f is a subset of $\mathbb{R}$ (with the induced natural topology).

Now suppose that (4.4) is satisfied and let $r \in \mathbb{R}^+$ be an arbitrarily chosen number. By the hypothesis we can write

$$\exists_\delta[\delta \in {}^*\mathbb{R}^+ \wedge \forall_{x,y}[x, y \in {}^*A \Rightarrow [|x-y| < \delta \Rightarrow |{}^*f(x) - {}^*f(y)| < r]]]$$

(which is a true proposition when, for example, $\delta \approx 0$). Hence, by transfer, we obtain

$$\exists_d[d \in \mathbb{R}^+ \wedge \forall_{x,y}[x, y \in A \Rightarrow [|x-y| < d \Rightarrow |f(x) - f(y)| < r]]]$$

which shows that f is uniformly continuous on A.[5] •

Note the difference (apparently innocuous) between the nonstandard formulations of the two forms of continuity (pointwise and uniform). A function $f : A \to \mathbb{R}$ is (pointwise) continuous on A if for each $y \in A$ and $x \in {}^*A$ such that $x \approx y$ we have ${}^*f(x) \approx f(y)$; the function is uniformly continuous on A if the relation ${}^*f(x) \approx {}^*f(y)$ holds for *all* points $x, y \in {}^*A$ such that $x \approx y$. While the pointwise continuity preserves "infinite closeness" relative to the standard points of A, the uniform continuity preserves "infinite closeness" at all points (standard and nonstandard) of the nonstandard extension of A.

If, in particular, we take $x \in A$ and $y \in {}^*A$ in theorem 4.45, then we simply obtain (pointwise) continuity:

Corollary 4.45.1 Every function uniformly continuous on A is continuous on A.

It is now easy to see that the function $f(x) = x^2$, continuous at all points $x \in \mathbb{R}$, is not uniformly continuous on $\mathbb{R}$. In fact, we have

$${}^*f(x) - {}^*f(y) = x^2 - y^2 = (x+y)(x-y)$$

and therefore from the relation $x - y \approx 0$ it does not necessarily follow that we have ${}^*f(x) - {}^*f(y) \approx 0$: it is enough for $(x+y)$ to be a sufficiently large infinite number to make ${}^*f(x) - {}^*f(y)$ fail to be infinitesimal.

Similarly, the function

$$f(x) = \frac{1}{x}, \qquad x \in (0, +\infty)$$

which is continuous everywhere within its domain, is not uniformly continuous. In fact, taking $\nu \in {}^*\mathbb{N}_\infty$, consider the numbers

$$x = \frac{1}{\nu^2}, \qquad y = \frac{1}{\nu} \quad \text{belonging to} \quad {}^*(0, \infty).$$

[5]**Alternative proof** Suppose that $f : A \subset \mathbb{R} \to \mathbb{R}$ is uniformly continuous on A and let $x \equiv [(x_n)_{n \in \mathbb{N}}]$ and $y \equiv [(y_n)_{n \in \mathbb{N}}]$ be two points of ${}^*A$ such that $x \approx y$. From the uniform continuity of f it follows that for any real number $r > 0$ there exists a real number $s > 0$ such that $\{n \in \mathbb{N} : x_n, y_n \in A \wedge |x_n - y_n| < s\} \subset \{n \in \mathbb{N} : |f(x_n) - f(y_n)| < r\}$. Since, by hypothesis $\{n \in \mathbb{N} : x_n, y_n \in A \wedge |x_n - y_n| < s\} \in \mathcal{U}$ whatever the real number $s > 0$ may be, then it follows from definition 3.9 that the set defined by $\{n \in \mathbb{N} : |f(x) - f(y)| < r\}$ belongs to the ultrafilter $\mathcal{U}$, for any real number $r > 0$. Consequently, for any two numbers $x, y \in {}^*A$, if $x \approx y$ then ${}^*f(x) \approx {}^*f(y)$.

Conversely, suppose that f is not uniformly continuous on A. Then there exists $r > 0$ such that for every $s > 0$ we can find elements $u, v \in A$ such that $|u - v| < s \wedge |f(u) - f(v)| \geq r$. In particular for each $n \in \mathbb{N}$, let $x_n, y_n \in A$ be such that $|x_n - y_n| < 1/n$ and $|f(x_n) - f(y_n)| \geq r$. Then the hyperreal numbers $x \equiv [(x_n)_{n \in \mathbb{N}}]$ and $y \equiv [(y_n)_{n \in \mathbb{N}}]$ belonging to ${}^*A$ are such that $x \approx y$ but ${}^*f(x) \not\approx {}^*f(y)$, which proves that the condition is also sufficient.

In this case we have $x \approx y$, but

$$^*f(x) = \nu^2, \quad \text{and} \quad ^*f(y) = \nu$$

from which it follows that $^*f(x) \not\approx {^*f(y)}$.

Note 4.46 A function f which satisfies the (nonstandard) condition given in theorem 4.45 is said to be **SU-continuous** on A.

For functions defined on certain specific types of domain the two forms of continuity (pointwise and uniform) simply coincide.

Theorem 4.47 If f is continuous on a compact $K \subset \mathbb{R}$ then f is uniformly continuous on K.

Proof Let $x, y \in {^*K}$ and $x \approx y$. Since $x, y \in \mathbf{ns}(^*K)$, then there exists $a \in K$ such that $x \approx a \approx y$. From the continuity of f we have

$$^*f(x) \approx f(a) \approx {^*f(y)}$$

and so f is uniformly continuous on K. •

Theorem 4.48 If $K \subset \mathbb{R}$ is compact and if $f : K \to \mathbb{R}$ is continuous than $f(K)$ is compact.

Proof Let $y \in {^*(f(K))} = {^*f(^*K)}$. Then there exists $x \in {^*K}$ such that $^*f(x) = y$. Since K is compact there exists $a \in K$ such that $x \approx a$, and from the continuity of f it follows that $y = {^*f(x)} \approx f(a)$. Thus, $y \in \mathbf{ns}(^*f(^*K))$ which means that $\mathbf{ns}(^*f(K)) = {^*f(^*K)}$, or, equivalently, that $f(K)$ is compact. •

Exercise 4.49 *Let $f : \mathbb{R} \to \mathbb{R}$ be a uniformly continuous function and $(x_n)_{n\in\mathbf{N}}$ a sequence of real numbers. Show that if $(x_n)_{n\in\mathbf{N}}$ is a Cauchy sequence then $(f(x_n))_{n\in\mathbf{N}}$ is also a Cauchy sequence.*

Note 4.50 Recall once more the nonstandard definition of continuity of a function at a point $a \in \mathbb{R}$:

$$\forall_x [x \in {^*A} \Rightarrow [x \approx a \Rightarrow {^*f(x)} \approx f(a)]] \ . \tag{4.5}$$

Note that this is a proposition of $^*\mathbb{R}$ which could not have been obtained by the transfer of a formally identical proposition relative to $\mathbb{R}$. Proposition (4.5) is, however, equivalent to the following one:

$$\forall_r [r \in \mathbb{R}^+ \Rightarrow \exists_s [s \in \mathbb{R}^+ \wedge$$

$$\forall_x [x \in A \wedge |x - a| < s \Rightarrow |f(x) - f(a)| < r]]] \ . \tag{4.6}$$

which is a proposition relative to $\mathbb{R}$. Now it does make sense to apply the Transfer principle to (4.6) so as to get

$$\forall_\epsilon[\epsilon \in {}^*\mathbb{R}^+ \Rightarrow \exists_\delta[\delta \in {}^*\mathbb{R}^+ \wedge$$
$$\forall_x[x \in {}^*A \wedge |x-a| < \delta \Rightarrow |f(x) - f(a)| < \epsilon]]] \quad . \tag{4.7}$$

In particular the numbers ϵ, δ occurring in expression (4.7) are positive hyperreals which may be infinitesimals.

A function ${}^*f$, the nonstandard extension of $f : A \subset \mathbb{R} \to \mathbb{R}$, which satisfies the condition (4.7) at a point $a \in A$, is said to be ***-continuous** at A. This shows that, in ${}^*\mathbb{R}$, it is possible to formulate two distinct concepts of continuity, namely S-continuity and what we have now defined as *-continuity. It is immediately clear that the following statement is true:

*A (standard) function $f : A \subset \mathbb{R} \to \mathbb{R}$ is S-continuous at a point $a \in A$ if and only if the function ${}^*f$ is *-continuous at the same point.*

Within the context of $\mathbb{R}$ then, the concepts of S-continuity and *-continuity are equivalent. Nevertheless, as will be seen later, if we consider other types of hyperreal functions of a hyperreal variable, which are not of the form ${}^*f$ for some standard function f, then this may not be the case. The two concepts are not always equivalent in the context of ${}^*\mathbb{R}$, and they do correspond there to distinct forms of continuity, with implications which are also distinct. This fact allows us to make a much deeper examination of the significance of "continuity" than is possible when it is simply considered from the standard point of view.

Exercise 4.51 *Using nonstandard definitions examine the continuity of each of the following functions:*

$$f_1(x) = \begin{cases} \sin\frac{1}{x}, & \text{if } x \neq 0 \\ 0 & \text{if } x = 0 \end{cases}$$

$$f_2(x) = \begin{cases} x\sin\frac{1}{x}, & \text{if } x \neq 0 \\ 0 & \text{if } x = 0 \end{cases}$$

$$f_3(x) = \begin{cases} x, & \text{if } x \text{ rational} \\ -x & \text{if } x \text{ irrational} \end{cases}$$

$$f_4(x) = \begin{cases} 0, & \text{if } x \text{ irrational or } x = 0 \\ \frac{1}{n} & \text{if } x = \frac{m}{n}, n \geq 1, \ (x \text{ in irreducible form}). \end{cases}$$

Continuity on intervals Continuous functions exhibit special behaviour relative to certain subsets of their domains. In particular, a continuous function transforms *intervals* into *intervals*. This is the essential content of what is usually called the *intermediate value theorem.*[6]

[6]This result can be proved more easily using the fact that every subset of ${}^*\mathbb{R}$ with a hypernatural number of elements always has a maximum and a minimum. (This will be proved later in full generality.)

Theorem 4.52 (Bolzano) If $f : [a, b] \to \mathbb{R}$ is a continuous function then for every value $\gamma \in \mathbb{R}$ lying strictly between $f(a)$ and $f(b)$ (with $f(a) \neq f(b)$) there exists $c \in (a, b)$ such that $f(c) = \gamma$.

Proof Without loss of generality, we may suppose that we have $f(a) < \gamma < f(b)$; the case in which $f(b) < \gamma < f(a)$ can obviously be treated similarly.

For each $n \in \mathbb{N}$ let $\pi_n = \{a \equiv x_0 < x_1 < \ldots < x_n \equiv b\}$ be the uniform partition of the interval $[a, b]$ into n equal subintervals of length equal to $(b - a)/n$. The set

$$\{x \in \pi_n : \quad f(x) < \gamma\}$$

is not empty ($f(x_0) \equiv f(a) < \gamma$) and is finite. Therefore it possesses a maximum s_n. Hence we can write

$$\forall_n [n \in \mathbb{N} \Rightarrow [a \leq s_n < b \wedge f(s_n) \leq \gamma \leq f(s_n + (b - a)/n)]]$$

from which, by transfer, we get

$$\forall_n [n \in {}^*\mathbb{N} \Rightarrow [a \leq s_n < b \wedge {}^*f(s_n) \leq \gamma \leq {}^*f(s_n + (b - a)/n)]] \quad .$$

Let $\nu \in {}^*\mathbb{N}$ be an infinite number and let π_ν be the partition of infinitesimal diameter

$$\delta_\nu = \frac{b - a}{\nu} \approx 0 \quad .$$

Since $a \leq s_\nu < b$ then s_ν is a finite number possessing a standard part, $c = \mathbf{st}(s_\nu) \in [a, b]$. As $\delta_\nu \approx 0$ so $s_\nu \approx c$ and $s_\nu + \delta_\nu \approx c$. Therefore from the continuity of f at the point c it follows that ${}^*f(s_\nu) \approx f(c) \approx {}^*f(s_\nu + \delta_\nu)$. Again taking into account the fact that ${}^*f(s_\nu) < \gamma \leq {}^*f(s_\nu + \delta_\nu)$, we have $f(c) \approx \gamma$. Since $f(c)$ and γ are real numbers we must finally have $f(c) = \gamma$, as it was required to show. •

Exercise 4.53 *By a process similar to that used in the intermediate value theorem, show that if $f : [a, b] \to \mathbb{R}$ is a continuous function then f attains its maximum and minimum values on $[a, b]$.*

Limits In many cases the study of the limit of a function at a point can be reduced to the study of pointwise continuity as presented above. Given a function $f : A \subset \mathbb{R} \to \mathbb{R}$ suppose that we have

$$\lim_{x \to a} f(x) = b$$

where b is a real number and $a \in \mathbb{R}$ is an accumulation point of A. Considering the extension of f to the set $A \cup \{a\}$, defined by

$$f_a(x) = \begin{cases} f(x) & \text{if } x \in A \backslash \{a\} \\ b & \text{if } x = a \end{cases}$$

then the limit of $f(x)$ when x tends towards a (through points belonging to A) is equal to $b \in \mathbb{R}$ if and only if the function f_a is continuous at the point a; that is, if and only if it satisfies the condition

$$\forall_x [x \in {}^*A \cup \{a\} \Rightarrow [x \approx a \Rightarrow {}^*f_a(x) \approx f_a(a)]] \quad .$$

Since for $x \in {}^*A \backslash \{a\}$ we have ${}^*f_a(x) = {}^*f(x)$ while, on the other hand, $f_a(a) = b$, then this condition is equivalent to the following

$$\forall_x[x \in {}^*A \Rightarrow [x \approx a \Rightarrow {}^*f_a(x) \approx b]] \ .$$

Note then that in this case we have

$$b = \mathbf{st}({}^*f(x))$$

for any $x \in {}^*A$ such that $x \approx a$.

Exercise 4.54 *Establish the equivalence*

$$\lim_{x \to a} f(x) = b \text{ if and only if } \forall_x[x \in {}^*A \Rightarrow [x \approx a \Rightarrow {}^*f(x) \approx b]]$$

where a and b are real numbers (and a is an accumulation point of the domain of f).

Exercise 4.55 *Obtain nonstandard conditions which translate the usual definitions of limit, upper limit and lower limit of a function at an accumulation point of its domain.*

4.3 Infinitesimal Calculus

The concepts and techniques of NSA introduced so far now allow us to give an account of the development of differential and integral calculus which returns formally to the original principles of Leibniz.

4.3.1 Differentiation

Let $a, b,$ $(a < b)$ be two elements of the extended real line $\bar{\mathbb{R}} \equiv \mathbb{R} \cup \{-\infty, +\infty\}$, and let $f : (a, b) \to \mathbb{R}$ be a function differentiable at the point $c \in (a, b)$. Then there exists a real number, denoted by $f'(c)$, which satisfies the equation

$$f(c+h) = f(c) + hf'(c) + hr(h)$$

where $h \in \mathbb{R}$ is such that $c + h \in (a, b)$ and $r(h)$, the remainder function, tends to zero when $h \to 0$. To put this another way: f is differentiable at c if there exists a real number, denoted by $f'(c)$, such that

$$\lim_{h \to 0} \left\{ \frac{f(c+h) - f(c)}{h} - f'(c) \right\} = 0 \ .$$

Using the nonstandard notion of limit given earlier, it is easy to verify that the (standard) definition of differentiability at a point c implies that for all infinitesimal $\delta \approx 0$ we have the relation

$$\frac{{}^*f(c+\delta) - f(c)}{\delta} \approx f'(c)$$

or, equivalently, that we have the equality

$${}^*f(c+\delta) = f(c) + \delta f'(c) + \delta\epsilon(\delta)$$

with $\epsilon(\delta) \approx 0$ for every $\delta \approx 0$. To see that the converse is also true we first introduce an (equivalent) nonstandard definition for the concept of differentiability.

Definition 4.56 A function $f : (a,b) \to \mathbb{R}$ is said to be **S-differentiable** at a point $c \in (a,b)$ if and only if for every $\delta \approx 0$, $\delta \neq 0$, the quotient

$$\frac{{}^*f(c+\delta) - f(c)}{\delta}$$

is finite, and has a standard part which does not depend on the particular infinitesimal $\delta \neq 0$.

As might be expected, the concepts of (standard) differentiability and of S-differentiability are equivalent.

Theorem 4.57 The function $f : (a,b) \to \mathbb{R}$ is differentiable at the point $c \in (a,b)$ if and only if it is S-differentiable at the same point, it then being possible to write

$$f'(c) = \mathbf{st}\left(\frac{{}^*f(c+\delta) - f(c)}{\delta}\right)$$

for any $\delta \approx 0$, $\delta \neq 0$, whatever.

Proof Having already dealt with the implication

differentiability ⇒ S-differentiability

it remains to show also that S-differentiability implies differentiability. Suppose then that f is S-differentiable at the point c, that is, there exists a real number $f'(c)$ for which the relation referred to in the theorem is satisfied and which for convenience, we can put in the following form

$$\forall_\delta \left[\delta \approx 0, \delta \neq 0 \Rightarrow \frac{{}^*f(c+\delta) - f(c)}{\delta} - f'(c) \approx 0\right].$$

The proof that this condition implies differentiability can be made by transfer. Let $r > 0$ be any real number whatever. From the S-differentiability of f at c it follows that

$$\exists_\gamma[\gamma \in {}^*\mathbb{R}^+ \wedge \forall_\delta[\delta \in {}^*\mathbb{R} \Rightarrow$$
$$[0 < |\delta| < \gamma \Rightarrow \left|\tfrac{{}^*f(c+\delta)-f(c)}{\delta} - f'(c)\right| < r]]]$$

is a true proposition in ${}^*\mathbb{R}$. (It is enough to take γ to be an arbitrary positive infinitesimal for this.) Applying the transfer principle we obtain

$$\exists_s[s \in \mathbb{R}^+ \wedge \forall_h[h \in \mathbb{R} \Rightarrow$$
$$[0 < |h| < s \Rightarrow \left|\tfrac{f(c+h)-f(c)}{h} - f'(c)\right| < r]]]$$

which is a true proposition in $\mathbb{R}$. But this proposition, true for every $r \in \mathbb{R}^+$, simply means that f is differentiable (in the usual sense) at c. •

Exercise 4.58 *Using infinitesimal arguments (based on theorem 4.57) derive the rules for the algebra of differentiable functions. That is to say, prove that if f and g are* *two functions differentiable at a point c in the interior of their domain, then $f+g$ and fg are differentiable at c, the same being true for the quotient f/g in the case when $g(c)$ differs from zero. Further, show that we have*

1. $(f \pm g)'(c) = f'(c) \pm g'(c)$
2. $(fg)'(c) = f'(c)g(c) + f(c)g'(c)$
3. $\left(\frac{f}{g}\right)'(c) = \frac{f'(c)g(c) - f(c)g'(c)}{[g(c)]^2}$.

Theorem 4.59 If f is differentiable at c then f is continuous at c.

Proof From the equality

$$^*f(c+\delta) = f(c) + \delta f'(c) + \delta\epsilon(\delta)$$

(where $\epsilon(\delta) \approx 0$), and the fact that $f'(c) \in \mathbb{R}$, it follows that

$$^*f(c+\delta) \approx f(c)$$

for any $\delta \approx 0$ (including $\delta = 0$). Thus f is continuous at c. •

Theorem 4.60 Let g be differentiable at the point c and suppose that f is differentiable at $g(c)$. Then $f \circ g$ is differentiable at c and

$$[f \circ g]'(c) = [f' \circ g](c).g'(c) \quad .$$

Proof By the preceding theorem, g is a function continuous at the point c and, therefore, $^*g(c+\delta) - g(c)$ is infinitesimal for any $\delta \approx 0$ (zero or non-zero). If, for some $\epsilon \approx 0$, we were to have $^*g(c+\epsilon) - g(c) = 0$ then, for any $\delta \approx 0, \delta \neq 0$ we would get

$$\frac{^*g(c+\delta) - g(c)}{\delta} \approx \frac{^*g(c+\epsilon) - g(c)}{\epsilon} = 0$$

and, therefore, in such a case we would get the result $g'(c) = 0$. Then we would have

$$^*g(c+\delta) = g(c) + \delta\eta(\delta)$$

where $\eta(\delta) \approx 0$ for any $\delta \approx 0$. If we have $\eta \neq 0$ then

$$\begin{aligned}
[f \circ g]'(c) &= \mathbf{st}\left(\frac{^*[f \circ g](c+\delta) - [f \circ g](c)}{\delta}\right) \\
&= \mathbf{st}\left(\frac{^*f[^*g(c+\delta)] - f[g(c)]}{\delta}\right) \\
&= \mathbf{st}\left(\frac{^*f[g(c) + \delta\eta] - [f[g(c)]}{\delta\eta}\eta\right) = 0 = f'[g(c)].g'(c)
\end{aligned}$$

which shows that the formula is satisfied in this case. The same result is also valid if we have $\eta = 0$ since then we would have $^*g(c+\delta) = g(c)$ and $^*f[^*g(c+\delta)] - f[g(c)] = 0$.

If $^*g(c+\delta) \neq g(c)$ for all $\delta \approx 0, \delta \neq 0$, then we can write

$$\frac{^*[f \circ g](c+\delta) - [f \circ g](c)}{\delta} = \frac{^*f[^*g(c+\delta)] - f[g(c)]}{\delta}$$

$$= \frac{^*f[^*g(c+\delta)] - f[g(c)]}{^*g(c+\delta) - g(c)} \frac{^*g(c+\delta) - g(c)}{\delta} .$$

As $^*g(c+\delta) = g(c) + \delta[g'(c) + \epsilon(\delta)] = g(c) + \delta'$, where $\delta' = \delta[g'(c) + \epsilon(\delta)] \approx 0$, we have

$$\frac{^*[f \circ g](c+\delta) - [f \circ g](c)}{\delta} = \frac{^*f[^*g(c+\delta)] - f[g(c)]}{^*g(c+\delta) - g(c)} . \frac{^*g(c+\delta) - g(c)}{\delta}$$

$$= \frac{^*f[g(c)+\delta'] - f[g(c)]}{\delta'} . \frac{^*g(c+\delta) - g(c)}{\delta}$$

$$\approx f'[g(c)].g'(c) = [f' \circ g].g'(c)$$

which completes the proof. •

Infinitesimal microscopes

As has now been shown, the use of infinite numbers and infinitesimals in NSA legitimises (*a posteriori*) the Leibnizian concepts for the determination of tangents of plane curves (see chap. 1). This would be more readily appreciated if we could amplify the graph of a function in an infinitesimal neighbourhood of a given point by means of a hypothetical *microscope of infinite power of resolution.*

Let σ be an arbitrary positive infinite number. An infinitesimal microscope of power $\sigma \in {}^*\mathbb{R}^+_\infty$ focussing on a point (a, b) of the hyperreal plane, (standard or nonstandard), is supposed to effect in mathematical terms the change of variables

$$M : \mathcal{B}_{1/\sigma}(a,b) \subset \mathbb{R}^2 \rightarrow \mathcal{B}_1(0,0) \subset {}^*\mathbb{R}^2$$

defined by

$$\begin{cases} u \equiv u(x,y) & = & \sigma(x-a) \\ v \equiv v(x,y) & = & \sigma(y-b) \end{cases}$$

where $\mathcal{B}_{1/\sigma}(a, b)$ denotes the euclidean ball of infinitesimal radius $1/\sigma$ centred at the point (a, b), and similarly $\mathcal{B}_1(0,0)$ is the open ball of unit radius centred at the origin.

Example 4.61 Let $f(x) = x^2, \quad x \in \mathbb{R}$.

1. Infinitesimal microscope focussed on the standard point $(0,0)$.
The change of coordinates is defined by

$$\begin{cases} u & = & \sigma x \\ v & = & \sigma y \end{cases}$$

giving the field of vision of the microscope

$$v = \sigma \left(\frac{1}{\sigma} u \right)^2 = \frac{u^2}{\sigma} .$$

As $|u| < 1$ then $0 \leq v \leq 1/\sigma \approx 0$ which means that in the infinitesimal neighbourhood of the origin, the amplified curve coincides, at least up to an infinitesimal, with the straight line $v = 0$.

This result corresponds to the fact that in this case we have $f'(0) = 0$.

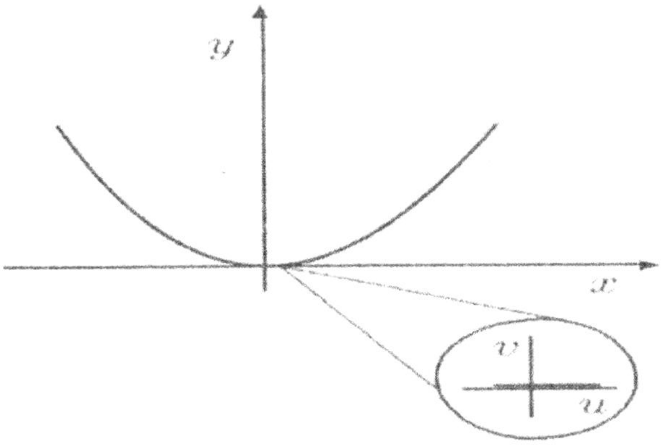

2. Infinitesimal microscope focussed on the standard point $(1,1)$. In this case the change of variables is defined by

$$\begin{cases} u = \sigma(x-1) \\ v = \sigma(y-1) \end{cases} .$$

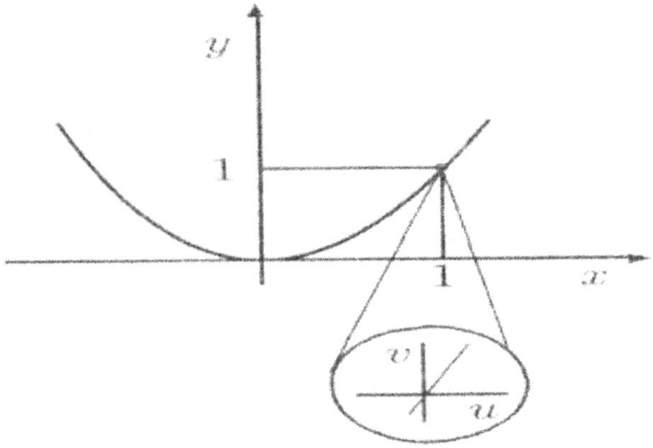

This gives

$$\frac{v}{\sigma}+1=\left(\frac{u}{\sigma}+1\right)^2$$

or, equivalently,

$$v=\frac{u^2}{\sigma}+2u\approx 2u\quad .$$

Just as before, taking account of the fact that $|u|<1$, we have $v\approx 2u$ which means that, in the infinitesimal neighbourhood of the point $(1,1)$, the microscopic graph coincides, up to an infinitesimal, with the straight line of slope 2: the derivative of f at the point $(1,1)$ is $s'(1)=2$.

3. Infinitesimal microscope focussed on the nonstandard point (ϵ,ϵ^2), where $\epsilon\approx 0$. Now focussing the microscope on a point infinitely close to the origin, but distinct from it, we have

$$\begin{cases} u &= \sigma(x-\epsilon) \\ v &= \sigma(y-\epsilon^2) \end{cases}$$

whence we get

$$v=2\epsilon u+\frac{u^2}{\sigma}\approx 0\quad .$$

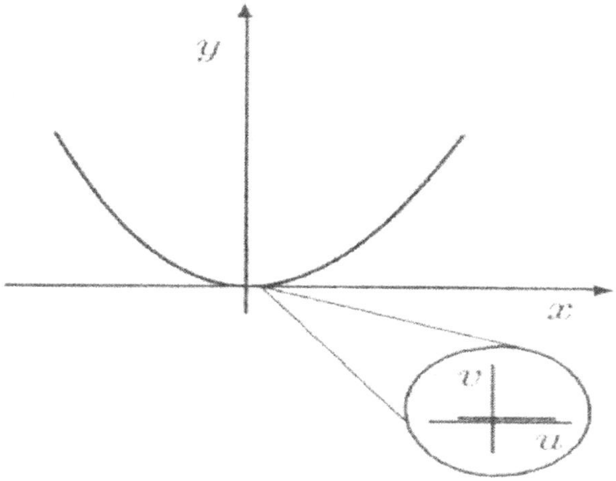

The microscopic graph coincides, up to an infinitesimal, with the axis $v=0$.

Example 4.62 Now consider the function defined by

$$f(x)=\begin{cases} x^2\sin\frac{1}{x} & \text{if } x\neq 0 \\ 0 & \text{if } x=0 \end{cases}\quad .$$

1. Infinitesimal microscope focussed on the standard point $(0,0)$.
The change of variables being defined by

$$\begin{cases} u &= \sigma x \\ v &= \sigma y \end{cases}$$

we have

$$v = \begin{cases} \frac{u^2}{\sigma} \sin \frac{\sigma}{u} & \text{for } u \neq 0 \\ 0 & \text{for } u = 0 \end{cases}$$

and, therefore, taking account of the fact that $|u| < 1$, we have $v \approx 0$ which means that f is differentiable at the origin and that $f'(0) = 0$.

2. Infinitesimal microscope focussed on the nonstandard point $(1/k\pi, 0)$ with $k \in {}^*\mathbb{N}_\infty$ and odd.
We will have

$$\begin{cases} u &= \sigma(x - 1/k\pi) \\ v &= \sigma y \end{cases}$$

whence we get

$$\begin{cases} x &= \frac{u}{\sigma} + \frac{1}{k\pi} \\ y &= \frac{v}{\sigma} \end{cases}$$

Taking, in particular, a microscope of power equal to $\sigma = k^2\pi^2$ we would have

$$\begin{cases} x &= \frac{u}{k^2\pi^2} + \frac{1}{k\pi} \\ y &= \frac{v}{k^2\pi^2} \end{cases}$$

and, therefore, for $|u| < 1$ we would always have $x \neq 0$. In the infinitesimal neighbourhood of the point in question we would get

$$\begin{aligned} v &= \frac{(u + k\pi)^2}{k^2\pi^2} \sin \frac{k^2\pi^2}{u + k\pi} \\ &= \left(\frac{u^2 + 2k\pi u}{k^2\pi^2} + 1 \right) \sin \left(k\pi - \frac{k\pi u}{u + k\pi} \right) \end{aligned}$$

Since

$$\frac{u^2 + 2k\pi u}{k^2\pi^2} \approx 0$$

$$\sin \left(k\pi - \frac{k\pi u}{u + k\pi} \right) = \sin \left(\frac{k\pi u}{u + k\pi} \right)$$

$$\frac{k\pi}{u + k\pi} \approx 1$$

then it follows finally that

$$v \approx \sin u$$

which shows that the microscopic image of the graph of f about the point $1/k\pi \approx 0$ is "approximately" sinusoidal.

In this example, as opposed to the preceding one, the microscopic image of the graph of the function at a point infinitely close to the origin, but distinct from it, differs from the graph of the function at the origin. This follows from the fact that the function is not continuously differentiable at the origin, in contrast to what happens in example 4.61.

Theorem 4.63 Let $f : (a, b) \to \mathbb{R}$ be a function with a local extremum at the point $c \in (a, b)$. If f is differentiable at c then $f'(c) = 0$.

Proof Suppose that f has a maximum at $c \in (a, b)$. Then, applying the transfer principle, we would have

$$^*f(c + \epsilon) \leq f(c)$$

for any $\epsilon \approx 0$. δ and δ' being infinitesimals, the first positive and the second negative, we have successively

$$f'(c) \approx \frac{^*f(c + \delta) - f(c)}{\delta} \leq 0 \leq \frac{^*f(c + \delta') - f(c)}{\delta'} \approx f'(c)$$

which, noting that $f'(c)$ is a real number, gives the equality $f'(c) = 0$. (The case when $f'(c)$ has a minimum at $c \in (a, b)$ can be treated similarly.) •

Exercise 4.64 *Let $f : [a, b] \to \mathbb{R}$ be a function continuous on $[a, b]$ and differentiable on (a, b). Prove*

1. Rolle's Theorem: If $f(a) = 0 = f(b)$ then there exists $x \in (a, b)$ such that $f'(x) = 0$.

2. The Mean Value Theorem: There exists $x \in (a, b)$ such that

$$f'(x) = \frac{f(b) - f(a)}{b - a} .$$

3. Corollaries:
If $f'(x) = 0$ in (a, b) then $f(x)$ is constant on $[a, b]$.
If $f'(x) > 0$ in (a, b) then $f(x)$ is monotone increasing on $[a, b]$.
If $f'(x) < 0$ in (a, b) then $f(x)$ is monotone decreasing on $[a, b]$.

4.3.2 Integration of continuous functions

We consider here only the Riemann integral of continuous functions in order to verify that the method of the calculation of plane areas developed in the 17th century can be founded on a secure base using hyperreal numbers. The method extends easily to the class of sectionally continuous functions. The integration of other functions and the study of other more advanced types of integration will only be considered in later chapters in the context of the "theory of Loeb measures".

Hyperfinite Darboux sums Let $f : [a, b] \to \mathbb{R}$ be a continuous function. For $n \in \mathbb{N}$ let $\Delta_n = (b - a)/n$ and consider a partition $\pi_n = \{s + j\Delta_n : j = 0, 1, \ldots, n\}$. Based on this partition we define the *lower* and *upper* Darboux sums in the usual way

$$S_n^L(f) = \sum_{j=1}^{n} m_j \Delta_n$$

$$S_n^U(f) = \sum_{j=1}^{n} M_j \Delta_n$$

where, for each $j = 1, 2, \dots, n$, m_j and M_j are respectively the minimum and maximum values of the function f in the interval $[a + (j-1)\Delta_n, a + j\Delta_n]$. It is clear that we have

$$S_n^L(f) \leq S_n^U(f)$$

and, by considering common refinements, we can establish the inequality

$$S_n^L(f) \leq S_{n'}^U(f)$$

whatever the numbers $n, n' \in \mathbb{N}$.

f being an arbitrarily fixed function, both $S_n^L(f)$ and $S_n^U(f)$ will actually be functions of the variable $n \in \mathbb{N}$. It is possible, therefore, to extend them in the usual way to ${}^*\mathbb{N}$, thereby obtaining ${}^*S_\nu^L(f)$ and ${}^*S_\nu^U(f)$ respectively, for each hypernatural number ν (finite or infinite).

Theorem 4.65 If $f : [a, b] \to \mathbb{R}$ is a continuous function, the relation

$${}^*S_\nu^L(f) \approx {}^*S_\nu^U(f)$$

is satisfied for every hypernatural number $\nu \in {}^*\mathbb{N}_\infty$.

Proof For any $n \in \mathbb{N}$ we have

$$0 \leq S_n^U(f) - S_n^L(f) \leq \lambda(b-a)$$

where $\lambda \equiv \lambda(n) = \max\{M_j - m_j : j = 1, 2, \dots, n\}$. Since λ is the maximum element of a finite set, then for each $n \in \mathbb{N}$ there exists k, $(1 \leq k \leq n)$, such that $\lambda(n) = M_k - m_k$, where M_k and m_k are respectively the maximum and minimum of f in the interval $I_k \equiv [a + (k-1)\Delta_n, a + k\Delta_n]$. Since f is continuous then M_k and m_k are attained in I_k and so we can state that for each $n \in \mathbb{N}$ there will exist $x, y \in [a, b]$ such that $|x - y| < \Delta_n$ and $\lambda(n) = |f(x) - f(y)|$. Formally

$$\forall_n [n \in \mathbb{N} \Rightarrow \exists_{x,y} [x, y \in [a, b] \wedge$$
$$|x - y| < \Delta_n \wedge 0 \leq S_n^U(f) - S_n^L(f) \leq (b-a)|f(x) - f(y)|]]$$

This proposition, which is relative to $\mathbb{R}$, can be transferred to ${}^*\mathbb{R}$ giving

$$\forall_\nu [\nu \in {}^*\mathbb{N} \Rightarrow \exists_{x,y} [x, y \in {}^*[a, b] \wedge$$
$$|x - y| < \Delta_\nu \wedge 0 \leq {}^*S_\nu^U(f) - {}^*S_\nu^L(f) \leq (b-a)|{}^*f(x) - {}^*f(y)|]]$$

which is a proposition true in ${}^*\mathbb{R}$.

Taking, in particular, $\nu \in {}^*\mathbb{N}_\infty$ we have $\Delta_\nu \equiv (b-a)/\nu \approx 0$ and, therefore, we will have $x \approx y$. From the continuity of f in $[a,b]$ (actually, in accordance with theorem 4.47, the *uniform* continuity of f) it follows that ${}^*f(x) \approx {}^*f(y)$ and consequently that ${}^*S^U_\nu(f) - {}^*S^L_\nu(f) \approx 0$. That is,

$$\forall_\nu [\nu \in {}^*\mathbb{N}_\infty \Rightarrow {}^*S^U_\nu(f) - {}^*S^L_\nu(f)]$$

as it was required to show. •

Theorem 4.66 If $f : [a,b] \to \mathbb{R}$ is a continuous function then

$${}^*S^U_\nu(f),\ {}^*S\nu^L(f) \in {}^*\mathbb{R}_b$$

$${}^*S^U_\nu(f) \approx {}^*S^U_\sigma(f) \quad \text{and} \quad {}^*S^L_\nu(f) \approx {}^*S^L_\sigma(f) \in {}^*\mathbb{R}_b$$

for any infinite hypernatural numbers ν and σ.

Proof Let ν be any number in ${}^*\mathbb{N}_\infty$. Then we have

$$|{}^*S^U_\nu(f)| = \left|\sum_{j=1}^{\nu} {}^*M_j \frac{b-a}{\nu}\right|$$

$$\leq \max_{1\leq j\leq \nu} |{}^*M_j| \sum_{j=1}^{\nu} \frac{b-a}{\nu} = (b-a).\ \max_{1\leq j\leq \nu} |{}^*M_j|$$

and so, since $|{}^*M_n| \leq \max_{a\leq x\leq b} |f(x)|$, $n = 1, 2, \ldots, \nu$, and since the number $\max_{1\leq j\leq \nu} |{}^*M_j|$ is finite, it follows that $|{}^*\bar{S}_\nu(f)|$ is also a finite number. Similarly we can show that $|{}^*S^L_\nu(f)|$ is a finite number.

To prove the second part of the theorem, let $\nu, \sigma \in {}^*\mathbb{N}_\infty$ be two arbitrarily chosen infinite numbers. Using suitable common hyperfinite refinements we obtain successively

$${}^*S^L_\nu(f) \leq {}^*S^U_\sigma(f) \approx {}^*S^L_\sigma(f) \leq {}^*S^U_\nu(f)$$

and, as ${}^*S^L_\nu(f) \approx {}^*S^U_\nu(f)$ we must get

$${}^*S^L_\nu(f) \approx {}^*S^L_\sigma(f) \approx {}^*S^U_\sigma(f) \approx {}^*S^U_\nu(f)$$

as it was required to show. •

From this there follows immediately the nonstandard definition of the Riemann integral of a continuous function f over the interval $[a,b]$

$$\int_a^b f(x)dx = \mathbf{st}({}^*S^L_\nu(f)) = \mathbf{st}({}^*S^U_\nu(f)), \tag{4.8}$$

where ν denotes any infinite hypernatural number.

Exercise 4.67 *Let f and g be two functions continuous on the interval $[a,b]$. Then, for any $\lambda \in \mathbb{R}$ and any $c \in [a,b]$,*

$$\text{(a)} \qquad \int_a^b \lambda f(x)dx = \lambda \int_a^b f(x)dx$$

(b) $\displaystyle\int_a^b [f(x)+g(x)]dx = \int_a^b f(x)dx + \int_a^b g(x)dx$

(c) $\displaystyle\int_a^b f(x)dx = \int_a^c f(x)dx \int_c^b f(x)dx$

(d) $\displaystyle f \leq g \Rightarrow \int_a^b f(x)dx \leq \int_a^b g(x)dx$

Verify these properties of the Riemann integral using the given nonstandard definition.

Chapter 5

Further Developments

5.1 Internal and External Sets and Functions

The nonstandard extensions of sets of real numbers and real-valued functions which we have considered up to now are merely particular examples of nonstandard mathematical objects in general. We can actually define arbitrary nonstandard sets and functions which need not be related to any previously given standard sets and functions. Indeed, it is perfectly possible to study in the context of NSA any type of mathematical object which might be encountered in real analysis. However, $\mathbb{R}$ and $^*\mathbb{R}$ are distinct sets and it is therefore not surprising that while $^*\mathbb{R}$ has *some* of the properties of $\mathbb{R}$ it does not possess *all* of them. As a particularly important example we can see immediately that it is impossible to transfer directly the **least upper bound axiom** of $\mathbb{R}$ to $^*\mathbb{R}$: it is enough to recall that $\mathbb{R}$ is a non-empty subset of $^*\mathbb{R}$ which is bounded above (by any positive infinite number) but which, nevertheless, has no supremum. On the other hand it is also easy to see that there are certain subsets of $^*\mathbb{R}$ to which this axiom does apply. For example the interval (x, y) with $x, y \in {}^*\mathbb{R}$, $(x < y)$ always has a least upper bound in $^*\mathbb{R}$, whatever the value of y.

Thus there are certain nonstandard objects which do possess the same formal properties as objects of similar type in real analysis, although this will not be the case for all nonstandard objects. It is therefore desirable to divide the objects of nonstandard analysis into two distinct classes: these comprise the so-called *internal* and *external* objects respectively. In this chapter we consider such a classification, not for all the possible types of nonstandard objects but at least in so far as sets of hyperreal numbers and numerical-valued functions are concerned.

Note 5.1 It was this distinction of mathematical objects into 'internal' or 'external' which first let Robinson see how the systematic use of infinitesimals proposed by Leibniz might be carried out. What is more, it allows us to justify the modern use of infinitesimals in contexts which go beyond the scope of elementary calculus to a still unknown extent.

5.1.1 Internal sets

Definition 5.2 A subset $\mathcal{A}$ of $^*\mathbb{R}$ is said to be **internal** if there exists a sequence $(A_n)_{n\in\mathbb{N}}$ of subsets of $\mathbb{R}$ such that for each hyperreal number $x = [(x_n)_{n\in\mathbb{N}}]$ we have

$$x \in \mathcal{A} \text{ if and only if } \{n \in \mathbb{N} : x_n \in A_n\} \in \mathcal{U} \ .$$

The subsets of $^*\mathbb{R}$ which are not internal are said to be **external**.

It is convenient at this stage to write

$$\mathcal{A} \equiv [(A_n)_{n\in\mathbb{N}}]$$

although this represents a simplification of the proper definition of the internal set $\mathcal{A}$ in terms of the equivalence class, modulo $\mathcal{U}$, $[(A_n)_{n\in\mathbb{N}}]$. It is useful in practice, but not entirely rigorous, and the relation between these two objects $\mathcal{A}$ and $[(A_n)_{n\in\mathbb{N}}]$ will be fully clarified later on in chapter 6. For the moment, however, the definition given above is good enough to allow a relatively elementary development of NSA.

If, in particular, we were to have $A_n = A \subset \mathbb{R}$ for $\mathcal{U}$-nearly all $n \in \mathbb{N}$, then

$$\mathcal{A} \equiv [(A_n)_{n\in\mathbb{N}}] = {}^*A$$

where $^*A$ is the nonstandard extension of the (standard) set A, defined earlier. Recall that the set $^*\mathbb{R}$ is itself just the nonstandard extension of $\mathbb{R}$. Another important nonstandard extension, also introduced earlier, is the set $^*\mathbb{N}$ of all hypernatural numbers (example 3.31).

The internal sets of $^*\mathbb{R}$ are the natural analogues of the subsets of $\mathbb{R}$, in that they possess the same formal properties. As a general rule, which will be fully justified later on, we can say that

> properties of subsets of $\mathbb{R}$ transfer directly to *internal subsets* of $^*\mathbb{R}$ although not to all arbitrary subsets of $^*R$.

The process of transfer can be realised term by term, modulo the ultrafilter $\mathcal{U}$ used in the construction of $^*\mathbb{R}$. For example, the fact that $\mathbb{N}$ is a well-ordered set means that

> every non-empty subset of $\mathbb{N}$ possesses a first element.

This property does not transfer directly to arbitrary subsets of $^*\mathbb{N}$, the set of the infinite hypernatural numbers: the set $^*\mathbb{N}_\infty = {}^*\mathbb{N}\backslash\mathbb{N}$, for example, is a non-empty subset of $^*\mathbb{N}$ which, nevertheless, does not possess a first element. However, the well-ordering property does transfer to the set of hypernatural numbers in the following, modified, sense:

Theorem 5.3 Every **internal** non-empty subset of $^*\mathbb{N}$ possesses a first element.

Proof Let $\mathcal{M} \subset {}^*\mathbb{N}$ be a non-empty internal subset. Then $\mathcal{M} \equiv [(M_n)_{n\in\mathbb{N}}]$ where $M_n \subset \mathbb{N}$, $M_n \neq \emptyset$ for $\mathcal{U}$-nearly all $n \in \mathbb{N}$. It is clear that, for $\mathcal{U}$-nearly all $n \in \mathbb{N}$, the set M_n possesses a first element, which we may write as m_n. Then $\mu \equiv [(m_n)_{n\in\mathbb{N}}]$ belongs to $\mathcal{M}$ and is its first element. •

Since there is no smallest infinite hypernatural number it follows at once from this theorem that ${}^*\mathbb{N}_\infty$ cannot be an internal set, and is therefore an external subset of ${}^*\mathbb{N}$.

For another example we return once again to the least upper bound property of ${}^*\mathbb{R}$ and show that this also transfers from subsets of $\mathbb{R}$ to *internal* subsets of ${}^*\mathbb{R}$:

Theorem 5.4 Every non-empty **internal** subset of ${}^*\mathbb{R}$ which is bounded above has a least upper bound.

Proof Let $\mathcal{A} \equiv [(A_n)_{n\in\mathbf{N}}]$ be a non-empty subset of ${}^*\mathbb{R}$ which is bounded above by $\alpha \equiv [(a_n)_{n\in\mathbf{N}}] \in {}^*\mathbb{R}$. Then for $\mathcal{U}$-nearly all $n \in \mathbb{N}$ the set $A_n \subset \mathbb{R}$ is not empty and is bounded above by $a_n \in \mathbb{R}$. Consequently $\sup(A_n)$ exists for $\mathcal{U}$-nearly all $n \in \mathbb{N}$, and it follows that the hyperreal number

$$\sup(\mathcal{A}) \equiv [(\sup(A_n))_{n\in\mathbf{N}}$$

is the least upper bound of $\mathcal{A} \subset {}^*\mathbb{R}$. •

Exercise 5.5 *Confirm that the number* $\sup(\mathcal{A})$, *as defined in the proof of of the preceding theorem, really is the least upper bound of the internal set* $\mathcal{A}$.

Exercise 5.6 *Show that boolean operations on internal sets generate internal sets, so that the family of internal sets is closed under (finitely many) boolean operations on sets.*

Working with an explicit sequential model for NSA, as we have done so far, does make concepts and methods of NSA easy to understand, at least to begin with. However the continual use of equivalence classes of sequences becomes increasingly tedious and, to some extent, limits both creativity and clarity. As one becomes more familiar with the methods of NSA, it is much easier to appeal to more elaborate analytic tools which allow substantial simplification of arguments. Examples of such tools are provided by the following consequences of Theorem 5.4.

Theorem 5.7 Overflow Let $\mathcal{A} \equiv [(A_n)_{n\in\mathbf{N}}]$ be an internal subset of ${}^*\mathbb{R}$. If $\mathcal{A}$ contains arbitrarily large finite elements then $\mathcal{A}$ contains an infinite element.

Proof If $\mathcal{A}$ were unbounded then there would be nothing to prove. Suppose then that $\mathcal{A}$ is bounded above. Let α be the least upper bound of $\mathcal{A}$. It is clear that α cannot be finite, and it follows from the definition of least upper bound that there must exist $x \in \mathcal{A}$ such that $\alpha/2 \leq x < \alpha$. Thus $\mathcal{A}$ contains an infinite element. •

Theorem 5.8 Underflow Let $\mathcal{A} \equiv [(A_n)_{n\in\mathbf{N}}]$ be an internal subset of ${}^*\mathbb{R}$. If $\mathcal{A}$ contains arbitrarily small positive infinite elements then $\mathcal{A}$ contains a finite element.

Proof Let $\mathcal{A}^+$ be the set formed of all the positive elements of $\mathcal{A}$ and let $b = \inf(\mathcal{A}^+)$. From the hypothesis it follows that $b \in {}^*\mathbb{R}_b$, and from the definition of greatest lower bound there must exist $x \in \mathcal{A}^+$ such that $b \leq x < b+1$ and therefore such that $x \in {}^*\mathbb{R}_b$. Thus $\mathcal{A}$ contains a finite element. •

There is an important corollary:

Corollary 5.8.1 Infinitesimal overflow Let $\mathcal{S}$ be an internal set of hyperreal numbers containing all numbers $x > 0$ which are such that $x \approx 0$. Then there exists a real number $r > 0$ such that $(0, r) \subset \mathcal{S}$.

Proof The set $\mathcal{M}$ defined by

$$\mathcal{M} = \{\alpha \in {}^*\mathbb{R}^+ : \quad (0, 1/\alpha) \subset \mathcal{S}\}$$

is an internal subset[1] of ${}^*\mathbb{R}$ which contains arbitrarily small positive infinite hyperreal numbers. By theorem 5.8 there exists $a \in \mathbb{R}^+$ such that $a \in \mathcal{M}$. Taking $r = 1/a$ completes the proof. •

This corollary shows in particular that $mon(0)$ is not an internal set. Further, theorem 5.8 together with thisorollary shows that it is not possible to separate *internally* the finite numbers from the infinite numbers or the infinitesimal numbers from the standard numbers.

Example 5.9 We now use theorem 5.8 to solve exercise 4.12 by purely nonstandard arguments.

Let $(a_n)_{n\in\mathbf{N}}$ be a Cauchy sequence and suppose if possible that it is not bounded. Then there exists $\sigma \in {}^*\mathbb{N}_\infty$ such that $a_\sigma \notin {}^*\mathbb{R}_b$. Consider the set defined by

$$\mathcal{M} = \{n \in {}^*\mathbb{N} : \quad |a_\sigma - a_n| < 1\} \ .$$

From the nonstandard characterisation of a Cauchy sequence we can conclude that $\mathcal{M}$ contains all the infinite numbers and therefore, by underflow, that it must contain a finite number $n_1 \in \mathbb{N}$. This means that $a_{n_1} \notin {}^*\mathbb{R}_b$, which is absurd since all the terms of the sequence $(a_n)_{n\in\mathbf{N}}$ are real numbers.

Another result of great importance and one whose generalisation will be seen to be crucial for the treatment of many questions of mathematical analysis is the so-called **principle of $\aleph_1$-saturation** or **countable saturation**:

Theorem 5.10 (Countable saturation) Let $(\mathcal{A}_r)_{r\in\mathbf{N}}$ be a sequence of internal sets with the **finite intersection property** [2] : that is, a sequence of sets such that $\bigcap_{r=1}^{i} \mathcal{A}_r \neq \emptyset$ for each $i \in \mathbb{N}$. Then we must have

$$\bigcap_{r\in\mathbf{N}} \mathcal{A}_r \neq \emptyset \ .$$

Proof Since each element $\mathcal{A}_r$ of the sequence is internal, we must be able to write

$$\mathcal{A}_r \equiv [(A_{r,n})_{n\in\mathbf{N}}]$$

where, for each $n = 1, 2, 3, \ldots$, $A_{r,n}$ is a standard set.

---

[1] Taking $\mathcal{S} \equiv [(s_n)_{n\in\mathbf{N}}]$ and $\alpha \equiv [(a_n)_{n\in\mathbf{N}}]$ and then defining, for each $n \in \mathcal{N}$,

$$M_n = \{a_n \in \mathcal{R}^+ : \quad (0, 1/a_n) \subset S_n\}$$

we obtain $\mathcal{M} \equiv [(M_n)_{n\in\mathcal{N}}]$ which shows that $\mathcal{M}$ is internal.

[2] This property is not generally valid in $\mathbb{R}$. The sequence of sets $(0, 1/n) \subset \mathbb{R}$, $n = 1, 2, \ldots$, for example, has the finite intersection property but $\bigcap_{n\in\mathbf{N}}(0, 1/n) = \emptyset$.

Since the given sequence has the finite intersection property, it follows that $\mathcal{A}_r \neq \emptyset$ for any $r \in \mathbb{N}$ whatsoever. In particular, $\mathcal{A}_1 \neq \emptyset$ and so without any loss of generality we may suppose that we have $A_{1,n} \neq \emptyset$ for all $n \in \mathbb{N}$. Since[3]

$$\bigcap_{r=1}^{i} \mathcal{A}_r \equiv \bigcap_{r=1}^{i} [(A_{r,n})_{n\in\mathbb{N}}] = \left[\left(\bigcap_{i=1}^{r} A_{r,n}\right)_{n\in\mathbb{N}}\right]$$

then from the finite intersection property it follows that

$$\{n \in \mathbb{N} : \bigcap_{r=1}^{i} A_{r,n} \neq \emptyset\} \in \mathcal{U}$$

for any $i = 1, 2, \ldots$.

$$\begin{array}{ccccc}
\mathcal{A}_1 & \mathcal{A}_2 & \ldots & \mathcal{A}_i & \ldots \\
A_{1,1} & A_{2,1} & \ldots & A_{i,1} & \ldots \\
A_{1,2} & A_{2,2} & \ldots & A_{i,2} & \ldots \\
\cdot & \cdot & & & \cdot \\
\cdot & \cdot & & & \cdot \\
\cdot & \cdot & & & \cdot \\
A_{1,n} & A_{2,n} & \ldots & A_{i,n} & \ldots \\
\cdot & \cdot & & & \cdot \\
\cdot & \cdot & & & \cdot
\end{array}$$

If we let

$$i_n = \max\left\{i \in \{1, 2, \ldots, n\} \quad : \bigcap_{r=1}^{i} A_{r,n} \neq \emptyset\right\}$$

then, for each $n = 1, 2, \ldots$, i_n is a well defined number since, by hypothesis, we have $A_{1,n} \neq \emptyset$. For each $n = 1, 2, \ldots$, the intersection $\bigcap_{r=1}^{i_n} A_{r,n}$ is not empty, and so we can choose $t_n \in \bigcap_{r=1}^{i_n} A_{r,n}$.

Setting $t \equiv [(t_n)_{n\in\mathbb{N}}] \in {}^*\mathbb{R}$ we now show that t belongs to the intersection of all the $\mathcal{A}_r$. For this it is enough to confirm that we have $t \in \mathcal{A}_r$ for every $r \in \mathbb{N}$. In fact, if i is any natural number whatever, then for each $n \in \mathbb{N}$ for which $i_n \geq i$ we must have $t_n \in A_{i,n}$ and therefore

$$\{n \in \mathbb{N} : t_n \in A_{i,n}\} \supset \{n \in \mathbb{N} : \quad i_n \geq i\} \ .$$

[3]Let $x \equiv [(x_n)_{n\in\mathbb{N}}] \in {}^*\mathbb{R}$. Then

$$x \in \bigcap_{r=1}^{i} \mathcal{A}_i \Rightarrow x \in \mathcal{A}_r, \forall_{r=1,2,\ldots,i} \Rightarrow \{n \in \mathbb{N} : x_n \in A_{r,n}\} \in \mathcal{U}, \forall_{r=1,2,\ldots,i}$$

$$\Rightarrow \{n \in \mathbb{N} : x_n \in \bigcap_{r=1}^{i} A_{r,n}\} \in \mathcal{U} \Rightarrow x \in \left[\left(\bigcap_{r=1}^{i} A_{r,n}\right)_{n\in\mathbb{N}}\right].$$

Since

$$\{n \in \mathbb{N} : i_n \geq i\} = \{n \in \mathbb{N} : n \geq i\} \cap \{n \in \mathbb{N} : \bigcap_{r=1}^{i} \mathcal{A}_{r,n} \neq \emptyset\},$$

the set $\{n \in \mathbb{N} : i_n \geq i\}$ belongs to $\mathcal{U}$, the same being therefore true of the set $\{n \in \mathbb{N} : t_n \in \mathcal{A}_{i,n}\}$. Thus $t \in \mathcal{A}_i$, and the result follows since i is arbitrary. •

Exercise 5.11 *Use countable saturation to show that there must exist positive infinite elements in* $^*\mathbb{R}$.

Exercise 5.12 *Let* $(\epsilon_n)_{n\in\mathbb{N}}$ *be a sequence of positive infinitesimal numbers. Use countable saturation to show that the least upper bound of the sequence is an infinitesimal number.* **Hint***: Note that the maximal element of the set* $\{\epsilon_{n_1}, \epsilon_{n_2}, \ldots, \epsilon_{n_k}\}$ *belongs to the set defined by*

$$[\epsilon_{n_1}, \frac{1}{n_1}) \cap [\epsilon_{n_2}, \frac{1}{n_2}) \cap \ldots \cap [\epsilon_{n_k}, \frac{1}{n_k}) \ .$$

Dually, show also that every sequence of positive infinite hyperreal numbers must have a positive infinite greatest lower bound.

Countable saturation has an immediate consequence with regard to the (external, Cantorian) cardinality of an internal set: there exist no internal sets which are countably infinite:

Theorem 5.13 An internal set is either finite or else uncountably infinite.

Proof Suppose, if possible, that the set $\mathcal{A} \equiv \{a_n : n \in \mathbb{N}\}$, with $a_i \neq a_j$ if $i \neq j$, is internal. Defining for each $n \in \mathbb{N}$

$$\mathcal{A}_n = \mathcal{A} \backslash \{a_1, a_2, \ldots, a_n\}$$

we have a sequence $(\mathcal{A}_n)_{n\in\mathbb{N}}$ of internal sets with the finite intersection property. Consequently, by countable saturation, the intersection $\bigcap_{n\in\mathbb{N}} \mathcal{A}_n$ would have to be different from $\emptyset$ which is, however, absurd. It follows at once that $\mathcal{A}$ cannot be an internal set. •

Corollary 5.13.1 If $(\mathcal{A}_r)_{r\in\mathbb{N}}$ is a sequence of internal sets then the union $\bigcup_{r\in\mathbb{N}} \mathcal{A}_r$ is internal if and only if it is equal to a finite union $\bigcup_{r=1}^{k} \mathcal{A}_r$ for some $k \in \mathbb{N}$.

Proof Every finite union of internal sets is an internal set (see Exercise 5.6). It remains to prove the converse. Suppose that $\mathcal{A} = \bigcup_{r\in\mathbb{N}} \mathcal{A}_r$ is internal. Then all the sets $\mathcal{A}\backslash\mathcal{A}_r$ will be internal and $\bigcap_{r=1}^{\infty}(\mathcal{A}\backslash\mathcal{A}_r) = \emptyset$. By $\aleph_1$-saturation there must exist $k \in \mathbb{N}$ such that $\bigcap_{r=1}^{k}(\mathcal{A}\backslash\mathcal{A}_r) = \emptyset$, or equivalently such that

$$\bigcap_{r=1}^{k}(\mathcal{A} \cap \mathcal{A}_r^c) = \mathcal{A} \cap (\bigcap_{r=1}^{k} \mathcal{A}_r^c) = \mathcal{A} \cap (\bigcup_{r=1}^{k} \mathcal{A}_r)^c = \mathcal{A}\backslash(\bigcup_{r=1}^{k} \mathcal{A}_r) = \emptyset \ .$$

On the one hand this shows that $\mathcal{A} \subset \bigcup_{r=1}^{k} \mathcal{A}_r$, while on the other hand from the definition of $\mathcal{A}$ we have $\bigcup_{r=1}^{k} \mathcal{A}_r \subset \mathcal{A}$. Hence $\mathcal{A} = \bigcup_{r=1}^{k} \mathcal{A}_r$. •

$\aleph_1$-saturation is an extremely important property of sets in the universe of NSA. In particular if for each internal set $\mathcal{A} \subset {}^*\mathbb{R}$ we define its standard part by

$$\mathbf{st}(\mathcal{A}) = \{\mathbf{st}(x) : \quad x \in \mathcal{A} \cap {}^*\mathbb{R}_b\},$$

then from $\aleph_1$-saturation we get the following result:

Theorem 5.14 If $\mathcal{A}$ is an internal subset of ${}^*\mathbb{R}$ then $\mathbf{st}(\mathcal{A})$ is closed in the usual topology of $\mathbb{R}$.

Proof Denoting the set $\mathbf{st}(\mathcal{A})$ more simply by A, we need to prove that A coincides with its closure: $\bar{A} = A$. Since $A \subset \bar{A}$ it is enough to show that $\bar{A} \subset A$. Accordingly, let a be any point of $\bar{A}$ whatsoever. For each $n \in \mathbb{N}$ the set defined by

$$\mathcal{A}_n = \mathcal{A} \cap \{x \in {}^*\mathbb{R} : |a - x| < 1/n\}$$

is internal, since it is the intersection of two internal sets. Further, these sets are not empty. For, since a is, by hypothesis, a point of closure of A, there must exist $a' \in A$ such that $|a' - a| < 1/n$; further, since $A = \mathbf{st}(\mathcal{A})$ then $a' = \mathbf{st}(x)$ for some $x \in \mathcal{A}$ and therefore $x \in \mathcal{A}_n$, and this shows that $\mathcal{A}_n$ is not empty. On the other hand, since $\bigcap_{j=1}^{n} \mathcal{A}_j = \mathcal{A}_n$ then every finite intersection of the sets $\mathcal{A}_n$ is not empty whence, by $\aleph_1$-saturation, it follows that

$$\bigcap_{n \in \mathbf{N}} \mathcal{A}_n \neq \emptyset \ .$$

Then, for any $x \in \bigcap_{n\in\mathbf{N}} \mathcal{A}_n$ we have $x \in \mathcal{A}$ and, since $|x - a| < 1/n$ for every $n \in \mathbb{N}$, $x \approx a$. In consequence, $\mathbf{st}(x) = a \in A \equiv \mathbf{st}(\mathcal{A})$, and it follows immediately that $\mathbf{st}(\mathcal{A})$ is closed. •

Corollary 5.14.1 Let A be any subset of $\mathbb{R}$. Then $\mathbf{st}({}^*A) = \bar{A}$.

Proof Since ${}^*A \subset {}^*\mathbb{R}$ is internal, then from theorem 5.14 we know that $\mathbf{st}({}^*A)$ is closed. We prove the equality $\mathbf{st}({}^*A) = \bar{A}$ by showing that $\bar{A} \subset \mathbf{st}({}^*A)$ and also that $\mathbf{st}({}^*A) \subset \bar{A}$.

Let a be any point of A. Then $a \in {}^*A \cap {}^*\mathbb{R}_b$ and, therefore, $a \in \mathbf{st}({}^*A)$ from which it follows that $A \subset \mathbf{st}({}^*A)$. Since $\bar{A}$ is the smallest closed set which contains A then we must have $\bar{A} \subset \mathbf{st}({}^*A)$.

To prove the other inclusion let a be an arbitrary point of $\mathbf{st}({}^*A)$. By the definition of $\mathbf{st}({}^*A)$ there exists $x \in {}^*A \cap {}^*\mathbb{R}_b$ such that $a = \mathbf{st}(x)$ and, therefore, $mon(a) \cap {}^*A \neq \emptyset$ which means that a is a point of closure of A. Since a is arbitrary we can write $\mathbf{st}({}^*A) \subset \bar{A}$, thus completing the proof. •

Hyperfinite sets The family of subsets of hypernatural numbers contains a subfamily of sets which are particularly important and interesting from the point of view of analysis. These are the so-called *hyperfinite sets.*

Definition 5.15 A subset $\mathcal{T} \equiv [(T_n)_{n\in\mathbf{N}}]$ is said to be **hyperfinite** (or ***-finite**) if for $\mathcal{U}$-nearly all $n \in \mathbb{N}$ the sets T_n are finite. If $\mathbf{card}(T_n)$ denotes the cardinality of T_n then the hypernatural number $\mathbf{card}(\mathcal{T}) \equiv [(\mathbf{card}(T_n))_{n\in\mathbf{N}}] \in {}^*\mathbb{N}$ denotes the **internal cardinality** of $\mathcal{T}$.

Example 5.16 Let $(m_n)_{n\in\mathbb{N}}$ be a sequence of arbitrary natural numbers and, for each $n \in \mathbb{N}$ let

$$T_n = \left\{0, \frac{1}{m_n}, \frac{2}{m_n}, \ldots, \frac{m_n - 1}{m_n}\right\} .$$

If ν denotes the hypernatural number defined by the sequence $(m_n)_{n\in\mathbb{N}}$ then,

$$\mathcal{T} \equiv [(T_n)_{n\in\mathbb{N}}] = \left\{0, \frac{1}{\nu}, \frac{2}{\nu}, \ldots, \frac{\nu - 1}{\nu}\right\}$$

is a hyperfinite set with internal cardinality equal to $\nu + 1$.

The set $\mathcal{T}$ constitutes a uniform hyperfinite partition of the interval ${}^*[0,1]$. In particular, if ν is an infinite number then $\mathcal{T}$ is a uniform infinite partition of that interval. Recall that it would have been partitions of this type which were considered for the proposed definition of the Riemann integral of continuous functions.

Example 5.17 Another example of a hyperfinite set is given by

$$\mathcal{M}_\omega = \{1, 2, \ldots, \omega\}$$

with $\omega \equiv [(n)_{n\in\mathbb{N}}]$. In the case of this set we have $\mathcal{M}_\omega \equiv [(M_n)_{n\in\mathbb{N}}]$ where, for $\mathcal{U}$-nearly all $n \in \mathbb{N}$, $M_n = \{1, 2, \ldots, n\} \subset \mathbb{N}$. Since for any $m \in \mathbb{N}$ whatsoever the set $\{n \in \mathbb{N} : m \in M_n\}$ is cofinite, it follows that $m \in \mathcal{M}_\omega$ and therefore that

$$\mathbb{N} \subset \mathcal{M}_\omega \subset {}^*\mathbb{N}$$

where all the inclusions are strict.

This result can be generalised: *every (standard) countably infinite set is a (proper) subset of a hyperfinite set.* Thus, let A be a countable subset of $\mathbb{R}$ such that $A = \{a_n \in \mathbb{R} : n \in \mathbb{N}\}$, and, for each $n \in \mathbb{N}$, set

$$A_n = \{a_1, a_2, \ldots, a_n\} .$$

Then the internal set $\mathcal{A} \equiv [(A_n)_{n\in\mathbb{N}}]$ is a hyperfinite set, of internal cardinality ω such that

$$A \subset \mathcal{A} \subset {}^*\mathbb{R}$$

which is itself a result of particular interest.

It is clear that every finite subset of ${}^*\mathbb{R}$ is hyperfinite but, as the previous examples have shown, there are hyperfinite sets which are not finite. The hyperfinite sets are of much importance in analysis because, although they may be infinite, they possess the more common properties of the finite sets. This turns out to be extremely attractive for the solution of many problems.

Note 5.18 Note that although a hyperfinite set with infinite internal cardinality will be a discrete set, it will nevertheless have external power (in the sense of Cantor) strictly greater than the power of a countable set. (v. theorem 5.13)

Theorem 5.19 Let $\mathcal{T}$ be a hyperfinite subset of hypernatural numbers. Then $\mathcal{T}$ will have both maximum and minimum elements.

Proof Suppose that we have $\mathcal{T} \equiv [(T_n)_{n\in\mathbb{N}}]$ where for $\mathcal{U}$-nearly all $n \in \mathbb{N}$ the set T_n is finite. Then for $\mathcal{U}$-nearly all $n \in \mathbb{N}$, the numbers $\min(T_n)$, $\max(T_n) \in \mathbb{R}$

would exist and, therefore, so would the hyperreal numbers $\min(\mathcal{T}) \equiv [(\min(T_n)_{n\in\mathbf{N}}]$, $\max(\mathcal{T}) \equiv [(\max(T_n)_{n\in\mathbf{N}}]$. These are, respectively, the minimum and maximum elements of the hyperfinite set $\mathcal{T}$. •

5.1.2 Internal functions

In a similar way we can distinguish between internal and external functions defined on $^*\mathbb{R}$ (or on a subset of $^*\mathbb{R}$).

Definition 5.20 For each $n \in \mathbb{N}$, let $f_n : A_n \to \mathbb{R}$ where $A_n \subset \mathbb{R}$. The function

$$F \equiv [(f_n)_{n\in\mathbf{N}}] : \mathcal{A} \equiv [(A_n)_{n\in\mathbf{N}}] \subset {}^*\mathbb{R} \to {}^*\mathbb{R}$$

defined for every $x \equiv [(x_n)_{n\in\mathbf{N}}] \in \mathcal{A}$ by

$$F(x) = [(f_n(x_n)_{n\in\mathbf{N}}]$$

is said to be an **internal function** defined on $\mathcal{A} \subset {}^*\mathbb{R}$.

Those functions defined on a proper or improper subset of $^*\mathbb{R}$ which are not internal are said to be **external functions**.

This definition is independent of the choice of the particular sequence $(f_n)_{n\in\mathbf{N}}$ to represent F, and also of the sequence $(x_n)_{n\in\mathbf{N}}$ which represents x.[4] If, in particular, we take $A_n = A \subset \mathbb{R}$ for $\mathcal{U}$-nearly all $n \in \mathbb{N}$, then $\mathcal{A} = {}^*A$; furthermore if, in addition, $f_n = f$ for $\mathcal{U}$-nearly all $n \in \mathbb{N}$, then the internal function

$$F \equiv [(f_n)_{n\in\mathbf{N}}] : {}^*A \to {}^*\mathbb{R}$$

is the nonstandard extension of the standard function $f : A \to \mathbb{R}$. In this case we write $F = {}^*f$.

Example 5.21 Let ω be the infinite number in $^*\mathbb{N}$ defined by $\omega \equiv [(n)_{n\in\mathbf{N}}]$. The functions $F, G : {}^*\mathbb{R} \to {}^*\mathbb{R}$ defined by

$$F(x) = \frac{1}{2} + \frac{1}{\pi}\arctan(\omega x)$$

$$G(x) = \frac{1}{\pi}\frac{\omega}{1+\omega^2 x^2}$$

are two examples of internal functions. In fact, we can write $F = [(f_n)_{n\in\mathbf{N}}]$ and $G = [(g_n)_{n\in\mathbf{N}}]$ where the (standard) functions

$$f_n(t) = \frac{1}{2} + \frac{1}{\pi}\arctan(nt)$$

$$g_n(t) = \frac{1}{\pi}\frac{n}{1+n^2 x^2}$$

are defined over the whole real line. Note that neither of the functions F or G is the

[4] Let $(f_{1n})_{n\in\mathbf{N}}$ and $(f_{2n})_{n\in\mathbf{N}}$ be two sequences belonging to $[(f_n)_{n\in\mathbf{N}}]$ and $(x_{1n})_{n\in\mathbf{N}}$ and $(x_{2n})_{n\in\mathbf{N}}$ two sequences belonging to $[(x_n)_{n\in\mathbf{N}}]$. Since we have $\{n \in \mathbb{N} : f_{1n}(x_{1n}) = f_{2n}(x_{2n})\} \supset \{n \in \mathbb{N} : x_{1n} = x_{2n}\} \cap \{n \in \mathbb{N} : f_{1n} = f_{2n}\}$ then, if $\{n \in \mathbb{N} : x_{1n} = x_{2n}\} \in \mathcal{U}$ and $\{n \in \mathbb{N} : f_{1n} = f_{2n}\} \in \mathcal{U}$, we must have $\{n \in \mathbb{N}: f_{1n}(x_{1n}) = f_{2n}(x_{2n})\} \in \mathcal{U}$ which proves the assertion.

nonstandard extension of a standard function. This shows that, while the class of the nonstandard extensions of standard functions is included in the class of the internal functions, the inclusion is strict.

Properties and concepts appropriate to standard functions can generally be transferred to internal functions. Thus, if each one of the functions f_n were differentiable in the standard sense at the point $x_n \in \mathbb{R}$,[5] then we could define the nonstandard ***-derivative** $^*DF$ of the internal function $F \equiv [(f_n)_{n\in\mathbf{N}}$ by writing

$$^*DF(x) = [(f'_n(x_n))_{n\in\mathbf{kN}}]$$

for $x \equiv [(x_n)_{n\in\mathbf{N}}] \in {}^*\mathbb{R}$.

The nonstandard differential operator $^*D$ possesses the same formal properties as the standard differential operator D. Thus if F and G are two internal functions for which the *-derivatives exist, then

$$^*D(F+G) = {}^*DF + {}^*DG$$

and

$$^*D(F.G) = ({}^*DF)G + F({}^*DG) \ .$$

In a similar way if, for each $n \in \mathbb{N}$, K_n is a compact subset of $\mathbb{R}$ and f_n is a function integrable over K_n, we can define the nonstandard ***-integral** of the internal function $F \equiv [(f_n)_{n\in\mathbf{N}}]$ over the hypercompact set $\mathcal{K} \equiv [(K_n)_{n\in\mathbf{N}}]$ of $^*\mathbb{R}$, by setting

$$^*\int_{\mathcal{K}} F(x)dx = \left[\left(\int_{K_n} f_n(t)dt\right)_{n\in\mathbf{N}}\right] \ .$$

Then all the usual properties of the standard integral (linearity, mean value theorems, etc.) transfer to this *integral.

Example 5.22 We consider again the internal function defined in example 5.21

$$F(x) = \frac{1}{2} + \frac{1}{\pi}\arctan(\omega x), \qquad x \in {}^*\mathbb{R}.$$

The *derivative of this function is the internal function defined by

$$G(x) = {}^*Df = \frac{1}{\pi}\frac{\omega}{1+\omega^2x^2}, \qquad x \in {}^*\mathbb{R}.$$

This function $G(x)$ is infinitesimal at every standard point x distinct from the origin: at some points within the monad of zero it will, however, assume infinite values. In addition, if α, β are any two non-infinitesimal numbers such that $\alpha < 0 < \beta$ then the nonstandard integral of $G(x)$ from α to β will be well defined and we will have

$$^*\int_\alpha^\beta G(x)dx \approx 1 \ .$$

[5]It is enough to consider this only for a set of indices belonging to $\mathcal{U}$.

Shadow of an internal function Let $F : {}^*\mathbb{R} \to {}^*\mathbb{R}$ be a given internal function and $\mathcal{A}$ an internal subset of ${}^*\mathbb{R}$. Recall that $\mathbf{st}(\mathcal{A}) = \{\mathbf{st}(x) : \quad x \in \mathcal{A} \cap {}^*\mathbb{R}_b\}$. Suppose that, for each $a \in \mathbf{st}(\mathcal{A})$, $F(a)$ is a finite hyperreal number. In this case it would make sense to define a (standard) function as follows:

$$f \equiv \mathbf{st}(F) : \mathbf{st}(\mathcal{A}) \to \mathbb{R}$$

$$x \mapsto f(x) = \mathbf{st}(F(x)) \quad .$$

The function $f \equiv \mathbf{st}(F)$ is called the **shadow** (or **trace**) of F on $\mathbf{st}(\mathcal{A}) \subset \mathbb{R}$. Note that if we were to have $F = {}^*f$ for some standard function f then the shadow of F would coincide with the function f of which it was the extension in the first place. However, there are internal functions which are not of this type but which, nevertheless, do possess a shadow on some subset of $\mathbb{R}$. In the case of the functions F and G of example 5.21, for example, we have

$$\mathbf{st}(F(x)) = \begin{cases} 0 & \text{if} \quad x < 0 \\ 1/2 & \text{if} \quad x = 0 \\ 1 & \text{if} \quad x > 0 \end{cases}$$

$$\mathbf{st}(G(x)) = 0 \qquad \text{if} \quad x \neq 0 \quad .$$

Exercise 5.23 *Let $x \approx 0$ and let $F_\epsilon : {}^*\mathbb{R}_0^+ \to {}^*\mathbb{R}$ be the function defined by*

$$F_\epsilon(x) = \left[\frac{x}{\epsilon}\right] \epsilon$$

(where $[.]$ denotes the integer part of a number). Show that the shadow of F_ϵ on $\mathbb{R}_0^+$ is the identity function.

5.1.2.1 Sequences of functions and internal functions

Pointwise convergence Let $(f_n)_{n \in \mathbf{N}}$ be a sequence of real functions of a real variable.[6] This sequence will converge (pointwise) to a function $f : \mathbb{R} \to \mathbb{R}$ if and only if for each $x \in \mathbb{R}$, given $r > 0$ there exists $n_0 \in \mathbb{N}$ (depending generally on x and on r) such that

$$\forall_n [n \in \mathbb{N} \Rightarrow [n \geq n_0 \Rightarrow |f_n(x) - f(x)| < r] \quad .$$

For each fixed $a \in \mathbb{R}$, the numerical sequence $(f_n(a))_{n \in \mathbf{N}}$ can be extended to a standard hypersequence[7]

$$(f_\nu(a))_{\nu \in {}^*\mathbf{N}} \quad .$$

From this, recalling the definition 4.1 of S-convergence of numerical sequences, we can define pointwise convergence of a sequence of functions as follows:

[6] If the functions f_n have not been defined on the entire real line we can, without loss of generality, extend them to the set $\mathbb{R}$ by setting $f_n(x) = 0$ at all points where they are not defined.

[7] The notation f_ν, for any given infinite hypernatural number ν, is a little ambiguous but is consistent with the notation for the terms of infinite order of a numerical standard hypersequence.

Definition 5.24 A (standard) sequence $(f_n)_{n\in\mathbf{N}}$ of functions defined on $\mathbb{R}$ converges pointwise to a function $f:\mathbb{R}\to\mathbb{R}$ if and only if the relation $f_\nu(a)\approx f(a)$ is satisfied for all hypernatural numbers $\nu\in {}^*\mathbb{N}_\infty$ and all $a\in\mathbb{R}$.

The sequence of functions $(f_n)_{n\in\mathbf{N}}$ naturally defines an internal function

$$F\equiv[(f_n)_{n\in\mathbf{N}}]: {}^*\mathbb{R}\to {}^*\mathbb{R}$$

which to each point $x\equiv[(x_n)_{n\in\mathbf{N}}]\in {}^*\mathbb{R}$ makes correspond the hyperreal number $F(x)\equiv[(f_n)_{n\in\mathbf{N}}]$. If ω denotes as usual the infinite hypernatural number $[(n)_{n\in\mathbf{N}}]$ then for any given $a\in\mathbb{R}\subset {}^*\mathbb{R}$ we have

$$F(a)=f_\omega(a) \quad .$$

Hence, if a sequence $(f_n)_{n\in\mathbf{N}}$ converges pointwise to a function f then we must have $f_\omega(a)\approx f(a)$ and, consequently, $F(a)\approx f(a)$ for all $a\in\mathbb{R}$. That is to say, if $(f_n)_{n\in\mathbf{N}}$ converges pointwise on $\mathbb{R}$ to a function f then $F\equiv[(f_n)_{n\in\mathbf{N}}]$ possesses a shadow on $\mathbb{R}$ which coincides with f.

Example 5.25 The sequence

$$f_n(x)=x^n,\qquad x\in[0,1];\qquad n=1,2,\ldots$$

defines the following internal function

$$F(x)=x^\omega,\qquad x\in {}^*[0,1]\subset {}^*\mathbb{R}.$$

This function is infinitesimal at all standard points $x\in[0,1)$ and is equal to 1 at the point $x=1$. The shadow of F is given by

$$\mathbf{st}F(x)=\begin{cases}0 & \text{if } 0\le x<1\\ 1 & \text{if } x=1\end{cases} \quad .$$

Since for each $x\in[0,1]$ and for every $\nu\in {}^*\mathbb{N}_\infty$ we have $f_\nu(x)\approx \mathbf{st}F(x)$ then $f(x)=\mathbf{st}F(x)$, $0\le x\le 1$, is the pointwise limit of the sequence of given functions. Note that nevertheless the fact that the relation $F(x)\approx f(x)$ is satisfied for all standard x does not itself guarantee the convergence of the sequence $(f_n)_{n\in\mathbf{N}}$ which generates F to a function $f\equiv\mathbf{st}F$. (Although it does indicate the existence of a subsequence which converges to this limit.) As an example, consider the following sequence of functions

$$\phi_n(x)=(-1)^n x^n,\qquad 0\le x\le 1$$

and let $\Phi\equiv[(\phi_n)_{n\in\mathbf{N}}]: {}^*[0,1]\to {}^*\mathbb{R}$. Assuming that the set of indices $\{2,4,6,\ldots\}$ belongs to the ultrafilter $\mathcal{U}$ then we actually have $\Phi=F$ and so

$$\Phi(x)\approx f(x)\equiv\mathbf{st}F(x),\qquad x\in[0,1] \quad .$$

However, the sequence $(\phi_n)_{n\in\mathbf{N}}$ does not converge to f at the point $x=1$. For each fixed $x\in[0,1)$ we have $\phi_\nu(x)\approx f(x)$ for any $\nu\in {}^*\mathbb{N}_\infty$, but for $x=1$ we have $\phi_{2\omega}(1)=1$ while $\phi_{2\omega+1}(1)=-1$. Consequently although in this case we would have

$\mathbf{st}\Phi = \mathbf{st}F = f$, the sequence $(\phi_n)_{n\in\mathbf{N}}$ converges pointwise to the function $\phi(x) = 0$ for $0 \le x < 1$ but does not converge at any other point.

On the other hand, the sequence

$$f_n(x) = \frac{2}{\pi} x \arctan(nx), \qquad x \in \mathbb{R}; \qquad n = 1, 2., \ldots$$

defines the function

$$F(x) = \frac{2}{\pi} x \arctan(\omega x), \qquad x \in {}^*\mathbb{R}$$

which, at standard points, is infinitely close to the (standard) function

$$f(x) = \begin{cases} -x & \text{if } x < 0 \\ +x & \text{if } x > 0 \end{cases} .$$

Consequently, if this sequence were to have a pointwise limit then it would coincide with $f(x)$.

Uniform convergence Let $(f_n)_{n\in\mathbf{N}}$ be a sequence of (standard) functions on a subset A of $\mathcal{R}$ which converges pointwise to a function $f : A \to \mathbb{R}$. In general, the limit function f would not necessarily possess the same properties as the elements of the sequence. For example, the terms of the sequence

$$f_n : [0,1] \to \mathbb{R}$$

$$x \mapsto f_n(x) = x^n, \qquad n = 1, 2, \ldots \tag{5.1}$$

are continuous functions: the pointwise limit of this sequence

$$f(x) = \begin{cases} 0 & \text{if } 0 \le x < 1 \\ 1 & \text{if } x = 1 \end{cases}$$

is, however, a discontinuous function. To study conditions under which the properties of elements of the sequence are preserved when we pass to the limit it is necessary to introduce a deeper notion of convergence, namely that of **uniform convergence**.

If A is any subset of $\mathbb{R}$ (possibly $\mathbb{R}$ itself) then a sequence of functions $(f_n)_{n\in\mathbf{N}}$ defined on A converges uniformly to a function $f : A \to \mathbb{R}$ if and only if, given any arbitrary real number $r > 0$, there always exists a natural number n_0 (depending only on r) such that

$$\forall_n [n \in \mathbb{N} \Rightarrow n \ge n_0 \Rightarrow \forall_x [x \in A \Rightarrow |f(x) - f_n(x)| < r]]] . \tag{5.2}$$

Before passing on to the nonstandard characterisation of uniform convergence, it is useful to examine the nonstandard extension of a sequence of functions in rather more detail. For a fixed point $x \in A \subset \mathbb{R}$ the nonstandard extension of $(f_n(x))_{n\in\mathbf{N}}$, which is simply a numerical sequence, is obtained by passing to a standard hypersequence as shown previously. However, we may need to consider the value of a term of nonstandard order of the hypersequence at points $x \in {}^*A$, which are perhaps themselves nonstandard. In such a situation we proceed as follows:

With the sequence $f_n(x), x \in A; n = 1, 2, 3, \ldots$ we associate a function of two variables defined by

$$\phi : \mathbb{N} \times A \to \mathbb{R}$$

$$(n, x) \mapsto \phi(n, x) = f_n(x) \quad .$$

This function possesses a nonstandard extension $^*\phi : {}^*\mathbb{N} \times {}^*A \to {}^*\mathbb{R}$ which is defined in the usual way. That is, for $\nu \equiv [(m_n)_{n \in \mathbf{N}}] \in {}^*\mathbb{N}$ and $x \equiv [(x_n)_{n \in \mathbf{N}}] \in {}^*A$ we put

$$^*\phi(\nu, x) \equiv [(\phi(m_n, x_n))_{n \in \mathbf{N}} \in {}^*\mathbb{R}]$$

and then of course we set

$$f_\nu(x) = {}^*\phi(\nu, x) \equiv [(\phi(m_n, x_n)_{n \in \mathbf{N}}] = [(f_{m_n}(x_n))_{n \in \mathbf{N}}] \quad .$$

The nonstandard characterisation of the notion of uniform convergence can now be formulated as follows:

Theorem 5.26 The sequence $(f_n)_{n \in \mathbf{N}}$ converges uniformly on $A \subset \mathbb{R}$ to a function $f : S \to \mathbb{R}$ if and only if for any point (standard or nonstandard) $x \in {}^*A$ we have

$$f_\nu(x) \approx {}^*f(x)$$

for every infinite hypernatural number $\nu \in {}^*\mathbb{N}_\infty$.

Proof Suppose that f_n tends to f uniformly on $A \subset \mathbb{R}$. Then, given any real number $r > 0$, there exists $n_0 \in \mathbb{N}$ such that (5.2) is satisfied. Using the relation $\phi(n, x) = f_n(x)$ this statement can be expressed in the form

$$\forall_n [n \in \mathbb{N} \Rightarrow [n \geq n_0 \Rightarrow \forall_x [x \in A \Rightarrow |f(x) - \phi(n, x)| < r]]]$$

from which by transfer we get

$$\forall_n [n \in {}^*\mathbb{N} \Rightarrow [n \geq n_0 \Rightarrow \forall_x [x \in {}^*A \Rightarrow |{}^*f(x) - {}^*\phi(n, x)| < r]]] \quad .$$

Let ν be an arbitrary member of $^*\mathbb{N}_\infty$. Since n_0 is finite, then for any real number $r > 0$ we must have $\nu > n_0$ and, therefore, $|{}^*f(x) - {}^*\phi(\nu, x)| < r$. Further, since the real number r is arbitrary and $^*\phi(\nu, x) = f_\nu(x)$, it follows that

$$\forall_\nu [\nu \in {}^*\mathbb{N}_\infty \Rightarrow \forall_x [x \in {}^*A \Rightarrow {}^*f(x) \approx f_\nu(x)]]] \quad . \tag{5.3}$$

Conversely, suppose that (5.3) is satisfied. Then for any given real number $r > 0$

$$\exists_p [p \in {}^*\mathbb{N} \wedge \forall_\nu [\nu \in {}^*\mathbb{N} \Rightarrow [\nu \geq p \Rightarrow \forall_x [x \in {}^*A \Rightarrow |{}^*f(x) - f_\nu(x)| < r]]]]$$

is a true proposition (it is enough to take $p \in {}^*\mathbb{N}$ infinite and the rest follows from the hypothesis). Since $f_\nu(x) = {}^*\phi(\nu, x)$ then transfer into $\mathbb{R}$ gives

$$\exists_p [p \in \mathbb{N} \wedge \forall_n [n \in \mathbb{N} \Rightarrow [n \geq p \Rightarrow \forall_x [x \in A \Rightarrow |f(x) - f_n(x)| < r]]]]$$

which is a true proposition for any $r \in \mathbb{R}^+$.

Thus $(f_n)_{n\in\mathcal{N}}$ converges uniformly to f on A.[8] •

Take, in particular, $\nu = \omega$ in theorem 5.26. If $(f_n)_{n\in\mathbb{N}}$ converges to f uniformly on A, then since $F \equiv [(f_n)_{n\in\mathbb{N}}]$, we obtain the relation

$$F(x) \approx {}^*f(x) \quad \text{for all} \quad x \in {}^*A \ .$$

It is now easy to see that the sequence of functions introduced in (5.1) does not converge uniformly on $[0,1]$ to a function f. In fact, taking, for example, the point $x \equiv [(x_n)_{n\in\mathbb{N}}] = [(1-n^{-1})_{n\in\mathbb{N}}] = 1-\omega^{-1} \in {}^*[0,1]$, we have ${}^*f(x) = 0$ while $F(x) = (1-\omega^{-1})^\omega \approx e^{-1}$; that is, $F(x) \not\approx {}^*f(x)$.

Considering theorem 5.26 just at standard points of A immediately gives the following result:

Corollary 5.26.1 If a sequence of functions $(f_n)_{n\in\mathbb{N}}$ converges uniformly to a function f on A then this sequence converges pointwise to the same function.

As one of the examples 5.25 shows clearly, a sequence of continuous functions may converge pointwise to a discontinuous function. If the convergence were uniform, however, this would not be the case: the uniform limit of a sequence of continuous functions is a continuous function.

To prove this result it is necessary to examine in a little more detail the behaviour of the internal functions of the form $f_\nu(x)$, $x \in {}^*A$, when ν is infinite. Such functions are not in general the nonstandard extensions of standard functions since they are not the result of applying the transfer principle. The values of $f_\nu(x)$, even for standard x, may not be real. Nevertheless we do have the following:

Theorem 5.27 If $(f_n)_{n\in\mathbb{N}}$ is a sequence of continuous functions on a set $A \subset \mathbb{R}$ then for any $\nu \in {}^*\mathbb{N}$ and any $y \in {}^*A$ there exists an infinitesimal $\delta > 0$ such that

$$\forall_x [x \in {}^*A \Rightarrow [|x-y| < \delta \Rightarrow f_\nu(x) \approx f_\nu(y)]] \ .$$

---

[8] **Alternative proof** Suppose that $f_n \to f$ uniformly on A, let $x \equiv [(x_n)_{n\in\mathbb{N}}]$ be any point in ${}^*A$ and let $r > 0$ and $n_0 \in \mathbb{N}$ be the numbers with the significance already attributed in (5.2). For any real $r > 0$ we have the inclusion

$$\{n \in \mathbb{N} : m_n \geq n_0\} \subset \{n \in \mathbb{N} : |f_{m_n}(x_n) - f(x_n)| < r\}$$

and, therefore, given that $\{n \in \mathbb{N} : m_n \geq n_0\} \in \mathcal{U}$ (since ν is an infinite hypernatural number), it follows from the definition of ultrafilter that we have $\{n \in \mathbb{N} : |f_{m_n}(x_n) - f(x_n)| < r\} \in \mathcal{U}$. We can then write

$$\forall_{r,\nu,x}[r \in \mathbb{R}^+ \wedge \nu \in {}^*\mathbb{N}_\infty \wedge x \in {}^*A \Rightarrow |f_\nu(x) - {}^*f(x)| < r]$$

whence

$$\forall_{\nu,x}[\nu \in {}^*\mathbb{N}_\infty \wedge x \in {}^*A \Rightarrow f_\nu(x) \approx {}^*f(x)].$$

To prove the converse suppose that $(f_n)_{n\in\mathbb{N}}$ does not converge uniformly on A to the function $f : A \to \mathbb{R}$.Then we have

$$\exists_r[r \in \mathbb{R}^+ \wedge \forall_m[m \in \mathbb{N} \Rightarrow$$

$$\exists_n[n \in \mathbb{N} \wedge n \geq m \wedge \exists_a[a \in A \wedge |f_n(a) - f(a)| \geq r]]]]$$

For each $m \in \mathbb{N}$ let n_m be a value of n and let x_m be a value of a which corresponds to it in the preceding formula. Then, setting $\eta \equiv [(n_m)_{n\in\mathbb{N}}]$ and $x \equiv [(x_m)_{m\in\mathbb{N}}]$ we must have $\eta \in {}^*\mathbb{N}_\infty$ and, therefore, $f_\eta(x) \not\approx f(x)$.

This theorem shows that under the stated conditions, although the transformation by f_ν of all the monad to which y belongs may not be completely included in the monad to which $f_\nu(y)$ belongs, nevertheless we do have this type of behaviour in a sufficiently small infinitesimal neighbourhood of y.

Proof The theorem is clearly satisfied for all standard n; it remains to prove its validity for $\nu \in {}^*\mathbb{N}_\infty$. The fact that $(f_n)_{n\in\mathbb{N}}$ is a sequence of functions continuous on A can be expressed in the following form:

$$\forall_n[n \in \mathbb{N} \Rightarrow \forall_y[y \in A \Rightarrow \forall_r[r \in \mathbb{R}^+ \Rightarrow \exists_s[s \in \mathbb{R}^+ \wedge$$
$$\forall_x[x \in A \Rightarrow [|x-y| < s \Rightarrow [|\phi(n,x) - \phi(n,y)| < r]]]]]]$$

Transferring to ${}^*\mathbb{R}$,

$$\forall_\nu[\nu \in {}^*\mathbb{N} \Rightarrow \forall_y[y \in {}^*A \Rightarrow \forall_\epsilon[\epsilon \in {}^*\mathbb{R}^+ \Rightarrow \exists_\eta[\eta \in {}^*\mathbb{R}^+ \wedge$$
$$\forall_x[x \in {}^*A \Rightarrow [|x-y| < \eta \Rightarrow [|{}^*\phi(\nu,x) - {}^*\phi(\nu,y)| < \epsilon]]]]]] \ .$$

Taking $\nu \in {}^*\mathbb{N}_\infty$, $y \in {}^*A$ and $\epsilon > 0$ infinitesimal, there exists $\eta \in {}^*\mathbb{R}^+$ such that for all $x \in {}^*A$

$$|x-y| < \eta \Rightarrow |f_\nu(x) - f_\nu(y)| < \epsilon \ .$$

If now δ is an infinitesimal such that $0 < \delta < \eta$, then since ϵ is also an infinitesimal, we can finally write

$$\forall_x[x \in {}^*A \Rightarrow [|x-y| < \delta \Rightarrow f_\nu(x) \approx f_\nu(y)]]$$

as was to be proved. •

Theorem 5.28 Let $(f_n)_{n\in\mathbb{N}}$ be a sequence of continuous functions converging uniformly on $A \subseteq \mathbb{R}$ to a function $f : A \to \mathbb{R}$. Then f is a continuous function.

Proof Let a be a point of A. For any $x \in {}^*A$ and $\nu \in {}^*\mathbb{N}$ we have the following inequality:

$$|{}^*f(x) - f(a)| \leq |{}^*f(x) - f_\nu(x)| + |f_\nu(x) - f_\nu(a)| + |f_\nu(a) - f(a)| \ .$$

For ν infinite it follows from the uniformity of the convergence that $f_\nu(x) \approx {}^*f(x)$ and $f_\nu(a) \approx f(a)$. On the other hand from the preceding theorem there exists an infinitesimal number $\delta > 0$ such that

$$|x-a| < \delta \Rightarrow f_\nu(x) \approx f_\nu(a) \ .$$

Now let $r > 0$ be an arbitrary real number. From what has gone before we may write

$$\exists_\delta[\delta \in {}^*\mathbb{R}^+ \wedge \forall_x[x \in {}^*A \Rightarrow [|x-a| < \delta \Rightarrow |{}^*f(x) - f(a)| < r]]]$$

whence, by transfer,

$$\exists_d[d \in \mathbb{R}^+ \wedge \forall_x[x \in A \Rightarrow [|x-a| < d \Rightarrow |f(x) - f(a)| < r]]]$$

which means that f is a function continuous at the point a. •

5.1.2.2 S-continuity and *-continuity

As remarked earlier (chapter 4, note 4.50) the continuity of a function $f : \mathbb{R} \to \mathbb{R}$ at a point $a \in \mathbb{R}$ can be described formally by

$$\forall_r [r \in \mathbb{R}^+ \Rightarrow \exists_s [s \in \mathbb{R}^+ \wedge \forall_x [x \in \mathbb{R} \\ \Rightarrow [|x - a| < s \Rightarrow |f(x) - f(a)| < r]]]] \tag{5.4}$$

which is equivalent to the nonstandard definition given by

$$\forall_x [x \in {}^*\mathbb{R} \Rightarrow [x \approx a \Rightarrow {}^*f(x) \approx f(a)]] \ . \tag{5.5}$$

Transferring (5.4) to the nonstandard universe we have

$$\forall_\epsilon [\epsilon \in {}^*\mathbb{R}^+ \Rightarrow \exists_\delta [\delta \in {}^*\mathbb{R}^+ \wedge \forall_x [x \in {}^*\mathbb{R} \\ \Rightarrow [|x - a| < \delta \Rightarrow |{}^*f(x) - f(a)| < \epsilon]]]] \ . \tag{5.6}$$

The definitions (5.5) and (5.6) are equivalent in the case of internal functions which are the nonstandard extensions of standard functions. However, when applied to arbitrary internal functions the two definitions are generally *not* equivalent. Hence we are led to consider two distinct concepts which we call **S-continuity** and ***-continuity**.

Definition 5.29 An internal function $F : {}^*\mathbb{R} \to {}^*\mathbb{R}$ is said to be *-continuous at a point $a \in {}^*\mathbb{R}$ (standard or nonstandard) if and only if for each $\xi \in {}^*\mathbb{R}^+$ there exists $\eta \in {}^*\mathbb{R}^+$ such that

$$\forall_x [x \in {}^*\mathbb{R} \Rightarrow [|x - a| < \eta \Rightarrow |F(x) - F(a)| < \xi]] \ .$$

Example 5.30 Setting $\omega \equiv [(n)_{n\in\mathbf{N}}]$ the internal functions defined by

$${}^*g(x) = \frac{1}{x}, \qquad x \in {}^*(0,1)$$

$$\Delta(x) = \begin{cases} \omega^2 x, & \text{if } \ 0 < x \le 1/\omega \\ 2\omega - \omega^2 x, & \text{if } \ 1/\omega < x < 2/\omega \\ 0, & \text{if } \ x \notin (0, 2/\omega) \end{cases}$$

$$G(x) = \sin(\omega x), \qquad x \in {}^*\mathbb{R}$$

are each *-continuous at every point of their respective domains.

The notion of *-continuity allows us to re-interpret theorem 5.27 as follows:

> if $(f_n)_{n\in\mathbf{N}}$ is a sequence of functions continuous on a set $A \subseteq \mathbb{R}$ then for any $\nu \in {}^*\mathbb{N}$ the internal function f_ν is *-continuous on ${}^*A$.

If a sequence $(f_n)_{n\in\mathbf{N}}$ of standard functions is pointwise convergent then its pointwise limit in $\mathbb{R}$ will be the shadow, $\mathbf{st}(F)$, in $\mathbb{R}$ of the internal function $F \equiv [(f_n)_{n\in\mathbf{N}}]$. The pointwise limit of a sequence of functions continuous at a point is not necessarily itself a function continuous at that point. Hence it is easy to understand that the *-continuity of an internal function F at a given point does not necessarily imply that its shadow, $\mathbf{st}(F)$, even when well defined, will be a function continuous at this point. To obtain conditions under which this implication is satisfied it is necessary to consider the other type of continuity.

Definition 5.31 An internal function $F : {}^*\mathbb{R} \to {}^*\mathbb{R}$ is said to be S-continuous at a point $a \in {}^*\mathbb{R}$ (standard or nonstandard) if and only if

$$\forall_x [x \in {}^*\mathbb{R} \Rightarrow [x \approx a \Rightarrow F(x) \approx F(a)]] \quad .$$

Considering again the functions ${}^*g, \Delta$, and G of example 5.30, we conclude that

1. ${}^*g$ is S-continuous at every standard point of its domain, but is not S-continuous at every point of ${}^*(0,1)$. In fact, if ν is a given infinite number and $x = \nu^{-1}$ then, setting $y = (2\nu)^{-1}$ we have $x \approx y$ but, nevertheless, ${}^*g(x) \not\approx {}^*g(y)$. *[Note, however, that ${}^*g$ is *-continuous at the point $x = \nu^{-1}$.]*

2. Δ is S-continuous at every standard point except at the origin. Taking, for example, $x = \omega^{-2}$ we have $\Delta(x) = 1$ while $\Delta(0) = 0$ and therefore, although $x \approx 0$ we have $\Delta(x) \not\approx \Delta(0)$.

3. The function G is not S-continuous at any point of its domain.

Note 5.32 There is no relation between the two forms of continuity in the general case. There are internal functions which are *-continuous but not S-continuous and vice-versa. For example, the function $F(x) = (-1)^{[x]}.\epsilon$, $x \in {}^*\mathbb{R}$, where ϵ is a non-zero infinitesimal and $[x]$ denotes the the integral part of x, is not *-continuous although it is S-continuous; by contrast, the function defined by $G(x) = \sin(\omega x)$, $a \in {}^*\mathcal{R}$, as seen above, is *-continuous but not S-continuous.[9]

From the given nonstandard treatment of the (standard) notions of continuity and uniform continuity we have the following results for a standard function f and its nonstandard extension ${}^*f$:

Theorem 5.33 Given a function $f : I \subset \mathbb{R} \to \mathbb{R}$ then:

1. f is continuous at a point $a \in I$ if and only if its nonstandard extension ${}^*f : {}^*I \to {}^*\mathbb{R}$ is S-continuous at the same point.

2. if ${}^*f$ is S-continuous at each point (standard or nonstandard) of ${}^*I \subset {}^*\mathbb{R}$, then f is uniformly continuous on I.

For an arbitrary internal function F, with a shadow $f \equiv \mathbf{st}(F)$ we have instead the following:

[9] Although *-continuous over the entire monad of a given standard point an internal function F may vary appreciably (that is to say, not infinitesimally) within this monad, as a result of which its shadow in $\mathbb{R}$ will be a function discontinuous at this point. To put it another way: the internal function may enter a monad with a given value and then vary appreciably after entry before leaving the monad. This shows that to guarantee the continuity of the shadow function at a given point it is necessary for the variation of the internal function in the interior of the monad of that point to be imperceptible in standard terms (that is, the variation must be infinitesimal).

Theorem 5.34 Let $F : {}^*I \to {}^*\mathbb{R}$ be any internal function, where I is an interval of $\mathbb{R}$. Then

1. if F is S-continuous at a (standard) point of $I \subset {}^*I$ and $F(I) \subset {}^*\mathbb{R}_b$, the shadow $f \equiv \mathbf{st}(F)$ of F on I is a continuous function,

2. if, further to the conditions of the preceding paragraph, the function F is S-continuous at every point of ${}^*I$, then the function $f \equiv \mathbf{st}(F)$ is uniformly continuous on I.

Proof (1) Under the conditions of the theorem it is clear that the shadow $f \equiv \mathbf{st}(F)$ exists and is a function well defined for all $a \in \mathbb{R}$ by $f(a) = \mathbf{st}F(x)$ whenever $x \approx a$ (in particular we have $f(a) = \mathbf{st}F(a)$). It remains to prove that F is continuous. Let a be a given point of $\mathbb{R}$ and $r > 0$ an arbitrarily chosen real number. The set

$$S_{a,r} = \{\eta \in {}^*\mathbb{R}^+ : \forall_x[x \in {}^*\mathbb{R} \Rightarrow [|x-a| < \eta \Rightarrow |F(x) - F(a)| < r]]\}$$

contains all the positive infinitesimals and, therefore, by infinitesimal overflow, contains a real number $\alpha > 0$. Taking $s = \alpha/2$, for example, we have

$$\forall_x[x \in {}^*\mathbb{R} \Rightarrow [|a-x| < s \Rightarrow |F(a) - F(x)| < r]] \quad .$$

From this result, taking into account that for $x \in \mathbb{R}$ we have

$$f(a) - f(x) = [f(a) - F(a)] + [F(a) - F(x)] + [F(x) - f(x)]$$

and that $f(a) \approx F(a)$ and $f(x) \approx F(x)$, we obtain finally

$$\forall_r[r \in \mathbb{R}^+ \Rightarrow \exists_s[s \in \mathbb{R}^+ \wedge \forall_x[x \in \mathbb{R} \Rightarrow [|a-x| < s \Rightarrow |f(a) - f(x)| < r]]]]$$

which shows that $f \equiv \mathbf{st}(F)$ is a continuous function at $a \in \mathbb{R}$.

(2) The proof is similar to the preceding, now using the fact that a and x belong to ${}^*I$ and that F is S-continuous at all points (standard and nonstandard) of ${}^*I$. •

Corollary 5.34.1 Let $F : {}^*\mathbb{R} \to {}^*\mathbb{R}$ be an internal function which is S-continuous and finite for all $a \in \mathbb{R}$. Then, setting $f \equiv \mathbf{st}(F)$ we must have ${}^*f(x) \approx F(x)$ at each point $x \in {}^*\mathbb{R}_b$.

Proof Let x be any point in ${}^*\mathbb{R}_b$ and $a = \mathbf{st}(x)$. From the preceding theorem it follows that $f \equiv \mathbf{st}(F)$ is a function continuous at $a \in \mathbb{R}$ and so ${}^*f(x) \approx f(a)$. On the other hand, we also have $F(x) \approx F(a)$ by S-continuity, and $F(a) \approx f(a)$ since f is the shadow of F in $\mathbb{R}$. The corollary then follows immediately since ${}^*f(x) - F(x) = [{}^*f(x) - f(a)] + [f(a) - F(a)] + [F(a) - F(x)]$. •

Example 5.35 Let $\mathcal{H} : {}^*\mathbb{R} \to {}^*\mathbb{R}$ be the internal function defined by

$$\mathcal{H}(x) = \begin{cases} 0, & \text{if } x \leq 0, \\ \frac{1}{2}\omega^2 x^2, & \text{if } 0 < x \leq 1/\omega \\ 2\omega x - \frac{1}{2}\omega^2 x^2 - 1 & \text{if } 1/\omega < x < 2/\omega \\ 1 & \text{if } x \geq 2/\omega \ . \end{cases}$$

This function is finite for all $x \in {}^*\mathbb{R}_b$: the shadow of $\mathcal{H}$, $\mathbf{st}(\mathcal{H})$, is the (standard) function

$$H_0(x) = \begin{cases} 0, & \text{if } x \leq 0 \\ 1, & \text{if } x > 0, \end{cases}$$

but note that ${}^*H_0 \neq \mathcal{H}$. The internal function $\mathcal{H}$ is S-continuous over every interval not containing the origin, but $\mathcal{H}$ is not S-continuous everywhere.[10]

Moreover, the shadow H_0 of $\mathcal{H}$ is discontinuous at the origin. Nevertheless it would be wrong to infer from this that when S-continuity fails at a standard point of the domain of an internal function there must necessarily be a discontinuity of the shadow of that function. The function Δ, referred to in Example 5.30 for instance, is S-continuous everywhere except at the origin and, in spite of that, the shadow of Δ is the null function, which is continuous everywhere.

Note that S-continuity is a much stronger form than *-continuity. If an internal function F is S-continuous at a given point $a \in \mathbb{R}$ then F will not exhibit large fluctuations in the interior of the monad of a; such variation as we may have must be infinitesimal. On the other hand *-continuity at the same point is perfectly compatible with the existence of fluctuations which are not infinitesimal (and which may even be infinite) in the interior of that monad. Thus it is to be expected that whether an internal function is or is not S-continuous at a given point will have important implications for the behaviour of the shadow (or standard part) of such a function, when it exists.

Example 5.36 A final example of interest is the internal function defined by

$$R(x) = \begin{cases} 0, & \text{if } x \leq 0 \\ \frac{1}{6}\omega^2 x^3 & \text{if } 0 < x \leq 1/\omega \\ \omega x^2 - \frac{1}{6}\omega^2 x^3 - x + \frac{1}{3\omega} & \text{if } 1/\omega < x < 2/\omega \\ x - \frac{1}{\omega} & \text{if } x \geq 2/\omega \end{cases}$$

which is S-continuous everywhere and whose shadow is the **ramp function** x_+ defined by

$$x_+ = \begin{cases} 0, & \text{if } x \leq 0 \\ x, & \text{if } x > 0 \end{cases} .$$

Now for the operator ${}^*D$ it is easy to verify the following equalities

$${}^*DR = \mathcal{H} \quad \text{and} \quad {}^*D^2R = {}^*D\mathcal{H} = \Delta \ .$$

The repercussions of these results in the universe of *standard* mathematical analysis can only be interpreted within the context of the theory of distributions.

5.2 Delta Functions, Standard and Nonstandard

"[...] As succesivas generalizacoes do conceito de numero visaram tornar sempre possivis certas operacoes [...] *Do mesmo modo, a operacao de derivacao, que nao e possivel para todas as funcoes continuas, passou a se-lo com a introducao doe novos entes chamados*

[10]However, $\mathcal{H}$ is *-continuous everywhere.

'*distribuicoes*". Por exemplo, a funcao de Weierstrass, que noa tem derivada no sentidp usual, assou a ter derivadas de todas as ordens, que nao sao funcoes, mas sim distribucoes ."

J. Sebastiao e Silva, [4]

The term *function* refers to a concept which is now established in the literature without ny ambiguity: by "*a real function of a real variable*" we understand a triplet (E, F, f) where E and F are non-empty sets of real numbers and f denotes a process which to ach element of E makes correspond one and only one element of F. This triplet is ormally represented in the form

$$f : E \to \mathbb{R}$$

which has the significance ascribed above. As one would expect, this definition did ot appear instantaneously; on the contrary, it is the fruit of an evolution which, in ts most significant phase, stretched over the last 300 years. In Appendix A we give ome notes on the most significant steps in the evolution of the concept of function p to the arrival of the abstract formulation described above. This concept, despite ts generality, is still not adequate for some applications (particularly in physics, in ngineering and, of course, in mathematics itself).

As a result of the fundamental theorem of the calculus, a function continuous n an interval admits, in that interval, primitives of all orders. That is to say, a unction continuous on an interval is infinitely integrable there. The same cannot be aid with respect to differentiation: continuity does not imply differentiability. There re functions continuous on $\mathbb{R}$ which are not differentiable at any point - such as, for xample, the Weierstrassian functions. This breakdown in symmetry of the behaviour f continuous functions is at the base of many of the difficulties in mathematical analsis. Such difficulties have led to the creation of new mathematical objects which, in ne form or another, have generalised the concept of function. Of particular imporance among such generalisations are the so-called "*distributions*", created by Laurent chwartz. In the context of the theory of distributions, a function continuous on an nterval is not only infinitely integrable but is also infinitely differentiable in that nterval. It is clear, therefore, that the associated concept of *generalised derivative* loes not correspond exactly to the usual definition of the derivative of a function at a oint. In fact, what is generalised in the context of the theory of distributions is not he concept of the derivative of a function at a point, but rather that of the derived unction conceived as a whole. It is not the appropriate place here to attempt to give rigorous development of the general idea of a distribution. Instead we shall at the noment only present an elementary study of the particularly important case of the lelta distribution, together with its derivatives and primitives.

The **unit step** "function" of Heaviside, H, and the **delta** "function" δ, which (in sense which can be made explicit) plays the role of its derivative, are two of the most vell known and significant examples of Schwartz distributions.[11] Although Heaviside 1850-1925) himself made as much use of δ as of H in his works, the most characteristic roperties of the delta function as well as the now universally adopted notation,

[11]The symbols H and δ, when interpreted in a standard context, do not designate *functions* in he strict sense of that word. However since they are frequently referred to as such in the technical terature, we shall adopt the same terminology here even though this does constitute an abuse of nguage.

$\delta(x)$, only appeared explicitly 20 years later in the text on quantum mechanics of P.A.M.Dirac [5]. For this reason we adopt the usual practice of referring to δ as the **Dirac delta function.**

In elementary text books (of physics and engineering) the delta function is frequently introduced in the form of a pointwise definition, as follows:

$$\delta(x) = \begin{cases} 0, & \text{if } x \neq 0 \\ +\infty, & \text{if } x = 0 \end{cases} \tag{5.7}$$

with the additional stipulation that this should be a function satisfying the condition

$$\int_{-\infty}^{+\infty} \delta(x)dx = 1 \quad . \tag{5.8}$$

Moreover it is claimed that, for each function $\phi : \mathbb{R} \to \mathbf{C}$, continuous on a neighbourhood of the origin,

$$\int_{-\infty}^{+\infty} \phi(x)\delta(x)ds = \phi(0) \quad . \tag{5.9}$$

In point of fact there can exist no standard function which exhibits this behaviour (no matter what theory of integration might be considered). However, by using purely formal manipulations, (5.7) and (5.8) can be given apparent justification in the following way:

Let $H_c : \mathbb{R} \to \mathbb{R}$ be the function defined by

$$H_c(x) = \begin{cases} 0, & \text{if } x < 0 \\ c, & \text{if } x = 0 \\ 1, & \text{if } x > 0 \end{cases} \tag{5.10}$$

where $c \in (0, 1)$. The classical derivative of H_c coincides with the null function for all $x \neq 0$ but is not defined for $x = 0$. Nevertheless it could be argued that a meaning for $H_c'(0)$ is given in the following sense:

$$H_c'(0) = \lim_{x \to 0} \frac{H_c(x) - H_c(0)}{x} = +\infty \quad .$$

If we now let $\delta(x)$ denote a mathematical object which is assumed to behave in some *operational* sense as the derivative of $H_c(x)$, then this could serve to motivate, if not to justify, the anomalous definition offered as (5.7). With the same logic (or perhaps lack of it!) with which δ is considered to be the derivative of H_c we can regard H_c as a "primitive" of δ. Thus, carrying out a formal integration of the alleged function $\delta(x)$, we generate the result

$$\int_{-\infty}^{\infty} \delta(x)dx = \lim_{A \to +\infty} H_c(A) - \lim_{B \to -\infty} H_c(B) = 1$$

which serves as a motivation for (5.8).

Although purely formal and without any real mathematical justification, the fact remains that assuming the existence of the so-called "functions" $H(x)$ and $\delta(x)$ which satisfy, in some sense, the properties referred to above, does turn out to be of enormous value in a wide variety of applications[21]. This is certainly the case for physicists in quantum mechanics and for engineers in systems and signal theory. To justify the use of such symbols we need to look for a rigorous definition of them, but ideally one which will be as elementary as possible.

5.1.3 The Dirac delta "function"

Operational significance Given a function $\phi : \mathbb{R} \to \mathbb{C}$ we call the **support** of ϕ the closure of the set of points on which this function is not null. For any non-negative integer p we denote by $C_c^p \equiv C_c^p(\mathbb{R})$ the vector space of all those functions which are continuously differentiable up to at least order p and whose supports are compact. The space of all the functions belonging to C_c^p for some $p \in \mathbb{N}$ we generally denote by C_c^∞. That this space does have elements other than the null function is shown by the example of the function of Schwartz defined by

$$\phi(x) = \begin{cases} \exp\left(\frac{-1}{1-x^2}\right) & \text{if } |x| \leq 1 \\ 0 & \text{if } |x| > 1 \end{cases} .$$

With the function H_c defined as in (5.10), where c can now denote any real number whatever, and with ϕ being any function in $C_c \equiv C_c^0$, the integral

$$\int_{-\infty}^{+\infty} H_c(x)\phi(x)dx = \int_0^{+\infty} \phi(x)dx \tag{5.11}$$

is always well defined. The right hand side of (5.11) shows further that the value of the integral is independent of the constant c occurring in the definition of H_c. It follows that, so far as the value of the integral (5.11) is concerned, we can identify the whole family of locally integrable functions $\{H_c\}_{c\in\mathbb{R}}$ with a single *functional* $\mathcal{H} : C_c \to \mathbb{C}$ defined by

$$< \mathcal{H}, \phi > \equiv \mathcal{H}[\phi] = \int_0^{+\infty} \phi(x)dx \ .$$

This functional is easily seen to be linear; moreover since it is independent of the value of c it is appropriate to denote it simply by $\mathcal{H}$. We may often speak loosely of the "function" $\mathcal{H}$, but this is strictly an abuse of language unless the term is understood in some generalised sense. No matter which of the elements H_c may be chosen to represent $\mathcal{H}$, this symbol can at most only specify an *equivalence class* of functions.[12]

The identification of a family of functions with a functional is frequent in mathematical analysis. Any locally integrable function $f : \mathbb{R} \to \mathbb{C}$ (that is to say, any function integrable on compacts) can give rise to a generalised function F understood in this sense. That is to say, we can define a functional $F : C_c \to \mathbb{C}$ by setting

$$< F, \phi > \equiv F[\phi] = \int_{-\infty}^{+\infty} \phi(x)f(x)dx \ . \tag{5.12}$$

[12]Here we already have a certain implicit generalisation of the concept of function, which is now extended to refer to a whole class of "equivalent" functions (where equivalence is defined by some appropriate relation).

This symbol F also denotes an equivalence class of functions, since (interpreting the integration sign in the sense of Lebesgue) we can substitute for f any other function $f_1 : \mathbb{R} \to \mathbf{C}$ such that $f_1(x) = f(x)$ almost everywhere (that is, for all of $\mathbb{R}$ except at most on a set of measure zero).[13]

When f is continuous we can, without danger of confusion, legitimately identify the corresponding functional F with the function f since in this case the equivalence class f contains no other element with the distinguishing property of continuity. Suppose now that f belongs to C^1 and let ϕ be a function also continuously differentiable and with compact support contained, for example, in the interval $[a, b] \subset \mathbb{R}$. Integrating by parts and noting that $\phi(a) = \phi(b) = 0$, we have

$$\int_{-\infty}^{+\infty} f'(x)\phi(x)dx = \int_{\infty}^{+\infty} f(x)[-\phi'(x)]dx \quad . \tag{5.13}$$

The function f', being continuous, also generates a functional over the space C^1 and it is natural to call this functional the **derivative** of the functional F generated by f and to represent it by the symbol DF. From equation (5.13) it follows that the functional $DF : C_c^1 \to \mathbf{C}$ in this case can be defined for any $\phi \in C_c^1$ by

$$< DF, \phi >= \int_{-\infty}^{+\infty} f'(x)\phi(x)dx \tag{5.14}$$

or, equally well, by

$$< DF, \phi >= \int_{-\infty}^{+\infty} f(x)[-\phi'(x)]dx \quad . \tag{5.15}$$

Choosing the equation (5.15) to define DF, it is possible to extend the differential operator D to any functional F generated by a function f which is locally integrable but which is not necessarily differentiable. In this case, f' may not exist and therefore the right-hand side of (5.14) may not make sense. Nevertheless the integral on the right-hand side of (5.15) is always well defined. For example,

$$x_+^{\frac{1}{2}} = \begin{cases} +\sqrt{x} & \text{if } x \geq 0 \\ 0 & \text{if } x < 0 \end{cases}$$

is a function continuously differentiable in the usual sense only on $\mathbb{R}\backslash 0$. Its classical derivative is the function

$$\frac{1}{2}x_+^{-\frac{1}{2}} = \begin{cases} +1/(2\sqrt{x}) & \text{if } x > 0 \\ 0 & \text{if } x < 0 \end{cases}$$

which is unbounded in any neighbourhood of the origin. However, whatever function $\phi \in C_c^1$ we happen to choose, the functions $x_+^{\frac{1}{2}}$ and $\frac{1}{2}x_+^{-\frac{1}{2}}$ satisfy equation (5.13). Thus $\frac{1}{2}x_+^{-\frac{1}{2}}$ is a representative of the derivative DF of the functional F generated by the ordinary function $x_+^{\frac{1}{2}}$.

[13]Note that this new generalisation of the concept of function depends implicitly on the Lebesgue theory of integration. It is common practice nowadays to describe as a function f a whole class of functions (defined in the sense of Dirichlet) which differ at most only on a set of Lebesgue measure zero.

Returning once more to equation (5.13), suppose that we substitute for f an arbitrary representative H_c of the generalised function $\mathcal{H}$. The function $H_c(x)$ admits an ordinary derivative for any $c \in \mathbb{R}$, namely the function which is null at all points $x \neq 0$ and is not defined at the point $x = 0$. The left-hand side of (5.13) now gives, for any $\phi \in \mathcal{C}_c^1$,

$$\int_{-\infty}^{+\infty} \phi(x) H_c'(x) dx = \lim_{\epsilon \downarrow 0} \int_{-\infty}^{-\epsilon} \phi(x) H_c'(x) dx$$

$$+ \lim_{\epsilon \downarrow 0} \int_{\epsilon}^{+\infty} \phi(x) H_c'(x) dx = 0$$

while from the right-hand side of (5.13), supposing that ϕ has support contained in the interval $[a, b]$ we obtain

$$\int_{\infty}^{+\infty} H_c(x)[-\phi'(x)]dx = \int_0^b [-\phi'(x)]dx = [-\phi(x)]_0^b = \phi(0)$$

a result which is seen to be independent of the constant c. The integration by parts formula (5.13) is therefore not satisfied in this case. However we can apply the definition of the derivative of a functional over $\mathcal{C}_c$ given in (5.15) to the functional $\mathcal{H}$ so as to get

$$< D\mathcal{H}, \phi >= \int_{-\infty}^{+\infty} H_c(x)[-\phi'(x)]dx = \phi(0)$$

for any $\phi \in \mathcal{C}_c^1$. This functional, which carries out the so-called **sampling operation** $\phi \mapsto \phi(0)$, we now agree to denote by the symbol δ. Thus we have

$$\delta = D\mathcal{H}$$

which allows us to say, that *the δ "function" of Dirac is, in this sense, the (generalised) derivative of the unit step "function" of Heaviside.* If we interpret the expression

$$\int_{-\infty}^{+\infty} \delta(x)\phi(x)dx \left(\equiv \int_{-\infty}^{+\infty} DH(x)\phi(x)dx \right)$$

in a merely symbolic sense, attributing to it, by definition, the value which the functional derivative $D\mathcal{H}$ assumes at the point $\phi \in \mathcal{C}_c^1$, then we can restore the formal validity of the formula of integration by parts and write

$$\int_{-\infty}^{+\infty} \delta(x)\phi(x)dx = \int_{-\infty}^{+\infty} \mathcal{H}(x)[-\phi'(x)]dx \ .$$

The integral on the right-hand side can be understood in the usual sense once we substitute for $\mathcal{H}$ an arbitrary one of its representatives H_c. We then arrive at the familiar symbolic expression

$$\int_{-\infty}^{+\infty} \phi(x)\delta(x)dx = \phi(0) \ . \tag{5.16}$$

Note finally that the formula (5.16) can be given a more general significance if we allow ϕ to be any function continuous on some neighbourhood of the origin. Taking, in particular, the function $\phi(x) = 1$ for all $x \in \mathbb{R}$ we obtain the formula (5.8) proposed (heuristically) by Dirac.

Generalised differentiation Since $\mathcal{C}_c^\infty$ is contained in $\mathcal{C}_c^p$, for any $p \in \mathbb{N}_0$, then everything said above remains true if we confine attention to test functions ϕ belonging to $\mathcal{C}_c^\infty$. For the sake of simplicity, and with no loss of generality, we shall from now on take $\mathcal{C}_c^\infty$ as the domain of test functions under consideration, unless explicitly stated to the contrary. With (5.15) as motivation we can now introduce the following definition:

Definition 5.37 Given a functional F defined on $\mathcal{C}_c^\infty$ the **generalised derivative** of F is the functional DF defined by

$$< DF, \phi > \equiv DF[\phi] = F[-\phi'] \equiv < F, -\phi' >$$

for any $\phi \in \mathcal{C}_c^\infty$.

Let $x_+ : \mathbb{R} \to \mathbb{R}$ be the function defined by

$$x_+ = \begin{cases} x & \text{if } x \geq 0 \\ 0 & \text{if } x < 0 \end{cases} .$$

Since this is a function continuous on $\mathbb{R}$ we can, with no danger of confusion, identify it with the functional which it generates on $\mathcal{C}_c^\infty$. Denoting this functional by $\mathbf{x}_+$ we then have

$$< \mathbf{x}_+, \phi > = \int_0^{+\infty} x\phi(x)dx$$

for any $\phi \in \mathcal{C}_c^\infty$. The generalised derivative of $\mathbf{x}_+$ is the functional $D\mathbf{x}_+ : \mathcal{C}_c^\infty \to \mathbf{C}$ defined by

$$< D\mathbf{x}_+, \phi > \equiv D\mathbf{x}_+[\phi] = < \mathbf{x}_+, -\phi' >$$

$$= \int_0^{+\infty} x[-\phi'(x)]dx = \int_0^{+\infty} \phi(x)dx$$

which shows that $D\mathbf{x}_+$ coincides with the functional $\mathcal{H}$, the unit step function of Heaviside. Accordingly it makes sense to write

$$\mathcal{H} = D\mathbf{x}_+ .$$

The derivative of $D\mathbf{x}_+$ is the functional $D^2\mathbf{x}_+ : \mathcal{C}_c^\infty \to \mathbf{C}$ defined by

$$< D^2\mathbf{x}_+, \phi > \equiv D^2\mathbf{x}_+[\phi] = < D\mathbf{x}_+, -\phi' >$$

$$= < \mathbf{x}_+, \phi'' > = \int_0^{+\infty} x\phi''(x)dx = \phi(0)$$

which shows that $D^2\mathbf{x}_+$ coincides with the functional δ, that is with the so-called Dirac delta function. Thus we can also write

$$\delta \equiv D\mathcal{H} = D^2 x_+$$

which means therefore that *the Dirac delta "function" is equal to the second generalised derivative of the continuous function* x_+.

Applying now the definition 5.37 to the delta function itself we have

$$D\delta[\phi] =< D\delta, \phi >=< \delta, -\phi' >= -\phi'(0) \tag{5.17}$$

for any $\phi \in C_c^\infty$. This generalised derivative $D\delta$ of the functional δ is frequently written as δ', although this notation could lead to an unfortunate confusion with the usual operation of differentiation applied to functions differentiable in the ordinary sense. In practice, however, this usage is very common and does not really cause much confusion, since the context makes clear the sense that we are expected to give to the operation concerned. We can go on to define the generalised derivative of order $m \in \mathbb{N}_0$ of the functional δ by induction, so as to get

$$D^m\delta[\phi] \equiv \delta^m[\phi] =< \delta, (-1)^m\phi^{(m)} >= (-1)^m\phi^{(m)}(0) \tag{5.18}$$

for any $\phi \in C_c^\infty$. The formula (5.18) can be simply represented in the following, more familiar, way

$$\int_{-\infty}^{+\infty} \delta^{(m)}(x)\phi(x)dx = (-1)^m\phi^{(m)}(0)$$

which is a generalised formula also put forward by Dirac. We can therefore conclude that

the Dirac delta function is indefinitely differentiable, and the derivative $\delta^{(m)}$ of arbitrary order $m \in \mathbb{N}_0$ is the (generalised) derivative of order $m+2$ of the continuous function x_+.

5.1.4 Nonstandard representations

Consider, for example, the sequence of functions $d_n : \mathbb{R} \to \mathbb{R}$, $n = 1, 2, \ldots$, defined by

$$d_n(t) = \begin{cases} n^2 t & \text{if } 0 < t \leq 1/n \\ 2n - n^2 t & \text{if } 1/n < t < 2/n \\ 0 & \text{if } t \notin (0, 2/n) \end{cases} .$$

A sketch graph of the general term d_n is shown below.

The pointwise limit of this sequence of functions vanishes everywhere; that is to say, $\lim_{n\to\infty} d_n(t) = 0$, for all $t \in \mathbb{R}$. Nevertheless the sequence possesses some interesting properties which warrant further examination.

(1) Since, for each $n \in \mathbb{N}$, the integral of d_n over the whole of $\mathbb{R}$ is equal to 1, we have

$$\lim_{n\to\infty} \int_{-\infty}^{+\infty} d_n(t)dt = 1$$

although the integral of the pointwise limit of the d_n (as $n \to \infty$) is equal to zero. As is well known, the operations of integration and of passing to the limit are not, in general, commutative, but it always causes some surprise that the limit of the integrals of a sequence of functions which converges (pointwise) to zero *everywhere* can be different from zero.

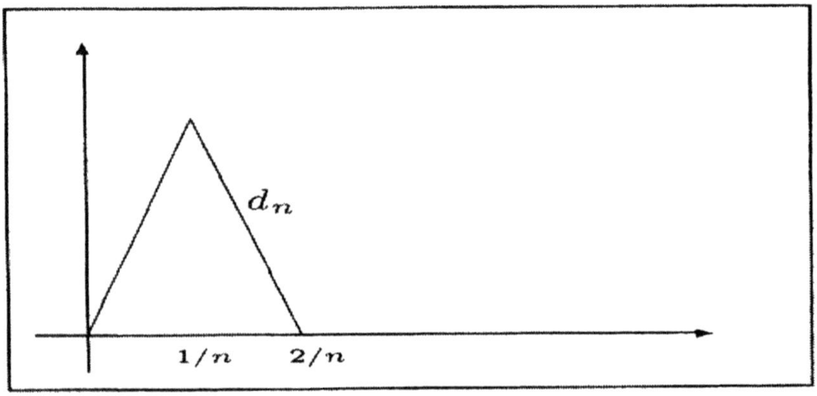

It is perhaps even more disconcerting in the case of a limit of the form

$$\lim_{n\to\infty} \int_{-\infty}^{+\infty} d_n(t)\phi(t)dt$$

where ϕ denotes a function continuous on a neighbouthood of the origin. We see that

$$\phi(0) - \int_{-\infty}^{+\infty} d_n(t)\phi(t)dt = \int_{-\infty}^{+\infty} [\phi(0) - \phi(t)]d_n(t)dt$$

$$= \int_0^{1/n} [\phi(0) - \phi(t)]n^2 dt + \int_{1/n}^{2/n} [\phi(0) - \phi(t)](2n - n^2 t)dt$$

and, since for n sufficiently large $\phi(0) - \phi(t)$ is a function continuous on the interval $(0, 2/n)$ then from the generalised mean value theorem we have

$$\phi(0) - \int_{\infty}^{+\infty} d_n(t)\phi(t)dt = \frac{1}{2}[\phi(0) - \phi(\xi_n)] + \frac{1}{2}[\phi(0) - \phi(\eta_n)]$$

where $0 < \xi_n < 1/n$ and $1/n < \eta_n < 2/n$. Letting n tend to infinity it follows that ξ_n and η_n tend to 0 and therefore that

$$\lim_{n\to\infty} \int_{-\infty}^{+\infty} d_n(t)\phi(t)dt = \phi(0) \quad .$$

This obliges us to conclude that, although the sequence $(d_n)_{n\in\mathbb{N}}$ itself tends to zero when n tends to infinity it is nevertheless effective in sampling values of functions ϕ which are continuous on a neighbourhood of the origin. We shall now look for a nonstandard interpretation of these facts.

Although the internal function $\Delta_d \equiv [(d_n)_{n\in\mathbb{N}}]$ has a shadow which is null in $\mathbb{R}$, it is *not* itself a null function in $^*\mathbb{R}$. In fact we have the following

$$^*\int_{-a}^{+a} \Delta_d(x)dx = 1 \quad \text{and} \quad ^*\int_{-a}^{+a} \Delta_d(x)\,^*\phi(x)dx \approx \phi(0)$$

where $a > 0$ is any real number and ϕ is any function continuous on a neighbourhood of the origin. These two properties indicate that the internal function Δ_d can be a representative, in the universe of NSA, of the standard object which is known as the Dirac delta function, in the sense that Δ_d satisfies some of the most characteristic properties of δ, suitably understood. If, however, we wish to work with derivatives of the δ function then Δ_d (which is not *differentiable) may not be the most convenient of representatives. In such a case, it would be better to represent δ by an internal function Δ such that, for every $m \in \mathbb{N}_0$, $\delta^{(m)}$ possesses a representative of the form $^*D\Delta^m$. In this way we could replace the *generalised differentiation of standard functionals* in universe of real analysis by the *ordinary *differentiation of internal functions* in the universe of NSA.

5.2.2.1 Elementary study of delta "functions"

Let $g(t)$, $t \in \mathbb{R}$ be a function belonging to $C^\infty(\mathbb{R})$ which is positive[14] , monotonely convergent to zero as $|t| \to \infty$, and such that

$$\int_{-\infty}^{+\infty} g(t)dt = 1 \ .$$

The internal function $\Delta_g : {^*\mathbb{R}} \to {^*\mathbb{R}}$ defined by $\Delta_g = \omega.{^*g}(\omega x)$, where $\omega \equiv [(n)_{n\in\mathbb{N}}] \in {^*\mathbb{N}_\infty}$, satisfies the condition[15]

$$\Delta_g(x) \approx 0, \qquad \text{for all} \quad x \notin mon(0) \ . \tag{5.19}$$

If $(a_n)_{n\in\mathbb{N}}$ is any given sequence of positive real numbers which tends to $+\infty$, then $\gamma = [(a_n)_{n\in\mathbb{N}}]$ is a positive infinite hyperreal number. Since

$$1 = \int_{-\infty}^{+\infty} g(t)dt = \lim_{n\to\infty} \int_{-na_n}^{+na_n} g(t)dt = \lim_{n\to\infty} \int_{-a_n}^{+a_n} ng(nt)dt$$

we can therefore write

$$^*\int_{-\gamma}^{+\gamma} \Delta_g(x)dx = \left[\left(\int_{-a_n}^{+a_n} ng(nt)dt\right)\right] \approx 1 \ . \tag{5.20}$$

[14]This condition can be relaxed, allowing g to be only conditionally integrable. This would allow us to obtain nonstandard representations of the delta function of arbitrary sign.

[15]That this condition follows from the restrictions imposed on the function g, can be shown by appeal to Pringsheim's theorem (see, for example, the *Curso de Algebra Superior* of J.Vincents Goncalves (2nd ed. p.117)).

Equations (5.19) and (5.20) show that, in the universe of nonstandard analysis, the properties (5.7) and (5.8) referred to in §5.2, are actually compatible with respect to the internal function Δ_g, provided we replace the equals sign $=$ by the sign $\approx$ and the infinity sign ∞ by the infinite positive number number γ.

Note 5.38 If, in particular, we choose for g a function infinitely differentiable on $\mathbb{R}$ and with support contained in an interval $(0, a)$, for some $a > 0$, then $\Delta_g(x)$, $x \in {}^*\mathbb{R}$ is a function vanishing at all standard points (including the origin) but is not the null function in the nonstandard universe.

Let $\nu \equiv [(m_n)_{n\in\mathbf{N}}]$ be an arbitrarily fixed positive infinite number (we may suppose that $m_n > 0$ for all $n \in \mathcal{N}$); since

$$\lim_{n\to\infty} \int_{-m_n}^{+m_n} g(t)dt = 1$$

then

$$1 \approx {}^*\int_{-\nu}^{+\nu} {}^*g(t)dt = {}^*\int_{-\nu/\omega}^{+\nu/\omega} \omega\, {}^*g(\omega x)dx = {}^*\int_{-\nu/\omega}^{+\nu/\omega} \Delta_g(x)dx$$

from which we can conclude that, for any infinite hyperreal number $\gamma \geq \nu/\omega$,

$${}^*\int_{-\gamma}^{-\nu/\omega} \Delta_g(x)dx \approx {}^*\int_{+\nu/\omega}^{+\gamma} \Delta_g(x)dx \approx 0 \ .$$

Taking, in particular, $\nu = \sqrt{\omega}$ we would have $\nu/\omega = 1/\sqrt{\omega} \equiv \epsilon \in mon(0)$, giving

$${}^*\int_{-\epsilon}^{+\epsilon} \Delta_g(x)dx \approx 1 \text{ and } {}^*\int_{-\gamma}^{-\epsilon} \Delta_g(x)dx \approx 0 \approx {}^*\int_{+\epsilon}^{+\gamma} \Delta_g(x)dx$$

which shows that *the *integral of the internal fumction Δ_g is infinitesimal outside an infinitesimal neighbourhood of the origin.*

Now let $f : \mathbb{R} \to \mathbf{C}$ be a bounded[16] function continuous at the origin; for any infinitesimal number $\epsilon \equiv [(e_n)_{n\in\mathbf{N}}] > 0$ such that $\omega\epsilon \in {}^*\mathbb{R}_\infty$ and for any infinite number $\gamma > \omega\epsilon$, we have

$${}^*\int_{-\gamma}^{+\gamma} \Delta_g(x)\, {}^*f(x)dx = \left\{ {}^*\int_{-\gamma}^{-\epsilon} + {}^*\int_{-\epsilon}^{+\epsilon} + {}^*\int_{+\epsilon}^{+\gamma} \right\} \Delta_g(x)\, {}^*f(x)dx \ .$$

Since f is bounded, we have

$$\left| {}^*\int_{+\epsilon}^{+\gamma} \Delta_g(x)\, {}^*f(x)dx \right| = \left| \left[\left(\int_{e_n}^{a_n} ng(nt)f(t)dt \right)_{n\in N} \right] \right|$$

$$\leq \|f\|_\infty \left[\left(\int_{ne_n}^{na_n} g(\tau)d\tau \right)_{n\in N} \right] \approx 0$$

[16] In the case of an unbounded function f we could get the same result by working with a kernel function g which has compact support. In this case Δ_g would be null (and not merely infinitesimal) outside the monad of the origin.

and, similarly, for the integral over the interval $[-\gamma, -\epsilon]$. We then get

$$^*\int_{-\gamma}^{+\gamma} \Delta_g(x)\,^*f(x)dx \approx\ ^*\int_{-\epsilon}^{+\epsilon} \Delta_g(x)\,^*f(x)dx$$

$$= {}^*f(\tau)\,^*\int_{-\epsilon}^{+\epsilon} \Delta_g(x)dx \approx f(0)$$

where $-\epsilon < \tau < +\epsilon$. From the continuity of f at the origin it follows that we have $^*f(\tau) \approx f(0)$ and therefore that

$$^*\int_{-\gamma}^{+\gamma} \Delta_g(x)\,^*f(x)dx \approx f(0) \ .$$

That is to say, Δ_g carries out for the function f the fundamental sampling operation which is the defining characteristic of the Dirac delta function. Accordingly we can write

$$< \delta, f >= \mathbf{st}\left(^*\int_{-\gamma}^{+\gamma} \Delta_g(x)\,^*f(x)dx\right) = f(0) \ . \tag{5.21}$$

Similarly we can derive the equality

$$< \delta^{(m)}, f >= \mathbf{st}\left(^*\int_{-\alpha}^{+\alpha} {}^*D^m\Delta_g(x)\,^*f(x)dx\right)$$

$$= (-1)^m f^m(0) \tag{5.22}$$

for any $m \in \mathbb{N}$ and any f bounded and sufficiently differentiable at the origin.

Denoting by $C_b^p\{0\}$ the vector space (over $\mathbb{C}$) of all bounded functions continuously differentiable up to at least order p in a neighbourhood of the origin, then although there is no trace of a delta function itself on $\mathbb{R}$, we can legitimately speak of the **shadow** of a delta function on $C_b^p\{0\}$, or of the trace of the *derivative of order p of a delta function on $C_b^p\{0\}$. In this way the spaces $C_b^p\{0\}$ behave as a species of *"filters"* of the standard universe which allow us to detect the presence of delta functions or their *derivatives in the nonstandard universe. All the linear properties associated with delta functions (sum, product by a finite number or *derivation) are detectable by these "filters". The same cannot be said for nonlinear operations. For example, the product of delta functions or of *derivatives of delta functions are not detectable through the "filters" $C_b^p\{0\}$, as the following discussion should make clear.

Example 5.39 Let θ be an indefinitely differentiable function with compact support contained in the interval $[a, b]$, and whose integral over $\mathbb{R}$ is equal to 1. Then $\Delta_\theta(x) = \omega\,^*\theta(\omega x)$, $x \in {}^*\mathbb{R}$, is a delta function which vanishes outside the infinitesimal interval $[\epsilon_a, \epsilon_b] \subset mon(0)$, where $\epsilon_a = a/\omega$ and $\epsilon_b = b/\omega$. Given a function $f \in C_b\{0\}$ we then have

$$^*\int_{-\gamma}^{+\gamma} \Delta_\theta(x)\,^*f(x)dx \approx f(0) \ .$$

The square of Δ_θ is an internal indefinitely *differentiable function which is easily seen not to be a delta function, in general. For any $f \in C_b\{0\}$ we will get

$$^*\int_{-\gamma}^{+\gamma} \Delta_\theta^2(x)\,^*f(x)dx =\ ^*\int_{\epsilon_a}^{\epsilon_b} \Delta_\theta^2(x)\,^*f(x)dx$$

$$= \Delta_\theta(\tau) \, {}^*\!\int_{\epsilon_a}^{\epsilon_b} \Delta_\theta(x) \, {}^* f(x) dx \approx \Delta_\theta(\tau) f(0)$$

where $\epsilon_a < \tau < \epsilon_b$. If $f(0) \neq 0$ then $\Delta_\theta(\tau) f(0)$ will generally be an infinite number, which shows that δ_θ^2 is not a delta function.

It may happen, however, that θ is a function for which $\Delta_\theta(\tau)$ takes, for each $f \in C_b\{0\}$, a bounded value; in this case we obtain

$${}^*\!\int_{-\gamma}^{+\gamma} \Delta_\theta^2(x) \, {}^* f(x) dx = c_j . {}^*\!\int_{-\gamma}^{+\gamma} \Delta_\theta(x) \, {}^* f(x) dx$$

an equality which could serve to justify the result $\delta^2 = c\delta$, with arbitrary constant c, which appears in the quoted literature.

Chapter 6

Foundations of Nonstandard Analysis

6.1 Standard and Nonstandard Models

The original treatment of Nonstandard Analysis given by Robinson in

A. Robinson *Nonstandard Analysis, Studies in Logic and the Foundations of Mathematics* (North-Holland, 1974)

is a comprehensive account of the general theory which nevertheless cannot be regarded as readily approachable. Accordingly there have since been several attempts to devise alternative treatments which are less demanding. Particularly valuable in this respect is the account given in the important article

W.A.J.Luxemburg "What is Nonstandard Analysis?", *Amer. Math. Monthly*, June-July (1972), pp. 38-67.

This contains a minutely detailed description of the construction of nonstandard models by means of ultraproducts. Such constructions probably offer the most intuitive and easily accessible way of making clear the foundations of NSA to those relatively unfamiliar with the specialised area of mathematical logic. It is just this kind of model, based on an ultrafilter over $\mathbb{N}$, which has been used in the preceding chapters, admittedly in a more or less *ad hoc* manner.

The present chapter is designed to present a more rigorous view of the foundations of NSA, in a form which could be further extended by using the techniques developed within it, especially that of the transfer principle. To begin with, however, it is necessary to give a brief account of model theory, with some indication of how it may be used to develop the modern form of infinitesimal analysis, to which Robinson himself gave the name of nonstandard analysis.

6.1.1 Formal languages and models

"Modern mathematics might be described as the science of abstract objects, be they real numbers, functions, surfaces, algebraic structures or whatever. Mathematical logic adds a new dimension to this science by paying attention to the language used in mathematics, to the way abstract objects are defined, and to the laws of logic which govern us as we reason about these objects."

J.Barwise, *Handbook of Mathematical Logic, Studies in Logic and the Foundations of Mathematics, Vol.90* (North-Holland, 1977).

Following Chang and Keisler, [4], we can describe model theory in the following terms:

> **Model theory** is that part of mathematical logic which treats the relations existing between a given *formal language* and its interpretations, that is, its *models*.

The theory of models has three fundamental components:

(i) a formal language $\mathcal{L}$ (a succinct description of which will follow),
(ii) a mathematical structure $\mathcal{M}$ (that is, a non-empty set equipped with an arbitrary number of operations, relations and, possibly, some special elements), and
(iii) an *interpretation* of $\mathcal{L}$ in $\mathcal{M}$, that is, a description of the form taken by the language $\mathcal{L}$ when its terms, sentences, etc, are applied to the objects of $\mathcal{M}$.

The properties of any mathematical structure are properly described in terms of *"sentences"* (or *"propositions"* , a concept which itself needs to be clarified within the context of the particular formal language[1] concerned. We shall only consider *first order* languages, that is, languages containing variables which range over individual terms of the universe in question but not over higher order objects. Thus, in such a language, it is not possible to make statements about all the subsets of the universe. A formal language $\mathcal{L}$ for a first order structure is constructed from a collection of symbols of the following types:

(a) a constant (name) c for each individual,
(b) a symbol for each function f of n variables,
(c) a symbol for each relation R of n variables,

$\mathcal{L}$ also contains the following logical symbols:[2]
(a) variables $v_1, v_2, \ldots, v_n, \ldots,$
(b) $=$, which denotes equality,
(c) logical connectives $\neg, \wedge, \vee, \Rightarrow, \Leftrightarrow$,
d) quantifiers $\forall, \exists$,
(e) parentheses and commas.

---

[1]To formulate the properties of the ordered field of real numbers, for example, it is necessary to have a formal language $\mathcal{L}(\mathbb{R})$ which contains a vocabulary rich enough to carry out the required task without ambiguity.

[2]The symbols $\neg, \wedge, \vee, \Rightarrow, \Leftrightarrow$ and $\forall, \exists$ can be expressed in terms of only two logical symbols and one quantifier. Hence, if desirable, we could reduce the alphabet of the language $\mathcal{L}$. In general there is usually little advantage to be gained in doing this and, on the contrary, it often makes reading of the text more difficult. In some situations, however, when it is necessary to make inductive proofs over complicated formulas it may be useful to work with a reduced alphabet.

With these basic symbols, which constitute the **alphabet** of the language,we can now form more complex syntactic entities:

Definition 6.1 The **terms** of the language $\mathcal{L}$ are sequences of symbols constructed in accordance with the following rules:

t1. every variable is a term;

t2. every constant is a term;

t3. if $\tau_1, \tau_2, \ldots, \tau_n$ are terms and if f is a function of n variables then $f(\tau_1, \tau_2, \ldots, \tau_n)$ is a term.

A sequence of symbols is a term if and only if it can be obtained by application of **t1**, **t2**, and **t3** a finite number of times.

If in a term $\tau \equiv \tau(v_1, v_2, \ldots, v_n)$ we replace each variable v_i with a constant c_i, we obtain a so-called **constant term**: this may be equal to some constant of the theory, but on the other hand may turn out to have no meaning. For example, "$(x+2)^2$" or "$\forall_x \phi$" are terms which have clear significance in an appropriate context: by contrast, "$x \inf \forall + \alpha\pi 4$" is an expression to which it would be difficult to attribute any significance whatever.

Given n terms $\tau_1, \tau_2, \ldots, \tau_n$ and an n-ary relation R, an **atomic formula** of the language $\mathcal{L}$ is an expression of the form [3]

$$R(\tau_1, \tau_2, \ldots, \tau_n).$$

Definition 6.2 The set of all **formulas** of the first order language $\mathcal{L}$ is defined as follows:

f1. every atomic formula is a formula of the language $\mathcal{L}$,

f2. if ϕ, ψ are formulas of $\mathcal{L}$ then so also are

$$(\neg\phi), (\phi \wedge \psi), (\phi \vee \psi), (\phi \Rightarrow \psi), (\phi \Leftrightarrow \psi);$$

f3. if ϕ is a formula of $\mathcal{L}$ and v is a variable then

$$(\forall_v \phi), \quad (\exists_v \phi)$$

are also formulas of $\mathcal{L}$.

A sequence of symbols is a formula if and only if it can be obtained by application of **f1**, **f2**, **f3** a finite number of times.

$\phi(v_1, v_2, \ldots, v_n)$ will denote a formula in which occurrences of variables are taken from the set $\{v_1, v_2, \ldots, v_n\}$, (compare with notation for terms). A variable which

[3] If a, b denote real numbers, the expressions $= (a, b)$ and $\leq (a, b)$ are simple examples of atomic formulas of the language $\mathcal{L}(\mathbb{R})$ of real analysis: in practice it is more usual, and more convenient, to write them as $a = b$ and $a \leq b$ respectively.

occurs in the range of a quantifier in a given formula is said to be a **bound variable** of the formula concerned; a variable which is not bound is said to be a **free variable.**

Definition 6.3 A formula which has no free variables is called a **sentence.**

Given any mathematical structure $\mathcal{M}$ let $\mathcal{L}$ be a formal language which contains sufficient constants to describe all the formal properties of $\mathcal{M}$. An **interpretation** of $\mathcal{M}$ is a mapping $\iota : \mathcal{L} \to \mathcal{M}$ such that

1. if c is a constant of $\mathcal{L}$ then $\iota(c) \in \mathcal{M}$;

2. if f is an n-ary function of $\mathcal{L}$ then $\iota(f)$ is a function of $\mathcal{M}^n$ in $\mathcal{M}$;

3. if R denotes an n-ary relation of $\mathcal{L}$ then $\iota(R)$ is an n-ary relation defined on $\mathcal{M}$.

Given a mathematical structure $\mathcal{M}$ and an appropriate formal language $\mathcal{L}$, we may naturally expect there to be more than one possible interpretation. However for the sake of simplicity here we shall suppose in each case that ι is a well determined interpretation which is fixed once and for all. Because of this no distinction is made between the *names* of constants, functions and relations and the *actual* constants, functions and relations themselves. That is to say, in all that follows, unless explicitly stated to the contrary, we always identify c with $\iota(c)$, f with $\iota(f)$ and R with $\iota(R)$.

If ϕ is an interpretation in $\mathcal{M}$ of a sentence of $\mathcal{L}$, then ϕ is either true or else false. The notion of **true sentence** is given formally in accordance with the following definition:

Definition 6.4 :

1. An atomic formula $R(\tau_1, \tau_2, \ldots, \tau_n)$ is true if each of the terms $\tau_1, \tau_2, \ldots, \tau_n$ is defined and, moreover, the n-tuple $(\tau_1, \tau_2, \ldots, \tau_n)$ belongs to R,

2. If ϕ and ψ are sentences the logical values (or truth values) of all possible logical combinations of ϕ and ψ are given in accordance with the following table:

ϕ	ψ	$\neg\phi$	$\phi \wedge \psi$	$\phi \vee \psi$	$\phi \Rightarrow \psi$	$\phi \Leftrightarrow \psi$
1	1	0	1	1	1	1
1	0	0	0	1	0	0
0	1	1	0	1	1	0
0	0	1	0	0	1	1

where 1 denotes truth and 0 denotes falsity.

3. The sentence $\forall_x \phi(x)$ is true if $\phi(c)$ is true for every possible constant c.

4. The sentence $\exists_x \phi(x)$ is true if $\phi(c)$ is true for some possible constant c.

Definition 6.5 A **theory T** of the language $\mathcal{L}$ is a subset of all the sentences of $\mathcal{L}$; thus

$$\mathbf{T} \subset \mathbf{Sent}(\mathcal{L}).$$

A mathematical structure $\mathcal{M}$ equipped with a language $\mathcal{L}$ is said to be a **model** of the theory **T** if all the sentences of **T** when interpreted in $\mathcal{M}$ are true. We then write $\mathcal{M} \models \mathbf{T}$.

Let Σ be a set (finite or infinite) of sentences; we shall refer to these as **hypotheses**. From the mathematical point of view it is the *consequences* of Σ which are of interest. Hence it is necessary to devise a deductive system which allows us to pass logically from one sentence (or group of sentences) to another (that is, to pass from premises to conclusions). In the first place, there are certain general rules or principles which should apply to any theory whatsoever: these are the so-called **logical axioms**. (By contrast, the hypotheses of a theory are extra-logical axioms). In the second place, together with the logical axioms, there must be a system of rules which allow us to draw conclusions from the set of hypotheses Σ: these are called the **rules of inference**. In order to keep the exposition short we shall not present either the logical axioms or the rules of inference explicitly here.

Given a set of hypotheses Σ, a **deduction of** Σ (in the language $\mathcal{L}$) is a finite sequence of sentences $S_1, S_2, \ldots, S_n$ such that for each S_i, $(1 \leq i \leq n)$ one of the following assertions is true:

1. S_i is a logical axiom,
2. S_i belongs to Σ, or
3. S_i is inferred from the preceding formulas by means of the rules of inference.

The final sentence of the deduction, S_n, is said to be a **theorem** of Σ.

A theory **T** (in the language $\mathcal{L}$) is said to be **consistent** if there exists no sentence ϕ of $\mathcal{L}$ such that $\phi \wedge \neg\phi$ will be a theorem of **T**. It is said to be **complete** if it is consistent and if, for every sentence ϕ of $\mathcal{L}$, it turns out that either ϕ is a theorem of **T** or else that $\neg\phi$ is a theorem of **T**.

Definition 6.6 Let **T** be a theory of the language $\mathcal{L}$ and Σ a subset of **T**. We say that Σ is a **set of axioms** of **T** if, for every sentence ϕ of $\mathcal{L}$, ϕ is a theorem of **T** if and only if ϕ is a theorem of Σ.

6.1.2 First order arithmetic

A theory may have several models. If all the models of a theory are isomorphic then the theory really only has one model. However, there are theories which do possess non-isomorphic models: it is then usual to distinguish between *standard* and *nonstandard* models. The first time the term *nonstandard model* appeared was in a celebrated article of the Norwegian mathematician T.Skolem, born in 1887:

Skolem, T. Uber die nicht-charakterisierberkeit der zahlenreihe miltels endlich oder unendlich vider aussagen mit aussenlich zahlenvariablen. *Fund. Math.*, **23** (1934), pp.159-161.

In this paper Skolem described a nonstandard model for the well-known axiomatic system of Dedekind-Peano[4] for the arithmetic of the natural numbers: in the next

[4]Here we follow A.F.Oliveira, "Logica e Arithmetica", *Gradiva*, 1991, p.128: although popularised by G.Peano, this axiomatisation was devised essentially by R.Dedekind, and is therefore assigned the names of both mathematicians.

section we shall compare the familiar (standard) model of natural number arithmetic with a nonstandard model of the same theory.

6.1.2.1 The Dedekind-Peano axioms

The arithmetic of the natural numbers can be derived from a fundamental set of axioms devised by R.Dedekind and G.Peano. As with any other mathematical theory it should be possible to formulate the axioms (as well as other sentences derived from them) in terms of an appropriate formal language with an adequate set of symbols. In the present case, apart from the logical symbols already referred to, and letters $n, m \ldots$ for numerical variables, it is necessary to have special symbols for certain specific constants. Typically these will be required for such concepts as zero, successor, addition, product, less than, etc. For convenience we shall use some abbreviations. In particular n^+ will denote the successor of n, $m \neq n$ will denote the sentence $\neg[m = n]$, and so on.

The Dedekind-Peano Axiomatisation

1. $\exists_n[n = 0]$
2. $\forall_n \exists_m [m = n^+]$
3. $\forall_n [n^+ \neq 0]$
4. $\forall_{n,m}[n^+ = m^+ \Rightarrow m = n]$
5. $\forall_n [n + 0 = n]$
6. $\forall_{n,m}[n + m^+ = (n + m)^+]$
7. $\forall_n [n.0 = 0]$
8. $\forall_{n,m}[n.m^+ = n.m + n]$
9. $[p(0) \wedge [\forall_n [p(n) \Rightarrow p(n+1)]]] \Rightarrow \forall_n p(n)$.

Note that the axioms **1.** to **8.** are sentences: they contain no free variables. In the case of axiom **9.**, however, we actually have an infinite number of possible sentences which can be generated when we substitute for p any specific formula (involving a natural number).

6.1.2.2 Models for first order arithmetic

The Dedekind-Peano axioms have a model which it is natural to describe from now on as the **standard model**. This is the usual system of natural numbers $\mathbf{N} = (\mathbb{N}_0, +, \leq)$, where $\mathbb{N}_0$ is the set

$$\mathbb{N}_0 = \{0, 1, 2, 3, 4, \ldots\}$$

and "0", "successor", "sum", "product" and "less than or equal" have their familiar interpretation. All models of the Dedekind-Peano axioms which are not isomorphic to $\mathbf{N}$ are described as **nonstandard models of arithmetic.**

The existence of such nonstandard models of the arithmetic of the natural numbers can be inferred from the *compactness theorem* of mathematical logic which, in its turn, is a consequence of the *completeness theorem* of Godel:

Theorem 6.7 (Completeness) A set Σ of sentences (axioms) in a given language $\mathcal{L}$ is consistent if and only if it possesses a model.

Theorem 6.8 (Compactness) If every finite subset of a set Σ of sentences of a given language possesses a model, then Σ has a model.[5]

The system $\mathbf{N} \equiv (\mathbb{N}_0, +, \cdot, \leq)$ of the natural numbers is, as stated earlier, a model (the standard model) of the Dedekind-Peano axioms. Now let Σ be the set formed of all the sentences which are true in $\mathbf{N}$ together with the countable set of sentences which affirm the existence of a number ν greater than all the natural numbers; that is to say a number which is such that

$$\nu > 0, \nu > 1, \nu > 2, \ldots, \nu > k, \ldots,$$

The system $\mathbf{N}$ is a model for each of the finite subsets S of Σ, since each sentence of S is either a true sentence of $\mathbf{N}$ or else a sentence of the form $\nu > k$ for some $k \in \mathbb{N}_0$; since S is finite it contains only a finite number of sentences of this latter type, true therefore in $\mathbf{N}$. From the compactness theorem the set Σ then possesses a model, denoted by $^*\mathbf{N} \equiv (^*\mathbb{N}_0, +, \cdot, \leq, \ldots)$, such that every arithmetic sentence (of the first order) is true in $\mathbf{N}$ and also true in $^*\mathbf{N}$.

To avoid confusion and to simplify the interpretation the elements of $^*\mathbb{N}_0$, which are the "natural numbers" of the nonstandard model, will be called **hypernatural numbers.** The set $^*\mathbb{N}_0$ of these hypernatural numbers contains copies of the standard natural numbers (which we denote by the same names as those assigned to them in the standard model), but also contains many other elements

$$^*\mathbb{N}_0 = \{0, 1, 2, \ldots, \nu, \nu + 1, \ldots, \nu^2, \nu^2 + 1, \ldots\} \ .$$

Although this argument using the compactness theorem is sufficient to ensure the existence of nonstandard models for the Dedekind-Peano axioms, it is always desirable in these circumstances to present an explicit example of such a model. Accordingly Skolem, in the article already referred to, constructed a model of the form described in chapter 3, (but using sequences of natural numbers rather than real numbers). This gives the set

$$^*\mathbb{N}_0 = (\,^*\mathbb{N}_0^{\mathbf{N}})/\mathcal{U}$$

which contains (a copy of) $\mathbb{N}_0$, since the equivalence classes of $\mathcal{U}$-nearly all constant sequences form a subset of $^*\mathbb{N}_0$. Moreover, the arithmetic operations and the order relation are of the form described in chapter 3. Thus we obtain a mathematical structure

$$^*\mathbf{N} = (\,^*\mathbb{N}_0, +, \cdot, \leq)$$

which is a nonstandard model of Dedekind-Peano arithmetic. This model satisfies the following properties, among others:

- $^*\mathbb{N}_0$ contains (an isomorphic copy of) $\mathbb{N}_0$
- $^*\mathbb{N}_0$ is a proper extension of $\mathbb{N}_0$
- Every sentence (of the first order) which involves elements $[(m_n)_{n\in\mathbf{N}}]$, $[(p_n)_{n\in\mathbf{N}}]$, ..., is true in $^*\mathbf{N}$ if and only if the sentences corresponding to the coordinates $m_n, p_n, \ldots$, are true in $\mathbf{N}$ for $\mathcal{U}$-nearly all natural numbers $n \in \mathbb{N}$.

[5]That the second theorem follows from the first can easily be seen: in fact if Σ has no model it will not be consistent, and it would therefore be possible to deduce a contradiction from this set of sentences. But a deduction uses only a finite number of sentences and therefore there would exist a finite subset of sentences of Σ which would have no model.

Once Dedekind-Peano arithmetic is seen to admit standard and nonstandard models which contain (isomorphic copies of) the natural numbers then it is natural to classify the elements of the nonstandard model in accordance with their position relative to the standard model. Thus, the elements of $^*\mathbb{N}_0$ which belong to the (copy of) $\mathbb{N}_0$ are said to be **finite** while those which do not belong to $\mathbb{N}_0$ are said to be **infinite**. The existence of infinite hypernatural numbers shows that the nonstandard model cannot be isomorphic to the standard model (whose language does not even make use of the terms *"finite"* and *"infinite"*.[6]

This brief account of a nonstandard model for Dedekind-Peano arithmetic may serve to show why Abraham Robinson chose the name *Nonstandard Analysis* for the theory he developed in the 1960s. However, despite its historical interest, it is not our object to develop any further study of standard and nonstandard models in general. In what follows we merely wish to present a fundamental account of Nonstandard Analysis in a context more general than that of the preceding chapters.

6.2 Proper Extensions of $\mathbb{R}$

In the 17th century Leibniz proposed the adoption of a numerical system

- which would contain infinite numbers and infinitesimals as well as the real numbers,
- and in which the usual rules of the calculus would continue to apply.

Interpreted strictly in the light of what was understood at the time, these two principles are contradictory; suitably interpreted in the context of Robinson's Nonstandard Analysis they do not clash. This has already been shown in the restricted situation considered in chapter 3 with respect to sentences relativised to $\mathbb{R}$.

The first principle can certainly be realised since it is possible to construct a non-Archimedian extension of $\mathbb{R}$, containing infinite numbers and infinitesimals. This has already been made clear by the construction of the field $^*\mathbb{R}$ of hyperreal numbers given in chapter 3. Note, however, that there is an immense number of possible models which extend $\mathbb{R}$. It is not merely the case that to each and every ultrafilter $\mathcal{U}_{\mathbb{N}}$ over $\mathbb{N}$ there corresponds an extension $^*\mathbb{R} \equiv \mathbb{R}^{\mathbb{N}}/\mathcal{U}_{\mathbb{N}}$ of $\mathbb{R}$. More importantly the same process of construction might be carried out with respect to index sets of higher cardinality than $\mathbb{N}$. In this way we could obtain nonstandard models of the first order theory of the real numbers which would not generally be isomorphic.

Accordingly we now consider, as briefly as possible, the construction of an extension $^*\mathbb{R}$ based on an *arbitrary* infinite set of indices, Γ. To begin with we need to introduce some general concepts concerning *filters* and *ultrafilters* over Γ.

Definition 6.9 A **filter** over Γ is a family $\mathcal{F}_\Gamma$ of subsets of Γ which satisfy the following properties:

[6] The term "finite" does not form part of the first order language appropriate to **N**: within this model an admissible way of stating that *"m is a (standard) finite number"* is given by the sentence *"m is an element of* $\mathbb{N}_0$*"*. Thus, the application of the terms "finite" and "infinite" only makes sense in the context of the nonstandard model $^*\mathbf{N}$.

(1) $\emptyset \notin \mathcal{F}_\Gamma$ and $\Gamma \in \mathcal{F}_\Gamma$;
(2) $[A \in \mathcal{F}_\Gamma \wedge B \in \mathcal{F}_\Gamma] \Rightarrow A \cap B \in \mathcal{F}_\Gamma$;
(3) $[A \in \mathcal{F}_\Gamma \wedge A \subset B \subset \Gamma] \Rightarrow B \in \mathcal{F}_\Gamma$.

A filter $\mathcal{F}_\Gamma$ is said to be **free** if it is such that $\bigcap\{F : F \in \mathcal{F}_\Gamma\} = \emptyset$.

The set $\mathcal{F}_{0\Gamma}$ of the cofinite subsets of Γ is, as is easily verified, an example of a filter over Γ, and is called the **Fréchet filter**. $\mathcal{F}_{0\Gamma}$ is a free filter since for any element $a \in \Gamma$ the set $\Gamma\backslash\{a\}$ belongs, by definition, to $\mathcal{F}_{0\Gamma}$ and therefore $a \notin \bigcap\{A : A \in \mathcal{F}_{0\Gamma}\}$. It follows at once that $\bigcap\{A : A \in \mathcal{F}_{0\Gamma}\} = \emptyset$. More generally we have

Theorem 6.10 $\mathcal{F}_\Gamma$ is a free filter over Γ if and only if it contains $\mathcal{F}_{0\Gamma}$.

Proof In the first case suppose that $\mathcal{F}_\Gamma$ is a free filter and therefore that,

$$\bigcap\{A : A \in \mathcal{F}_\Gamma\} = \emptyset \quad .$$

Given a set of the form $\Gamma\backslash\{a_1, \ldots, a_n\} \in \mathcal{F}_{0\Gamma}$, we show that this always belongs to $\mathcal{F}_\Gamma$. Since $\mathcal{F}G_\Gamma$ is a free filter the element a_1 cannot belong to all its sets because otherwise we would then have $\bigcap\{F : F \in \mathcal{F}_\Gamma\} \supset \{a_1\}$. Thus there exists $A \in \mathcal{F}_\Gamma$ such that $a_1 \notin A$; consequently we have $A \subset \Gamma\backslash\{a_1\} \equiv B_1$ and, therefore, $B_1 \in \mathcal{F}_\Gamma$. Similarly we can show that the set $B_2 \equiv \Gamma\backslash\{a_2\}$ belongs to $\mathcal{F}_\Gamma$ and therefore that $B_1 \cap B_2 \equiv \Gamma\backslash\{a_1, a_2\}$ also belongs to $\mathcal{F}_\Gamma$. Repeating this reasoning as many times as necessary we have $\Gamma\backslash\{a_1, \ldots, a_n\} \in \mathcal{F}_\Gamma$ from which there follows the inclusion $\mathcal{F}_{0\Gamma} \subset \mathcal{F}_\Gamma$.

Conversely suppose that $\mathcal{F}_{0\Gamma} \subset \mathcal{F}_\Gamma$. Then

$$\bigcap\{B : B \in \mathcal{F}_\Gamma\} \subset \bigcap\{A : A \in \mathcal{F}_{0\Gamma}\} = \emptyset$$

and therefore $\mathcal{F}_\Gamma$ is a free filter. •

Definition 6.11 A family $\mathcal{U}_\Gamma$ of subsets of Γ such that
(a) $\mathcal{U}_\Gamma$ is a free filter over Γ, and
(b) $\forall_A[A \subset \Gamma \Rightarrow [A \in \mathcal{U}_\Gamma \vee A^c \equiv \Gamma\backslash A \in \mathcal{U}_\Gamma]]$

is said to be a **free ultrafilter** over Γ.

The existence of free ultrafilters is guaranteed by the following (purely existential) result already referred to :

Theorem 6.12 (Tarski) Every free filter can be extended to a free ultrafilter.

Proof Let M be the set of all filters over Γ which contain the Fréchet filter $\mathcal{F}_{0\Gamma}$. M is not empty since it at least contains the Fréchet filter itself. The set M can be ordered with respect to the relation $\subset$. Let K be a chain[7] of M and let $S = \bigcup\{\mathcal{F} : \mathcal{F} \in K\}$. S is, as is easily seen, a filter over Γ which contains the Frechet filter $\mathcal{F}_{0\Gamma}$. Consequently $S \in M$. Applying Zorn's Lemma we can deduce that M possesses a maximal element $\mathcal{U}_\Gamma$. It remains to prove that $\mathcal{U}_\Gamma$ is an ultrafilter over Γ.

[7]A **chain** is any totally ordered subset.

It is clear that $\mathcal{U}_\Gamma$ is a free filter which contains the Fréchet filter. Now let A and B be two subsets of Γ such that $A \cup B = \Gamma$ and suppose that, for example, $A \notin \mathcal{U}_\Gamma$. Considering the auxiliary family $\mathcal{G}$ of subsets of Γ defined by

$$\mathcal{G} = \{G \subset \Gamma : A \cup G \in \mathcal{U}_\Gamma\}$$

we can verify that

- $\mathcal{G}$ is a filter over Γ
- $\mathcal{U}_\Gamma \subseteq \mathcal{G}$.

Since $A \cup B = \Gamma$ then $A \cup B \in \mathcal{U}_\Gamma$ and, therefore, $B \in \mathcal{G}$. Now taking account of the fact that $\mathcal{U}_\Gamma$ is a maximal filter it follows that we must have $\mathcal{G} \subset \mathcal{U}_\Gamma$ and therefore $G = \mathcal{U}_\Gamma$ whence $B \in \mathcal{U}_\Gamma$ or, equivalently, $\mathcal{U}_\Gamma$ is an ultrafilter as asserted. •

Given an infinite index set Γ, consider the set of all generalised sequences (functions from Γ into $\mathbb{R}$)

$$\mathbb{R}^\Gamma = \{\{a_\gamma\}_{\gamma \in \Gamma} : a_\gamma \in \mathbb{R}\} \quad .$$

Letting $\mathcal{U}_\Gamma$ be a free ultrafilter over Γ, the quotient set

$$\mathbb{R}^\Gamma / \mathcal{U}_\Gamma \equiv {}^*\mathbb{R}$$

can be equipped with the operations of addition and multiplication and an ordering relation defined componentwise, modulo $\mathcal{U}_\Gamma$. The result is a non-Archimedian ordered field which extends $\mathbb{R}$. All the definitions considered in the models based on sequences of real numbers are equally applicable (when suitably adapted) to such kinds of model. The mapping

$$\iota : \mathbb{R} \to {}^*\mathbb{R} \tag{6.1}$$

which makes each element $a \in \mathbb{R}$ correspond to the equivalence class represented by the constant generalised sequence $(a_\gamma)_{\gamma \in \Gamma}$ with $a_\gamma = a$ for all $\gamma \in \Gamma$, embeds $\mathbb{R}$ in ${}^*\mathbb{R}$. It can thus be said that $\mathbb{R}$ is a *subfield* of ${}^*\mathbb{R}$, indeed a proper subfield, thus implying that ${}^*\mathbb{R}$ must be non-Archimedian.

6.3 The (second) Leibniz Principle

Recall that Leibniz's proposal required not merely the existence of an enlarged number system, containing infinitely large and infinitely small elements as well as the real numbers themselves. It also demanded that this system should continue to enjoy the same properties as the original real number system. The realisation of this second requirement needs a considerably more delicate approach. The problem was solved by A.Robinson in a most ingenious way, using tools developed within modern logic, particularly model theory. The essential point was the need for an adequate interpretation of the Leibniz second principle which would not ultimately result in a contradiction. Now there are certainly some properties of the original system which do transfer to the new system. But it seems to be required that *all* properties valid in the first system should also be valid in the second system, and conversely: and yet, after all, the two systems are clearly distinct! What is really needed is the formulation of a general principle of transfer between the two systems which effectively delimits the largest possible set of those properties which can be transferred from one to the other.

In Chapter 5 we introduced the notions of internal set and internal function, in order that these internal objects should be, in the last analysis, the true inheritors of the properties valid for standard objects of the same type. However, in order to do this we had to to resort again to the use of the ultrafilter $\mathcal{U}$, since the Leibniz principle relativised to $\mathbb{R}$ is not applicable in general. By returning to the construction of the basic model, we now seek to establish a version of the Leibniz Principle which will be comprehensive and will allow the general development of nonstandard analysis, regardless of model. We shall then clarify the reasons for making a general distinction between internal and external objects, in terms of sequences of standard objects of the same type, modulo a given ultrafilter $\mathcal{U}$ over $\mathbb{N}$.

6.3.1 Superstructures

Let Γ be an infinite set of indices and $\mathcal{U} \equiv \mathcal{U}_\Gamma$ a free ultrafilter over Γ. We fix, once and for all, the ultrafilter $\mathcal{U}$ and we let ${}^*\mathbb{R}$ denote the quotient $\mathbb{R}^\Gamma/\mathcal{U}$, the **set of hyperreal numbers.**

To establish a transfer principle applicable to all the objects of mathematical analysis (numbers, sets, functions, functionals, spaces of functions, etc.) we must consider sentences where quantification can be made over the set of all such objects. That is, we must be able to allow quantification over the whole *universe of real analysis*, or over the whole *universe of nonstandard analysis*), and not just over the set $\mathbb{R}$ (or the set ${}^*\mathbb{R}$). We wish to transfer sentences of the type *"for all nonempty subsets of ..."* or, *"there exists a function defined on the set A ..."* and to do this, it is necessary to be able to quantify over a set which contains among its elements sets of numbers, functions, etc.

As is well-known, each and every mathematical theory can be formalised in terms of sets, the theory of sets thus being the *"primitive theory"* established axiomatically beforehand. One of the most frequently adopted axiomatisations of set theory is Zermelo-Fraenkel set theory, ZF. In the present context it is necessary to add one further axiom, the axiom of choice, to obtain the axiom system denoted by ZFC. All the objects of any mathematical theory formulated on this basis are ultimately defined as sets. Now, the fact that nothing but sets exist makes the development of any theory highly technical, extremely tedious and difficult to understand. For example, a fully rigorous treatment could oblige us to to consider the real numbers 1 and π as sets (or sets of sets, or sets of sets of sets, or ...), although in point of fact it may well be entirely irrelevant to specify the elements which constitute these objects. In most cases therefore, it is desirable to develop the theory based on sets of objects which, in the context of that particular theory, may be themselves regarded as *"non-sets"*. That is to say the theory should be based on *primitive* objects which, within the context of the theory concerned, are considered to have no elements and which therefore have no internal structure. The set of all such primitive elements we will call **the set of atoms** or **the set of individuals** of the theory. Over and above these atoms, all the other objects of the theory are sets.

In real analysis, the set of atoms is, naturally enough, chosen to be the set $\mathbb{R}$ of the real numbers. All the other objects of real analysis are sets of real numbers, sets of sets of real numbers, sets of such sets, and so on. That is to say, over and above the real numbers (the atoms of the theory), all other objects can be defined in terms of sets of real numbers, and supersets of these. In particular, a function or, more generally a relation, may be defined in terms of the notion of ordered pair, and this is itself a set of sets.[8] Thus, as is well known, we can define an ordered pair (in the sense of Kuratowski) by setting

$$(x, y) \equiv \{\{x\}, \{x, y\}\}$$

where the asymmetry of the definition makes clear which is the first and which is the second element of that ordered pair. And, of course, a function $f : \mathbb{R} \to \mathbb{R}$ can be identified as a certain specific set of such ordered pairs;

$$f \equiv \{(x, y) : y = f(x)\} \ .$$

Since real analysis can thus be considered as the theory of sets whose basic set of atoms is $\mathbb{R}$, it can be formally denoted as **ZFC**$\mathbb{R}$: Zermelo-Fraenkel + choice + $\mathbb{R}$.

What can be said for real analysis can also be said with greater generality for any other theory based on a set of atoms X (where we suppose that $\mathbb{R} \subset X$). In this way we obtain a theory **ZFCX** whose axioms include the following:

Axiom of Atoms

$$\forall_a [a \in \mathbf{X} \Leftrightarrow a \neq \emptyset \wedge \neg[(\exists_t [t \in a)]] \ .$$

This actually says that: *an atom contains no elements, but nevertheless is not the empty set.*[9]

The remainder of **ZFCX** is essentially identical with **ZFC**, and it can be shown that the consistency of **ZFCX** depends only on the consistency of **ZFC**.

Let X be an appropriate set of atoms and consider the family of sets $(X_n)_{n=0,1,2,\ldots}$ defined recursively by

$$X_0 = X : \qquad X_{n+1} = X_n \cup \mathcal{P}(X_n), \qquad n = 0, 1, 2, \ldots,$$

[8] In the context of the construction which will follow, it is usual to define functions and relations in a way slightly different from the usual one, though equivalent to it. Functions and relations will now be understood as appropriate subsets of ordered n-tuples for suitable values of n, in accordance with the following definition:

Definition For $n \geq 2$ we define an **ordered n-tuple** to be a set

$$(x_1, x_2, \ldots, x_n) \equiv \{(1, x_1), (2, x_2), \ldots, (n, x_n)\}$$

where $x_1, x_2, \ldots, x_n$ are objects of the theory.

Note that the concepts of ordered pair and 2-tuple are distinct. Here, the concept of ordered pair serves only to define the n-tuple. An object of the form (x_1, x_2) in this context will generally be interpreted as a 2-tuple and not as an ordered pair.

The definition of ordered n-tuple has the advantage of putting on an equal footing (relative to the construction which is to follow) all functions and relations, independently of the number of their variables.

[9] In **ZFC** the unique object which contains no elements is the set $\emptyset$.

where $\mathcal{P}(Y))$ denotes the power set of Y.

We define the **superstructure** over the set X of atoms to be the set given by

$$\mathcal{V}(X) = \bigcup_{n=0}^{\infty} X_n \quad .$$

The elements of X_0 are the atoms, while the elements of $\mathcal{V}(x)\backslash X_0$ are called the **entities** of the superstructure. We also refer to X_0 as the **base** of the superstructure; the set $X_n \backslash X_{n-1}$ is called the **level** of **order** n of the superstructure $\mathcal{V}(X)$.

The elements of $X_n \backslash X_{n-1}$ are said to be entities of order n. For an element α of $\mathcal{V}(X)$ we may also write $r(\alpha) = n \geq 1$ so as to signify that it is an entity of order n: if α denotes an atom then we write $r(\alpha) = 0$.

As can be seen from the definition the sets $X_n, n = 0, 1, 2, \ldots$ satisfy the following chains of relations:

$$X_0 \subset X_1 \subset X_2 \subset \ldots \subset X_n \subset \ldots$$

$$X_0 \in X_1 \in X_2 \in \ldots \in X_n \in \ldots$$

The superstructure $\mathcal{V}(X)$ contains all the entities which might appear in the development of any theory based on the elements of X, thereby satisfying various other properties which makes it reasonable to call it the **universe** over X.

Since the relation $X_{n-1} \subset X_n$ implies that $\mathcal{P}(X_{n-1}) \subset \mathcal{P}(X_n)$ it can be shown by induction that

$$X_{n+1} = X \cup \mathcal{P}(X_n), \qquad n = 0, 1, 2, \ldots \tag{6.2}$$

from which there follows the so-called **transitivity** of the set X_{n+1} (for $n = 0, 1, 2, \ldots$). This is to be understood in the following sense

$$\forall_{a,b}[a \in b \in X_{n+1} \Rightarrow a \in X_n \quad .$$

In fact, we have from (6.2) that if $b \in X_{n+1}$ then $b \in X$ or $b \subset X_n$; however b cannot belong to X (because the elements of X, by hypothesis, do not have elements). Consequently, $b \subset X_n$ and, therefore, $a \in X_n$.

Since every element of $\mathcal{V}(X)$ is either an atom or else belongs to X_{n+1} for some $n \in \mathbb{N}_0$ then $\mathcal{V}(X)$ is itself a *transitive set*

$$\forall_{a,b}[a \in b \in \mathcal{V}(X) \Rightarrow a \in \mathcal{V}(X)] \quad .$$

Exercise 6.13 *Show that*

(1) $X_{n+1} = X \cup \mathcal{V}(X_n), \quad n = 0, 1, 2, \ldots,$
(2) $a, b \in X_n \Rightarrow \{a, b\} \in X_{n+1} \wedge (a, b) \in X_{n+2},$
(3) $a \in X_n \Rightarrow \mathcal{P}(a) \in X_{n+2}.$

An arbitrary subset of elements of $\mathcal{V}(X)$ may not belong to $\mathcal{V}(X)$; that is to say from the relation $A \subset \mathcal{V}(X)$ it does not necessarily follow that $A \in \mathcal{V}(X)$. Thus although the inclusion $\mathcal{V}(X) \subset \mathcal{V}(X)$ holds, it is not true that $\mathcal{V}(X)$ is a member of $\mathcal{V}(X)$. Another example of an entity contained in $\mathcal{V}(X)$ which does not belong to $\mathcal{V}(X)$ is any subset of $\mathcal{V}(\mathcal{R})$ which contains an element of each level (with no upper limit). In the general case, from the way in which the superstructure $\mathcal{V}(X)$ is constructed, we have the following criterion:

A set $A \subset \mathcal{V}(X)$ belongs to $\mathcal{V}(X)$ if and only if there exists an upper bound of all the orders of the elements of A. that is, if and only if there exists $n \in \mathbb{N}$ such that

$$\forall_a [a \in A \Rightarrow r(a) \leq n] \quad . \tag{6.3}$$

In fact, if $A \in \mathcal{V}(X)$ then $A \in X_{n+1}$ for some $n \geq 0$ and, therefore, it satisfies condition (6.3). Conversely, suppose that there exists $n \geq 0$ such that for all $a \in A$ we have $r(a) \leq n$. Then $A \subset X_n$ and therefore $A \in \mathcal{V}(X)$.

6.3.1.1 Embedding of superstructures

Carrying out the construction described above with $X = \mathbb{R}$ and again with $X = {}^*\mathbb{R}$ we obtain the superstructures $\mathcal{V}(\mathbb{R})$ and $\mathcal{V}({}^*\mathbb{R})$. These constitute respectively **the universe of real analysis** and the **universe of nonstandard analysis**. We intend now to establish a general transfer principle which is an extension to $\mathcal{L}(\mathbb{R})$ of the transfer principle relativised to $\mathbb{R}$, already studied at the end of Chapter 3. Recall that the field $\mathbb{R}$ was *"embedded"* in the field ${}^*\mathbb{R}$, belonging to $\mathcal{V}({}^*\mathbb{R})$, by means of the mapping $a \mapsto {}^*a \equiv [(a, a, \ldots)]$. We naturally wish to preserve this embedding and extend it to the whole structure $\mathcal{V}(\mathbb{R})$. To do this it is necessary to construct an injective mapping ${}^* : \mathcal{V}(\mathbb{R}) \to \mathcal{V}({}^*\mathbb{R})$ which satisfies the conditions

(a) $\forall_a [a \in \mathbb{R} \Rightarrow {}^*a = a]$

(b) $\forall_{a,b} [a, b \in \mathcal{V}(\mathbb{R}) \Rightarrow [a \in b \Leftrightarrow {}^*a \in {}^*b]]$.

Note that in the subformula ${}^*a \in {}^*b$ of the condition (b) the membership symbol $\in$ is to be interpreted in the sense of the universe $\mathcal{V}({}^*\mathbb{R})$: this condition then affirms that the subformula $a \in b$ in $\mathcal{V}(\mathbb{R})$ must be true when and only when ${}^*a \in {}^*b$ is true in $\mathcal{V}({}^*\mathbb{R})$. It is therefore necessary to define what happens in passing from the interpretation of $\in$ in $\mathcal{V}(\mathbb{R})$ to the interpretation of $\in$ in $\mathcal{V}({}^*\mathbb{R})$ in a way which fully satisfies this condition. As we shall see, this can be achieved in stages: we define the mapping * by the composition of two other mappings - the constant mapping κ and a mapping known as the **Mostowski collapse**.

The constant mapping κ When we embed $\mathbb{R}$ in ${}^*\mathbb{R}$ we use equivalence classes of (generalised) constant sequences of the form $[(r)_{\gamma \in \Gamma}]$ with $r \in \mathbb{R}$. The constant mapping will be the extension to $\mathcal{V}(\mathbb{R})$ of the embedding referred to in (6.1). For this we begin by applying to the whole superstructure $\mathcal{V}(\mathbb{R})$ the same process which allows the extension of $\mathbb{R}$ to ${}^*\mathbb{R}$. Thus, we first define the ultrapower

$$\mathcal{V}_{\mathcal{U}}[\mathbb{R}] \equiv (\mathcal{V}(\mathbb{R}))^{\Gamma} / \mathcal{U}$$

which is a set formed of equivalence classes, modulo $\mathcal{U}$, of (generalised) sequences of objects belonging to $\mathcal{V}(\mathbb{R})$. The relations of equality, $=_{\mathcal{U}}$, and of membership $\in_{\mathcal{U}}$, between objects of $\mathcal{V}_{\mathcal{U}}(\mathbb{R})$ are then defined as follows: given any two elements $\alpha \equiv [\{a_\gamma\}_{\gamma \in \Gamma}]$ and $\beta \equiv [\{b_\gamma\}_{\gamma \in \Gamma}]$ of $\mathcal{V}_{\mathcal{U}}[\mathbb{R}]$,

$$\alpha =_{\mathcal{U}} \beta \Leftrightarrow \{\gamma \in \Gamma : a_\gamma = b_\gamma\} \in \mathcal{U}$$

$$\alpha \in_{\mathcal{U}} \beta \Leftrightarrow \{\gamma \in \Gamma : a_\gamma \in b_\gamma\} \in \mathcal{U}. \tag{6.4}$$

These relations $=_{\mathcal{U}}$ and $\in_{\mathcal{U}}$ are not the relations of *equality* and *membership* in the usual set-theoretic sense of the terms; in fact, given two classes $[\{a_\gamma\}_{\gamma\in\Gamma}]$ and $[\{b_\gamma\}_{\gamma\in\Gamma}]$ of $\mathcal{V}(\mathbb{R})$ an expression of the type $[\{a_\gamma\}_{\gamma\in\Gamma}] \in [\{b_\gamma\}_{\gamma\in\Gamma}]$ means that $[\{a_\gamma\}_{\gamma\in\Gamma}]$ is an element of $[\{b_\gamma\}_{\gamma\in\Gamma}]$, which would then have to be characterised as a set whose elements are of the form $[\{a_\gamma\}_{\gamma\in\Gamma}]$, and not an equivalence class of sequences $\{b\}_{\gamma\in\Gamma}$ of elements of $\mathcal{V}(\mathbb{R})$, as is really the case. Hence the necessity of the special definitions given for $=_{\mathcal{U}}$ and $\in_{\mathcal{U}}$.

We shall now embed $\mathcal{V}(\mathbb{R})$ in $\mathcal{V}({}^*\mathbb{R})$ by means of the constant (generalised) sequences, defining the mapping

$$\kappa \; : \; \mathcal{V}(\mathbb{R}) \to \mathcal{V}_{\mathcal{U}}[\mathbb{R}]$$

$$\alpha \mapsto \kappa(\alpha) = [(\alpha)_{\gamma\in\Gamma}] \; .$$

Theorem 6.14 If $a, b \in \mathcal{V}(\mathbb{R})$ are two arbitrary entities then
(1) $a \in b \Leftrightarrow \kappa(a) \in_{\mathcal{U}} \kappa(b)$
(2) $a = b \Leftrightarrow \kappa(a) =_{\mathcal{U}} \kappa(b)$.

Proof (1) If $a \in b$ then the set of indices $\{\gamma \in \Gamma : a_\gamma \in b_\gamma\}$ belongs to $\mathcal{U}$ already since for all $\gamma \in \Gamma$, $a_\gamma = a$ and $b_\gamma = b$. Consequently $\kappa(a) \equiv [(a)_{\gamma\in\Gamma}] \in_{\mathcal{U}} [(b_{\gamma\in\Gamma}] \equiv \kappa(b)$. Conversely, if $\kappa(a) \in_{\mathcal{U}} \kappa(b)$ then $\{\gamma \in \Gamma : a \in b\} \in \mathcal{U}$; since $\emptyset \notin \mathcal{U}$ then $a \in b$ for some index $\gamma \in \Gamma$ (and, therefore, for all those indices).

(2) The first part $a = b \Rightarrow [(a)_{\gamma\in\Gamma}] =_{\mathcal{U}} [(b)_{\gamma\in\Gamma}]$ is immediate as a result of the proper definition of the extension $\mathcal{V}_{\mathcal{U}}[\mathcal{R}]$. Conversely, if $\kappa(a) \equiv [(a)_{\gamma\in\Gamma}] =_{\mathcal{U}} [(b)_{\gamma\in\Gamma}] \equiv \kappa(b) \in \mathcal{U}$ then $\{\gamma \in \Gamma : a = b\} \in \mathcal{U}$ and since $\emptyset \notin \mathcal{U}$ it follows that we must have $a = b$. •

From the second part of the theorem it follows immediately that the *mapping* $\kappa : \mathcal{V}(\mathbb{R}) \to \mathcal{V}_{\mathcal{U}}[\mathbb{R}]$ *is injective*, thereby embedding $\mathcal{V}(\mathbb{R})$ in $\mathcal{V}_{\mathcal{U}}[\mathbb{R}]$.

In $\mathcal{V}_{\mathcal{U}}[\mathbb{R}]$ there exist elements representable by sequences of objects of $\mathcal{V}(\mathbb{R})$ whose order has no upper limit. However, for any $a \in \mathcal{V}(\mathbb{R})$ whatsoever, the element $\kappa \equiv [(a)_{\gamma\in\Gamma}] \in \mathcal{V}_{\mathcal{U}}[\mathbb{R}]]$ will be of a special type: either it will belong to $(\mathbb{R}_0)^\Gamma/\mathcal{U}$ or it will belong to $(\mathbb{R}_k \backslash \mathbb{R}_{k-1})^\Gamma/\mathcal{U}$ for some given $k \in \mathbb{N}$. These objects (of the codomain of κ) therefore belong to the so-called *"bounded ultrapower"* of $\mathcal{V}(\mathbb{R})$, which is defined as follows.

Definition 6.15 A generalised sequence $a \equiv \{a_\gamma\}_{\gamma\in\Gamma}$ of elements of $\mathcal{V}(\mathbb{R})$ is said to be **bounded** if there exists (a fixed) $i \in \mathbb{N}$ such that $r(a_\gamma) \leq i$ for any $\gamma \in \Gamma$ belonging to a subset which is in the ultrafilter $\mathcal{U}$. We call the **bounded ultrapower** of $\mathcal{V}(\mathbb{R})$ the part of $\mathcal{V}_{\mathcal{U}}[\mathbb{R}]$ consisting of equivalence classes representable by (generalised) bounded sequences. We denote the bounded ultrapower of $\mathcal{V}(\mathbb{R})$ by $\Pi_{\mathcal{U}}(\mathbb{R})$.

From this definition there follows immediately the equality

$$\Pi_{\mathcal{U}}(\mathbb{R}) = (\mathbb{R}_0)^\Gamma/\mathcal{U} \cup (\mathbb{R}_1 \backslash \mathbb{R}_0)^\Gamma/\mathcal{U} \cup \ldots \cup (\mathbb{R}_k \backslash \mathbb{R}_{k-1})^\Gamma/\mathcal{U} \cup \ldots$$

which then confirms the inclusion $\kappa(\mathcal{V}(\mathbb{R})) \subset \Pi_{\mathcal{U}}(\mathbb{R})$.

The Mostowski Collapse, μ

Recall that the ultimate objective here is the embedding of the superstructure $\mathcal{V}(\mathbb{R})$ in the superstructure $\mathcal{V}({}^*\mathbb{R})$. This requires first the identity mapping of the base, and second that the relations of *membership* and *equality* in $\mathcal{V}(\mathbb{R})$ should correspond respectively in $\mathcal{V}({}^*\mathbb{R})$ to *"valid"* relations of *membership* and *equality* (in a set-theoretic sense). Since we have already achieved an injective mapping which allows us to pass from $\mathcal{V}(\mathbb{R})$ to $\Pi_{\mathcal{U}}(\mathbb{R})$, we now define a new mapping which will allow us to pass from $\Pi_{\mathcal{U}}(\mathbb{R})$ to $\mathcal{V}({}^*\mathbb{R})$. As will be seen, this latter mapping then allows the realisation of the embedding *, by composition with the former.

Definition 6.16 The **Mostowski collapse** is defined to be the mapping

$$\mu : \Pi_{\mathcal{U}}(\mathbb{R}) \to \mathcal{V}({}^*\mathbb{R})$$

which satisfies the following conditions

(a) the restriction of μ to $(\mathbb{R}_0)^\Gamma/\mathcal{U}$ is the identity mapping;

(b) for any element $[\{\alpha_\gamma\}_{\gamma\in\Gamma}]$ of $\Pi_{\mathcal{U}}(\mathbb{R})\setminus\left((\mathbb{R}_0)^\Gamma/\mathcal{U}\right)$, the Mostowki transformation $\mu([\{\alpha_\gamma\}_{\gamma\in\Gamma}])$ is given inductively by

$$\mu([\{\alpha_\gamma\}_{\gamma\in\Gamma}]) = \{\mu([\{x_\gamma\}_{\gamma\in\Gamma}]) : [\{x_\gamma\}_{\gamma\in\Gamma}] \in_{\mathcal{U}} [\{\alpha_\gamma\}_{\gamma\in\Gamma}]\} \ .$$

Note that $\mathbb{R}_0 \equiv \mathbb{R}$ and, therefore, $(\mathbb{R}_0)^\Gamma/\mathcal{U} \equiv {}^*\mathbb{R}$, whence by (a) we have $\mu(x) = x$ for all $x \in {}^*\mathbb{R} \subset \Pi_{\mathcal{U}}(\mathbb{R})$. Suppose now that we have $A_\gamma \in \mathcal{P}(\mathbb{R})$ for $\mathcal{U}$-nearly all $\gamma \in \Gamma$. From the definition of the Mostowski collapse we have

$$\mu([\{A_\gamma\}_{\gamma\in\Gamma}]) = \{\mu([\{a_\gamma\}_{\gamma\in\Gamma}]) \in \mathcal{V}({}^*\mathbb{R}) : [\{a_\gamma\}_{\gamma\in\Gamma}] \in_{\mathcal{U}} [\{A_\gamma\}_{\gamma\in\Gamma}]\}$$

where $[\{a_\gamma\}_{\gamma\in\Gamma}]$ certainly belongs to the zeroth level. Then $\mu([\{a_\gamma\}_{\gamma\in\Gamma}]) \equiv [\{a_\gamma\}_{\gamma\in\Gamma}] \in {}^*\mathbb{R}$ and, therefore

$$\mu([\{A_\gamma\}_{\gamma\in\Gamma}]) = \{[\{a_\gamma\}_{\gamma\in\Gamma}] \in {}^*\mathbb{R} : [\{a_\gamma\}_{\gamma\in\Gamma}] \in_{\mathcal{U}} [\{A_\gamma\}_{\gamma\in\Gamma}]\}$$

that is, $\mu([\{A_\gamma\}_{\gamma\in\Gamma}])$ is a subset of ${}^*\mathbb{R}$ formed of all the numbers $[\{a_\gamma\}_{\gamma\in\Gamma}] \in {}^*\mathbb{R}$ such that the set $\{\gamma \in \Gamma : a_\gamma \in A_\gamma\}$ belongs to the ultrafilter $\mathcal{U}$. Similarly we can verify that if $\{f_\gamma\}_{\gamma\in\Gamma}$ is a sequence of functions mapping $\mathbb{R}$ into $\mathbb{R}$, then $\mu([\{f_\gamma\}_{\gamma\in\Gamma}])$ can be identified as a function

$$F : {}^*\mathbb{R} \to {}^*\mathbb{R}$$

defined for $x \equiv [\{x_\gamma\}_{\gamma\in\Gamma}] \in {}^*\mathbb{R}$ by $F(x) \equiv [(f_\gamma(x_\gamma))_{\gamma\in\Gamma}] \in {}^*\mathbb{R}$.

In general let $[\{\alpha_\gamma\}_{\gamma\in\Gamma}]$ belong to $\Pi_{\mathcal{U}}(\mathbb{R})\setminus{}^*\mathbb{R}$. From the definition of $\Pi_{\mathcal{U}}(\mathbb{R})$ it follows that $[\{\alpha_\gamma\}_{\gamma\in\Gamma}] \in (\mathbb{R}_k\setminus\mathbb{R}_{k-1})^\Gamma/\mathcal{U}$ for some $k \geq 1$. Hence, if $[\{a_\gamma\}_{\gamma\in\Gamma}] \in_{\mathcal{U}} [\{\alpha_\gamma\}_{\gamma\in\Gamma}]$ then $[\{a_\gamma\}_{\gamma\in\Gamma}]$ belongs to some level of order inferior to k having already been defined by its reduction as an element of $\mathcal{V}({}^*\mathbb{R})$. The given definition for μ thus makes sense.

heorem 6.17 μ is an injective mapping.

roof Let $[\{\alpha_\gamma\}_{\gamma\in\Gamma}]$ and $[\{\beta_\gamma\}_{\gamma\in\Gamma}]$ be two elements of $\Pi_{\mathcal{U}}(\mathbb{R})$. If $[\{\alpha_\gamma\}_{\gamma\in\Gamma}]$ and $\{\beta_\gamma\}_{\gamma\in\Gamma}]$ belong to $(\mathbb{R}_0)^\Gamma/\mathcal{U}$ then from $[\{a_\gamma\}_{\gamma\in\Gamma}] \neq_{\mathcal{U}} [\{\beta_\gamma\}_{\gamma\in\Gamma}]$ it follows that $\gamma\in\Gamma : \alpha_\gamma \neq \beta_\gamma\} \in \mathcal{U}$ and, therefore, that $\mu([\{\alpha_\gamma\}_{\gamma\in\Gamma}]) \neq \mu([\{\beta_\gamma\}_{\gamma\in\Gamma}])$.

Suppose now that $[\{\alpha_\gamma\}_{\gamma\in\Gamma}], [\{\beta_\gamma\}_{\gamma\in\Gamma}] \in \Pi_{\mathcal{U}}(\mathbb{R})\backslash((\mathbb{R}_0)^\Gamma/\mathcal{U})$ are two elements uch that $[\{\alpha_\gamma\}_{\gamma\in\Gamma}] \neq_{\mathcal{U}} [\{\beta_\gamma\}_{\gamma\in\Gamma}]$. Without any loss of generality we may suppose hat $\alpha_\gamma \neq \beta_\gamma$ for all $\gamma \in \Gamma$. Since

$$\alpha_\gamma \neq \beta_\gamma \Leftrightarrow \alpha_\gamma \not\subset \beta_\gamma \vee \beta_\gamma \not\subset \alpha_\gamma$$

hen

$$\alpha_\gamma \neq \beta_\gamma \Leftrightarrow \alpha_\gamma\backslash\beta_\gamma \neq \emptyset \vee \beta_\gamma\backslash\alpha_\gamma \neq \emptyset$$

nd therefore, since $\{\gamma\in\Gamma : \alpha_\gamma\backslash\beta_\gamma \neq \emptyset\} \cup \{\gamma\in\Gamma : \beta_\gamma\backslash\alpha_\gamma \neq \emptyset\} \equiv \Gamma$, one of the sets

$$\{\gamma\in\Gamma : \alpha_\gamma\backslash\beta_\gamma \neq \emptyset\} \quad \text{or} \quad \{\gamma\in\Gamma : \beta_\gamma\backslash\alpha_\gamma \neq \emptyset\}$$

elongs to the ultrafilter $\mathcal{U}$. Suppose, for example, that the set

$$A \equiv \{\gamma\in\Gamma : \alpha_\gamma\backslash\beta_\gamma \neq \emptyset\}$$

elongs to the ultrafilter $\mathcal{U}$. Then we construct an element $[\{\xi_\gamma\}_{\gamma\in\Gamma}] \in \Pi_{\mathcal{U}}(\mathbb{R})$ of the ollowing form: for each $\gamma\in A$ we choose $\xi_\gamma \in \alpha_\gamma\backslash\beta_\gamma$ and, for each $\gamma \notin A$ we choose $_\gamma$ arbitrarily. Then we have $[\{\xi_\gamma\}_{\gamma\in\Gamma}] \in_{\mathcal{U}} [\{\alpha_\gamma\}_{\gamma\in\Gamma}]$ but $[\{\xi_\gamma\}_{\gamma\in\Gamma}] \notin_{\mathcal{U}} [\{\beta_\gamma\}_{\gamma\in\Gamma}]$, and herefore the sets

$$\{[\{\eta_\gamma\}_{\gamma\in\Gamma}] : [\{\eta_\gamma\}_{\gamma\in\Gamma}] \in_{\mathcal{U}} [\{\alpha_\gamma\}_{\gamma\in\Gamma}]\}$$

$$\{[\{\eta_\gamma\}_{\gamma\in\Gamma}] : [\{\eta_\gamma\}_{\gamma\in\Gamma}] \in_{\mathcal{U}} [\{\beta_\gamma\}_{\gamma\in\Gamma}]\}$$

re necessarily distinct. Hence, $\mu([\{\alpha_\gamma\}_{\gamma\in\Gamma}]) \neq \mu([\{\beta_\gamma\}_{\gamma\in\Gamma}])$ which, together with he first part, completes the proof that μ is injective. •

he mapping *. We can now finally establish an embedding of the universe of real nalysis into the universe of nonstandard analysis.

heorem 6.18 The mapping

$$^* \equiv \mu\circ\kappa : \mathcal{V}(\mathbb{R}) \to \mathcal{V}({}^*\mathbb{R})$$

efined, for each $\alpha \in \mathcal{V}(\mathbb{R})$, by

$$^*\alpha = \mu\circ\kappa(\alpha) = \mu([(\alpha)_{\gamma\in\Gamma}])$$

an embedding of the superstructure $\mathcal{V}(\mathbb{R})$ in the superstructure $\mathcal{V}({}^*\mathbb{R})$.

roof Since κ and μ are injective mappings, * is also an injective mapping. From the efinition of κ and of μ it follows immediately that the *-transform of each element f $\mathbb{R}$ is an element of $^*\mathbb{R}$. On the other hand

$$\mu\circ\kappa(\mathbb{R}) = \mu([(\mathbb{R})_{\gamma\in\Gamma}]) = \{x \equiv [\{x_\gamma\}_{\gamma\in\Gamma}] : \{\gamma\in\Gamma : x_\gamma \in \mathbb{R}\} \in \mathcal{U}\}$$

which shows that $\mu \circ \kappa(\mathbb{R})$ coincides with the set ${}^*\mathbb{R}$ defined earlier. Consequently if for each $r \in \mathbb{R}$ we identify r with ${}^*r$ we have $\mathbb{R} \subset {}^*\mathbb{R}$, this being a proper inclusion. It remains to prove that the second condition necessary for the *-transformation to be an embedding of the the superstructures is satisfied. Let α, β be any elements of $\mathcal{V}(\mathbb{R})$ whatsoever. If $\alpha \in \beta$ then, from theorem 6.14, $\kappa(\alpha) \in_{\mathcal{U}} \kappa(\beta)$ and therefore, from the definition of μ it follows that $\mu \circ \kappa(\alpha) \in \mu \circ \kappa(\beta)$, or, equivalently, that ${}^*\alpha \in {}^*\beta$. Conversely, if ${}^*\alpha \in {}^*\beta$ then $\mu \circ \kappa(\alpha) \in \mu \circ \kappa(\beta)$ whence, taking into account the definition of $\mu(\kappa(\beta))$, we obtain the $\mathcal{U}$-relation $\kappa(\alpha) \in_{\mathcal{U}} \kappa(\beta)$. Appealing again to theorem 6.14, we have $\alpha \in \beta$. •

Theorem 6.19 The codomain of the Mostowski collapse is the following

$$\mu(\Pi_{\mathcal{U}}(\mathbb{R})) = {}^*\mathbb{R} \cup {}^*(\mathbb{R}_1) \cup \ldots \cup {}^*(\mathbb{R}_k) \cup \ldots$$

Proof Let α be any element of $\mu(\Pi_{\mathcal{U}}(\mathbb{R}))$. Then $\alpha = ([\{a_\gamma\}_{\gamma \in \Gamma}])$ with $[\{a_\gamma\}_{\gamma\in\Gamma}] \in \Pi_{\mathcal{U}}(\mathbb{R})$: that is, with $[\{a_\gamma\}_{\gamma\in\Gamma}] \in [\mathbb{R}_0]$ or with $[\{a_\gamma\}_{\gamma\in\Gamma}] \in [\mathbb{R}_k \backslash \mathbb{R}_{k-1}]$ for some $k \in \mathbb{N}$. It follows that $\{\gamma \in \Gamma : a_\gamma \in \mathbb{R}_k\} \in \mathcal{U}$, from which we may conclude that for some $k \in \mathbb{N}_0$,

$$\alpha \equiv \mu([\{a_\gamma\}_{\gamma\in\Gamma}]) \in \mu([(\mathbb{R}_k)_{\gamma\in\Gamma}]) = \mu \circ \kappa(\mathbb{R}_k) \equiv {}^*(\mathbb{R}_k) \quad .$$

Conversely, let α be an element of ${}^*\mathbb{R}_k$ for some $k \in \mathbb{N}_0$. Then $\alpha = \mu([\{a_\gamma\}_{\gamma\in\Gamma}])$ with $[\{a_\gamma\}_{\gamma\in\Gamma}] \in_{\mathcal{U}} [(\mathbb{R}_k)_{\gamma\in\Gamma}]$ for some $k \in \mathbb{N}_0$ and so $\alpha \in \mu(\Pi_{\mathcal{U}}(\mathbb{R})$. •

6.3.1.2 The Los-Mostowski Theorem and the Leibniz principle

Let $\mathcal{L}(\mathbb{R})$ and $\mathcal{L}({}^*\mathbb{R})$ be the formal languages appropriate to the superstructures $\mathcal{V}(\mathbb{R})$ and $\mathcal{V}({}^*\mathbb{R})$ for (standard) real analysis and nonstandard analysis respectively. Given any formula ϕ of the language $\mathcal{L}(\mathbb{R})$, we write $\mathcal{V}(\mathbb{R}) \models \phi$ to signify that ϕ is a true sentence in the universe $\mathcal{V}(\mathbb{R})$; similarly, we write $\mathcal{V}({}^*\mathbb{R}) \models \Phi$ to signify that the formula Φ of the language $\mathcal{L}({}^*\mathbb{R})$ is a true sentence in the universe $\mathcal{V}({}^*\mathbb{R})$.

Definition 6.20 Given a formula ϕ of $\mathcal{L}(\mathbb{R})$ the ***-transform** of ϕ is the formula $\Phi \equiv {}^*\phi$ obtained by substituting for each constant $\alpha \in \mathcal{V}(\mathbb{R})$ which occurs in it, its nonstandard extension ${}^*\alpha$.

For example, the formula $\phi(a)$ defined by

$$\begin{aligned}\phi(a) \quad : \quad & \forall_\epsilon[\epsilon \in \mathbb{R}^+ \Rightarrow \exists\delta[\delta \in \mathbb{R}^+ \wedge \\ & \forall_x[x \in \mathbb{R} \Rightarrow [|x-a| < \delta \Rightarrow |f(x) - f(a)| < \epsilon]]]]\end{aligned}$$

expresses the continuity of a function $f : \mathbb{R} \to \mathbb{R}$ at a given point $a \in \mathbb{R}$. Here the constants are the sets $\mathbb{R}^+$ and $\mathbb{R}$, the functions f and $|.|$ and the relation $<$. The variables ϵ, δ and x are **bounded variables** while a is a **free variable**. If, for the sake of simplicity of notation (which is always desirable), we denote the *-transforms of the relation $<$ and of the function f by the same symbols, then the *-transform of the given formula $\phi(a) \in \mathcal{L}(\mathbb{R})$ is the formula ${}^*\phi(a) \in \mathcal{L}({}^*\mathbb{R})$ defined by

$$\begin{aligned}{}^*\phi(a) \quad : \quad & \forall_\epsilon[\epsilon \in {}^*\mathbb{R}^+ \Rightarrow \exists\delta[\delta \in {}^*\mathbb{R}^+ \wedge \\ & \forall_x[x \in {}^*\mathbb{R} \Rightarrow [|x-a| < \delta \Rightarrow |{}^*f(x) - {}^*f(a)| < \epsilon]]]]\end{aligned}$$

(where δ and x now range over ${}^*\mathbb{R}^+$ and ${}^*\mathbb{R}$). This expresses the *-continuity of the nonstandard extension ${}^*f$ (of the function f) at the point $a \in {}^*\mathbb{R}$. On the other hand the relation $x \approx y$ which can be expressed in $\mathcal{L}({}^*\mathbb{R})$ by the formula

$$\forall_z[z \in \mathbb{R} \Rightarrow [z > 0 \Rightarrow |x - y| < z]]$$

involves the set $\mathbb{R}$ which is an external set of the universe of nonstandard analysis. Thus this formula is not the *-transform of any formula of $\mathcal{L}({}^*\mathbb{R})$. It is an example of an **external formula** of $\mathcal{L}({}^*\mathbb{R})$ whereas the *-transforms of formulas of $\mathcal{L}(\mathbb{R})$ are **internal formulas** of $\mathcal{L}({}^*\mathbb{R})$. The importance of this distinction between internal and external formulas of the language $\mathcal{L}({}^*\mathbb{R})$ resides in the following theorem whose proof is given in appendix B.

Theorem 6.21 (Los-Mostowski) Let $\phi(x_1, \ldots, x_k; b_1, \ldots, b_m)$, be a given formula of the language $\mathcal{L}(\mathbb{R})$, in which $x_1, \ldots, x_k$ are single free variables and $b_1, \ldots, b_m$ are constants, and let

$$\Phi(x_1, \ldots, x_k) \equiv {}^*\phi(x_1, \ldots, x_k; b_1, \ldots, b_m) \ .$$

If $\alpha^1 \equiv [(\alpha^1_\lambda)_{\gamma\in\Gamma}], \ldots, \alpha^k \equiv [(\alpha^k_\gamma)_{\gamma\in\Gamma}]$ are elements of $\Pi_{\mathcal{U}}(\mathbb{R})$ then the formula of the language $\mathcal{L}(\mathbb{R})$

$$\Phi(\mu(\alpha^1), \ldots, \mu(\alpha^k)) \equiv \Phi\left(\mu([(\alpha^1_\gamma)_{\gamma\in\Gamma}]), \ldots \mu([(\alpha^k_\gamma)_{\gamma\in\Gamma}])\right)$$

will be true in $\mathcal{V}({}^*\mathbb{R})$ if and only if the set

$$\{\gamma \in \Gamma : \mathcal{V}(\mathbb{R}) \models \phi(\alpha^1_\gamma, \ldots, \alpha^k_\gamma; b_1, \ldots, b_m)\}$$

belongs to the ultrafilter $\mathcal{U}$.

Taking into account that a proposition is a formula without free variables, and then applying theorem 6.21 to the case of formulas with no free variables we obtain:

Corollary 6.21.1 (Leibniz Principle) A sentence $\mathbf{p}$ is true in $\mathcal{V}(\mathbb{R})$ if and only if its transform ${}^*\mathbf{p}$ is true in $\mathcal{V}({}^*\mathbb{R})$; that is

$$\mathcal{V}(\mathbb{R}) \models \mathbf{p} \Leftrightarrow \mathcal{V}({}^*\mathbb{R}) \models {}^*\mathbf{p} \ .$$

The Leibniz principle can be used directly to obtain important properties of the *-mapping. As an example we give below a derivation of some of those properties.

Theorem 6.22 Let a, b be constants of $\mathcal{V}(\mathbb{R})$. Then

(1) ${}^*\emptyset = \emptyset$
(2) $a = b \Leftrightarrow {}^*a = {}^*b$ and $a \subset b \Leftrightarrow {}^*a \subset {}^*b$
(3) ${}^*\{a, b\} = \{{}^*a, {}^*b\}$ and ${}^*(a, b) = ({}^*a, {}^*b)$
(4) ${}^*(a \cup b) = {}^*a \cup {}^*b$, ${}^*(a \cap b) = {}^*a \cap {}^*b$ and ${}^*(a \backslash b) = {}^*a \backslash {}^*b$
(5) ${}^*(a \times b) = {}^*a \times {}^*b$.

Proof (1) If ${}^*\emptyset$ were different from $\emptyset$ then we would have

$$\exists_x[x \in {}^*\emptyset]$$

which, by transfer, would give

$$\exists_x [x \in \emptyset]$$

which is absurd. Hence, $^*\emptyset = \emptyset$.

(2) If $a, b \in \mathbb{R}$ the proposition will be valid by virtue of the definition of the *-mapping. Suppose that $a, b \in \mathcal{V}(\mathbb{R})\backslash\mathbb{R}$. Then the equality $a = b$ is equivalent to the following sentence in $\mathcal{V}(\mathbb{R})$

$$\forall_x [x \in a \Leftrightarrow x \in b]$$

whence, by transfer, we have

$$\forall_x [x \in {}^*a \Leftrightarrow x \in {}^*b]$$

which is a sentence valid in $^*\mathbb{R}$ equivalent to the equality $^*a = {}^*b$.

Similarly, $a \subset b$ is equivalent to the sentence valid in $\mathcal{V}(\mathbb{R})$,

$$\forall_x [x \in a \Rightarrow x \in b]$$

whence, by transfer, we obtain the sentence valid in $\mathcal{V}({}^*\mathbb{R})$,

$$\forall_x [x \in {}^*a \Rightarrow x \in {}^*b]$$

which is equivalent to $^*a \subset {}^*b$.

(3) Let $c = \{a, b\} \in \mathbb{R}_k$ for some $k \geq 1$. Then

$$\forall_x [x \in \mathbb{R}_{k-1} \Rightarrow x \in c \Leftrightarrow [x = a \vee x = b]]]$$

and therefore

$$\forall_x [x \in {}^*(\mathbb{R}_{k-1}) \Rightarrow x \in {}^*c \Leftrightarrow [x = {}^*a \vee x = {}^*b]]]$$

which means that $^*c = \{{}^*a, {}^*b\}$.

Taking into account the definition of ordered pair we have

$$^*(a, b) = {}^*\{\{a\}, \{b\}\}$$

$$\{{}^*\{a\}, {}^*\{a, b\}\} = \{\{{}^*a\}, \{{}^*a, {}^*b\}\} = ({}^*a, {}^*b)$$

thus obtaining the second part of point (3). •

Exercise 6.23 *Complete the proof of theorem 6.22.*

Note 6.24 (3) may be generalised, but only for finite sets. This is not, as might appear at first sight, a negative feature. In fact if for example the theorem were to apply to an arbitrary set $A \in \mathcal{V}(\mathbb{R})$, that is to say if the equality

$$^*A = \{{}^*a : a \in A\}$$

were always true, then the two models would be isomorphic, and we would not therefore have obtained any extension of the initial system.

Theorem 6.25 Let a, b be two arbitrary entities of $\mathcal{V}(\mathbb{R})$.
(a) If R is a relation in $a \times b$, then $^*R$ is a relation in $^*a \times {}^*b$ and $\mathbf{dom}({}^*R) = {}^*(\mathbf{dom}(R)$ and $\mathbf{cdom}({}^*R) = {}^*(\mathbf{cdom}(R))$.
(b) If f is a mapping of a into b then $^*f$ is a mapping of $^*a$ into $^*b$, with $^*(f(c)) = {}^*f({}^*c)$ for all $c \in a$.

Proof As $R \subset a \times b$ it follows that $^*R \subset {}^*a \times {}^*b$ from the preceding theorem. Letting $c = \mathbf{dom}(R)$, then

$$\forall_x[x \in c \Rightarrow \exists_y[y \in b \wedge (x, y) \in R]]$$

whence, by transfer, we have

$$\forall_x[x \in {}^*c \Rightarrow \exists_y[y \in {}^*b \wedge (x, y) \in {}^*R]]$$

which means that $^*c \subset \mathbf{dom}({}^*R)$. To prove the equality we have further

$$\forall_{x,y}[(x, y) \in (a \backslash c) \times b \Rightarrow (x, y) \notin R]$$

whence we obtain

$$\forall_{x,y}[(x, y) \in ({}^*a \backslash {}^*c) \times {}^*b \Rightarrow (x, y) \notin {}^*R]$$

which shows that if $x \notin {}^*c$ then $(x, y) \notin {}^*R$ or equivalently that $\mathbf{dom}({}^*R) \subset {}^*c$. Consequently, $^*c = \mathbf{dom}({}^*R)$.

The part relative to the codomain of R has a similar proof.

(b) If $f : a \to b$ then f is a subset of $a \times b$ such that

$$\forall_{x,y,y'}[x \in a \wedge y, y' \in b \Rightarrow [(x, y) \in f \wedge (x, y') \in f \Rightarrow y = y']] \ .$$

Then, on the one hand, we have $^*f \subset {}^*a \times {}^*b$ and, on the other hand, by transfer we have

$$\forall_{x,y,y'}[x \in {}^*a \wedge y, y' \in {}^*b \Rightarrow [(x, y) \in {}^*f \wedge (x, y') \in {}^*f \Rightarrow y = y']]$$

which shows that $^*f$ is a function from $^*a$ to $^*b$.

As $c \in a$ then $(c, f(c)) \in f$ and, therefore, $({}^*c, {}^*(f(c)) \in {}^*f$ whence we conclude that $^*f({}^*c) = {}^*(f(c))$, as was required to show. •

6.3.2 Internal and external entities

Since $^* \equiv \mu \circ \kappa$, the set defined by

$$\mu(\Pi_{\mathcal{U}}(\mathbb{R})) = {}^*\mathbb{R} \cup {}^*(\mathbb{R}_1) \cup \ldots \cup {}^*(\mathbb{R}_k) \cup \ldots \subset \mathcal{V}({}^*\mathbb{R})$$

contains the *transforms of all the standard objects of $\mathcal{V}(\mathbb{R})$. However the transformation is not surjective: as remarked earlier, there are objects in $\mathcal{V}({}^*\mathbb{R})$ which are not the result of the μ-transformation of any object of $\Pi_{\mathcal{U}}(\mathbb{R})$ and not therefore the *transforms of any standard objects. This fact, which is most important, shows that $\mathcal{V}({}^*\mathbb{R})$ is a universe which is much richer than the standard universe $\mathcal{V}(\mathbb{R})$, and leads to the following distinction between the objects of $\mathcal{V}({}^*\mathbb{R})$:

Definition 6.26 The objects of $\mathcal{V}({}^*\mathbb{R})$ which are of the form $\mu([\{\alpha_\gamma\}_{\gamma\in\Gamma}])$ and which, therefore, may be identified with $[\{\alpha_\gamma\}_{\gamma\in\Gamma}] \in \Pi_{\mathcal{U}}(\mathbb{R})$, are said to be **internal objects** of the universe of nonstandard analysis. All other objects are said to be **external objects**.

Definition 6.27 From definition 6.26 it follows that every element of $\mathcal{V}({}^*\mathbb{R})$ of the form ${}^*\alpha$ for some $\alpha \in \mathcal{V}(\mathbb{R})$ is internal: ${}^*\alpha$ is then described as the **nonstandard extension** of the corresponding (standard) object α.

Example 6.28 Let $[\{A_\gamma\}_{\gamma\in\Gamma}]$ be an element of $\Pi_{\mathcal{U}}(\mathbb{R})$ where, for $\mathcal{U}$-nearly all $\gamma \in \Gamma$, $A_\gamma \subset \mathbb{R}$. Then $\mathcal{A} = \mu([\{A_\gamma\}_{\gamma\in\Gamma}]) \subset {}^*\mathbb{R}$ is a subset of ${}^*\mathbb{R}$. Taking into account that the mapping μ is injective, it is usual to *"identify"* $\mathcal{A} \in \mathcal{V}({}^*\mathbb{R})$ with $[\{A_\gamma\}_{\gamma\in\Gamma}] \in \Pi_{\mathcal{U}}(\mathbb{R})] \subset \mathcal{V}_{\mathcal{U}}[\mathbb{R}]$ in order to emphasize the genesis of $\mathcal{A}$. This is the sense in which we write the equality $\mathcal{A} = [\{A_\gamma\}_{\gamma\in\Gamma}]$, as indicated in Chapter 5. In particular, if we have $A_\gamma = A \subset \mathbb{R}$ for $\mathcal{U}$-nearly all $\gamma \in \Gamma$ then $\mathcal{A} = {}^*A \equiv \mu([(A)_{\gamma\in\Gamma}])$. The same kind of identification can be made equally well between other objects of $\mathcal{V}(\mathbb{R})$ and $\Pi_{\mathcal{U}}(\mathbb{R})$, as was done in Chapter 5 with regard to the internal functions.

Theorem 6.29 An entity $\alpha \in \mathcal{V}({}^*\mathbb{R})$ is internal if and only if the sentence $\alpha \in {}^*a$ is true for some $a \in \mathcal{V}(\mathbb{R})$. That is to say, an entity is internal if and only if it belongs to the nonstandard extension of a standard entity.

Proof Let α be any standard entity in $\mathcal{V}(\mathbb{R})$. Then $\alpha = \mu([\{\xi_\gamma\}_{\gamma\in\Gamma}])$ with $[\{\xi_\gamma\}_{\gamma\in\Gamma}] \in \Pi_{\mathcal{U}}(\mathbb{R})$ or, equivalently, with $[\{\xi_\gamma\}_{\gamma\in\Gamma}] \in (\mathbb{R}_k\backslash\mathbb{R}_{k-1})^\Gamma/\mathcal{U}$ for some $k \in \mathbb{N}$. Then $\{\gamma \in \Gamma : \xi \in \mathbb{R}_k\} \in \mathcal{U}$ and, consequently, $\mu([\{\xi_\gamma\}_{\gamma\in\Gamma}]) \in ({}^*\mathbb{R}_k)$. The proposition is thus confirmed for $a = \mathbb{R}_k$.

Conversely, suppose that $\alpha \in {}^*a$ for some a belonging to $\mathcal{V}(\mathbb{R})$. Then $\alpha \in \mu([(a)_{\gamma\in\Gamma}])$; that is, α is of the form $\mu([\{\xi\}_{\gamma\in\Gamma}])$ where $[\{\xi_\gamma\}_{\gamma\in\Gamma}] \in_{\mathcal{U}} [(a)_{\gamma\in\Gamma}]$ and, therefore, α is an internal entity. •

Theorem 6.30 Let A be any object whatsoever of $\mathcal{V}(\mathbb{R})$. Then ${}^*\mathcal{P}(A) \equiv {}^*(\mathcal{P}(A))$ is the set of all the internal subsets of ${}^*A$.

Proof Let $B \in {}^*\mathcal{P}(A)$. Then B is internal; it remains to be shown that B belongs to $\mathcal{P}({}^*A)$. This can be achieved by transfer, as follows. Since we have

$$\forall_X[X \in \mathcal{P}(A) \Rightarrow X \subset A]$$

then by transfer we get

$$\forall_X[X \in {}^*\mathcal{P}(A) \Rightarrow X \subset {}^*A] \ .$$

Setting $X = B$ it follows that B is actually an internal subset of ${}^*A$.

Conversely, suppose that B is an internal subset of ${}^*A$. Then, from Theorem 6.29 it follows that there exists $C \in \mathcal{V}(\mathbb{R})$ for which we have $B \in {}^*C$. Since by transfer the sentence

$$\forall_X[X \in C \Rightarrow [X \subset A \Rightarrow X \in \mathcal{P}(A)]]$$

implies that

$$\forall_X[X \in {}^*C \Rightarrow [X \subset {}^*A \Rightarrow X \in {}^*\mathcal{P}(A)]]$$

then, setting $X = B$ we conclude that B belongs to ${}^*\mathcal{P}(A)$. •

We have thus proved that ${}^*\mathcal{P}(A)$ is the set of all internal subsets of ${}^*A$. $\mathcal{P}({}^*A)\backslash{}^*\mathcal{P}(A)$ is the set of all the external subsets of ${}^*A$.

Example 6.31 It has already been shown that the upper bound principle in $\mathbb{R}$ transfers to the internal sets of ${}^*\mathbb{R}$ but not to arbitrary sets. We now obtain the same result simply by transfer. The least upper bound axiom in $\mathbb{R}$ can be formulated as follows:

$$\forall_A[A \in \mathcal{P}(\mathbb{R}) \Rightarrow [A \neq \emptyset \wedge \exists_\alpha[\alpha \in \mathbb{R} \wedge$$
$$\forall_a[a \in A \Rightarrow a \leq \alpha]] \Rightarrow \exists_s[s \in \mathbb{R} \wedge s = \sup(A)]]]$$

which, by transfer, gives

$$\forall_A[A \in {}^*\mathcal{P}(\mathbb{R}) \Rightarrow [A \neq \emptyset \wedge \exists_\alpha[\alpha \in {}^*\mathbb{R} \wedge$$
$$\forall_a[a \in A \Rightarrow a \leq \alpha]] \Rightarrow \exists_s[s \in {}^*\mathbb{R} \wedge s = \sup(A)]]] \quad .$$

Hence the least upper bound axiom holds for *internal* subsets of ${}^*\mathbb{R}$.

Definition 6.32 A formula Φ of the language of Nonstandard Analysis $\mathcal{L}({}^*\mathbb{R})$ is said to be internal if its constants are all internal entities. Otherwise it is said to be external.

Theorem 6.33 (Principle of Internality) If $\Phi(x)$ is an internal formula of $\mathcal{L}({}^*\mathbb{R})$ and α denotes an internal set of $\mathcal{V}({}^*\mathbb{R})$ then the set $\{x \in \alpha : \Phi(x)\}$ is internal.

Proof Since α and $\Phi(x)$ are internal then, for some $m \in \mathbb{N}$, α and the constants of Φ will all belong to ${}^*(\mathbb{R}_m)$. Let ϕ be the formula of $\mathcal{L}(\mathbb{R})$ for which ${}^*\phi = \Phi$. The standard sentence

$$\forall_y[y \in \mathbb{R}_m \Rightarrow \exists_z[z \in \mathbb{R}_{m+1} \wedge z = \{x \in y : \phi(x)\}]]$$

is valid in $\mathcal{V}(\mathbb{R})$ and therefore, by transfer, the sentence

$$\forall_y[y \in {}^*(\mathbb{R}_m) \Rightarrow \exists_z[z \in {}^*(\mathbb{R}_{m+1}) \wedge z = \{x \in y : \Phi(x)\}]]$$

is valid in $\mathcal{V}({}^*\mathbb{R})$. The result follows when we take $y = \alpha$. •

Definition 6.34 Let $A \in \mathcal{V}(\mathbb{R})\backslash\mathbb{R}$. The **standard copy** of A, denoted by ${}^\sigma A$, is defined to be the entity of $\mathcal{V}({}^*\mathbb{R})$ given by

$${}^\sigma A = \{{}^*a : \quad a \in A\} \quad .$$

6.3.3 Complementary results

Comprehensiveness As we have seen in Chapter 4, given a sequence of real numbers $(a_n)_{n\in\mathbb{N}}$ it is always possible to extend it to a hypersequence $(a_n)_{n\in{}^*\mathbb{N}}$, which is an internal object. The technique applied does not work, however, if instead of a sequence of real numbers we have a sequence of hyperreal numbers, that is a function

$\alpha : {}^\sigma\mathbb{N} \to {}^*\mathbb{R}$. In certain problems it is important to obtain an (internal) extension of ${}^*\mathbb{N}$ in ${}^*\mathbb{R}$ which coincides with α in ${}^\sigma\mathbb{N} \subset {}^*\mathbb{N}$. As will be shown in the theorem which follows such an extension actually exists. This constitutes a most important property of the embedding ${}^*\mathcal{V}(\mathbb{R}) \to \mathcal{V}({}^*\mathbb{R})$ which we call **comprehensiveness.**

Theorem 6.35 (Comprehensiveness) Given a mapping $f : {}^\sigma A \to {}^*B$ where A and B denote two standard entities, $A, B \in \mathcal{V}(\mathbb{R})$, there exists an internal mapping $F : {}^*A \to {}^*B$ whose restriction to ${}^\sigma A$ coincides with f.

Proof Since for each ${}^*a \in {}^\sigma A$ we have $f({}^*a) \in {}^*B$ then $f({}^*a) = \mu([\{b_\gamma\}_{\gamma\in\Gamma}])$ with $[\{b_\gamma\}_{\gamma\in\Gamma}] \in_{\mathcal{U}} [(B)_{\gamma\in\Gamma}]$. Without any loss of generality we may assume that $b_\gamma \in B$ for all $\gamma \in \Gamma$. We then define a generalised sequence of standard functions

$$f_\gamma \ : \ A \to B$$

$$a \mapsto f_\gamma(a) = b_\gamma, \quad \gamma \in \Gamma \ .$$

For each $\gamma \in \Gamma$, $f_\gamma \in \mathcal{P}(A \times B)$, from which it follows that $\{\gamma \in \Gamma : f_\gamma \in \mathbb{R}_k\} \in \mathcal{U}$ for some $k \in \mathbb{N}$. Therefore there exists $j \leq k$ such that $\{\gamma \in \Gamma : f_\gamma \in \mathbb{R}_j \backslash \mathbb{R}_{j-1}\} \in \mathcal{U}$. Consequently $[\{f_\gamma\}_{\gamma\in\Gamma}] \in \Pi_{\mathcal{U}}(\mathbb{R})$ and therefore the definition

$$F = \mu([\{f_\gamma\}_{\gamma\in\Gamma}])$$

makes sense. It remains to confirm that F satisfies the conditions of the proposition. On the one hand F is an internal entity and on the other hand, noting that for all $x = \mu([\{x_\gamma\}_{\gamma\in\Gamma}]) \in {}^*A$ we have $F(x) = \mu([(f_\gamma(x_\gamma))_{\gamma\in\Gamma}])$, then F is well defined on ${}^*A$. For ${}^*a \in {}^\sigma A \subseteq {}^*A$ we get $F({}^*a) = \mu([f_\gamma(a))_{\gamma\in\Gamma}]) = \mu([\{b_\gamma\}_{\gamma\in\Gamma}]) = f({}^*a)$ and hence $F_{|{}^\sigma A} = f$. •

From the comprehensiveness of the model of nonstandard analysis it follows immediately that if $(A_n)_{n\in\mathbb{N}}$ is a sequence of entities of ${}^*(\mathbb{R}_k)$, for some $k \in \mathbb{N}_0$, then there exists an internal hypersequence of entities $(B_\nu)_{\nu\in{}^*\mathbb{N}}$ such that $A_n = B_n$ for all $n \in \mathbb{N}$.

Saturation Countable saturation of the model of **NSA** constructed by means of an ultrafilter over $\mathbb{N}$ implies that:

"If $\mathcal{A} \subset {}^*\mathbb{R}$ is an internal set of $\mathcal{V}({}^*\mathbb{R})$ then $\mathbf{st}(\mathcal{A}) = \{\mathbf{st}(x) \in \mathbb{R} : x \in \mathcal{A} \cap {}^*\mathbb{R}_b\}$ is a closed subset of $\mathcal{V}(\mathbb{R})$."

which is actually a restatement of Theorem 5.14. This result, which can be extended to any metric space, becomes invalid in the case of an arbitrary topological space (which may not necessarily satisfy the first axiom of countability). However, if instead of working with a model based on an ultrafilter over $\mathbb{N}$ we work with a model based on an ultrafilter over an index set Γ which is richer than $\mathbb{N}$, the validity of this proposition can be recovered.

First note that any given family $\{A_\alpha\}_{\alpha\in\Gamma}$ whatever of subsets, is said to possess the **finite intersection property** if and only if for every (finite) sequence of indices $\{\alpha_1, \alpha_2, \ldots, \alpha_k\} \subset \Gamma$

$$A_{\alpha_1} \cap A_{\alpha_2} \cap \ldots A_{\alpha_k} \neq \emptyset \ .$$

Definition 6.36 Let κ be a cardinal. The nonstandard model $\mathcal{V}({}^*S)$ is said to be **κ-saturated** if for any set of indices Γ with cardinality $< \kappa$ and any family $\{A_\alpha\}_{\alpha\in\Gamma}$ of internal sets with the finite intersection property we have $\bigcap_{\alpha\in\Gamma} A_\alpha \neq \emptyset$.

The κ-saturated models form the appropriate tools for the study of the topological spaces with bases of cardinality less than κ. A question which might be posed is whether it is known that κ-saturated models exist for a given cardinal κ. The reply has been given by Keisler, and subsequently improved by Kunen, from whom the following result derives:

Theorem 6.37 Given a superstructure $\mathcal{V}(\mathbb{R})$ and a cardinal κ there exists a κ-saturated superstructure $\mathcal{V}({}^*\mathbb{R})$ and a monomorphism ${}^* : \mathcal{V}(\mathbb{R}) \to \mathcal{V}({}^*\mathbb{R})$.

In the case when the cardinal κ is greater than or equal to the cardinal of $\mathcal{V}(\mathbb{R})$ the superstructure $\mathcal{V}({}^*\mathbb{R})$ obtained is said to be **polysaturated**.

Theorem 6.38 An internal infinite set in a κ-saturated model has cardinality greater than or equal to κ.

Proof If $\mathcal{A}$ were to be an infinite internal set of cardinal less than κ then using the saturation property of the family $\{\mathcal{A}\backslash\{a\}_{a\in\mathcal{A}}\}$ we would get

$$\bigcap_{a\in\mathcal{A}} (\mathcal{A}\backslash\{a\}) \neq \emptyset$$

which is absurd. •

Enlargements We begin with the following definition:

Definition 6.39 Let $\mathcal{A}$ be any set of $\mathcal{V}(\mathbb{R})$ (actually an element of $\mathbb{R}_n\backslash\mathbb{R}_0$ for some $n \in \mathbb{N}$) and denote by $\mathcal{P}_F(\mathcal{A})$ the set of all finite subsets of $\mathcal{A}$. The elements of ${}^*(\mathcal{P}_F(\mathcal{A})) \in \mathcal{V}({}^*\mathbb{R})$ are called **hyperfinite sets** or ***-finite sets**.

To put it another way, an internal set $\mathcal{A}$ is said to be hyperfinite if we have $\mathcal{A} \equiv \mu([\{\mathcal{A}_\gamma\}_{\gamma\in\Gamma}])$ where for $\mathcal{U}$-nearly all $\gamma \in \Gamma$, $\mathcal{A}_\gamma$ is a finite set.

Example 6.40 The hyperfinite sets behave formally as finite sets. For example we can show, by transfer, that every hyperfinite set of real numbers has a maximum and a minimum. Thus the sentence

$$\forall_A[A \in \mathcal{P}_F(\mathbb{R}) \Rightarrow \exists_M[M \in A \wedge \forall_a[a \in A \Rightarrow a \leq M]]]$$

which expresses formally that every finite set of real numbers has a maximum, is valid in $\mathcal{V}(\mathbb{R})$. By transfer we obtain the sentence

$$\forall_{\mathcal{A}}[\mathcal{A} \in {}^*\mathcal{P}_F(\mathbb{R}) \Rightarrow \exists_M[M \in \mathcal{A} \wedge \forall_a[a \in \mathcal{A} \Rightarrow a \leq M]]]$$

which is valid in $\mathcal{V}({}^*\mathbb{R})$ and which expresses precisely that every hyperfinite set of hyperreal numbers has a maximum. (Similarly, the corresponding result for the minimum may be seen to hold.)

Using the notion of internal bijection, we can give an alternative characterisation for hyperfinite sets which has the advantage of not depending on the particular model of NSA being considered.

Theorem 6.41 **T** is a hyperfinite set with internal cardinality $\nu \in {}^*\mathbb{N}$ if and only if there exists a bijection

$$\Phi : \{1, 2, \ldots, \nu\} \to \mathbf{T} \quad .$$

Proof Let $\mathbf{T} \equiv [(T_n)_{n\in\mathbb{N}}]$. If **T** is a hyperfinite set with internal cardinality $\nu \equiv [(m_n)_{n\in\mathbb{N}}]$, then without loss of generality we may suppose that for each $n \in \mathbb{N}$, T_n is a finite set of cardinality m_n. Then, for each $n \in \mathbb{N}$, there exists a bijection $\phi : \{1, 2, \ldots, m_n\} \to T_n$ and, therefore, $\Phi \equiv [(\phi_n)_{n\in\mathbb{N}}]$ is a bijection of $\{1, 2, \ldots, \nu\}$ into **T**.

Conversely, let $\Phi \equiv [(\phi_n)_{n\in\mathbb{N}}]$ be a bijection of $\{1, 2, \ldots, \nu\}$ into **T**. Then,

$$\mathbf{dom}(\Phi) = \{1, 2, \ldots, \nu\} \equiv [(\{1, 2, \ldots, m_n\})_{n\in\mathbb{N}}]$$

and therefore for $\mathcal{U}$-nearly all $n \in \mathbb{N}$

$$\mathbf{dom}(\phi_n) = \{1, 2, \ldots, m_n\} \quad . \tag{6.5}$$

On the other hand, since **T** is the image of $\{1, 2, \ldots, \nu\}$ by $\Phi \equiv [(\phi_n)_{n\in\mathbb{N}}]$, then $\mathbf{T} = [(\phi_n(\{1, 2, \ldots, m_n\}))_{n\in\mathbb{N}}]$ and therefore for $\mathcal{U}$-nearly all $n \in \mathbb{N}$,

$$\phi_n(\{1, 2, \ldots, m_n\}) = T_n. \tag{6.6}$$

From the injectivity of Φ it follows that for $\mathcal{U}$-nearly all $n \in \mathbb{N}$,

$$\phi_n \text{ is injective.} \tag{6.7}$$

Then the set **U** of all $n \in \mathbb{N}$ for which (6.5), (6.6) and (6.7) are simultaneously valid belongs to the ultrafilter $\mathcal{U}$. But for $n \in \mathbf{U}$, T_n has finite cardinality equal to m_n and hence **T** is a hyperfinite set with internal cardinality $\nu \equiv [(m_n)_{n\in\mathbb{N}}]$. •

A hyperfinite set $\mathcal{A}$ can be written in the form

$$\mathcal{A} = \{a_1, a_2, \ldots, a_\nu\}$$

(where the dots here may represent many more elements than would be encountered in analogous standard situations).

Definition 6.42 The superstructure $\mathcal{V}({}^*\mathbb{R})$ is said to be an **enlargement** of $\mathcal{V}(\mathbb{R})$ if for each set $a \in \mathcal{V}(\mathbb{R})$ there exists a *finite set $\mathcal{A} \in {}^*\mathcal{P}_F(A)$ such that $A \subset \mathcal{A} \subset {}^*A$ (that is, such that ${}^*a \in \mathcal{A}$ for each $a \in A$).

It can be shown that for a given superstructure $\mathcal{V}(\mathbb{R})$ it is possible to choose an index set Γ and an ultrafilter $\mathcal{U}$ over Γ such that the superstructure $\mathcal{V}({}^*\mathbb{R})$ obtained as indicated above is an enlargement of $\mathcal{V}(\mathbb{R})$.

Chapter 7

Hyperfinite Analysis

Hyperfinite sets are particularly important. Using them allows much of mathematical analysis to be developed by essentially finitistic processes. In this chapter and those which follow we will give various examples of this kind of approach to higher analysis.

7.1 Introduction

We assume a fixed model of NSA with a sufficiently high order of saturation. Let κ be an infinite hypernatural even number and let $\epsilon \equiv \epsilon_\kappa \approx 0$ be the positive infinitesimal defined by $\epsilon = \kappa^{-1}$. Then

$$\Pi \equiv \Pi_\kappa = \{-\kappa/2, -\kappa/2 + \epsilon, \ldots, 0, \ldots, \kappa/2 - \epsilon\}$$

$$= \{-\kappa/2 + j\epsilon : j = 0, 1, 2, \ldots, \kappa^2 - 1\} \subset {}^*\mathbb{R} \qquad (7.1)$$

is an internal hyperfinite set of hyperreal numbers of internal cardinality κ^2. It is usually referred to as the **hyperfinite real line.**

Note 7.1 If in particular we take $\kappa = \nu!$ for some $\nu \in {}^*\mathbb{N}_\infty$, then Π will contain all the finite rational numbers: that is to say, the inclusion $\mathbb{Q} \subset \Pi$ will be satisfied. In fact, let $q = m/n \in \mathbb{Q}$ with $hcf(m, n) = 1$ and, without loss of generality, suppose that $n > 0$. Then we have

$$q \equiv \frac{m}{n} = \frac{m.\nu!/n}{\nu!} = \frac{m.1.2.\ldots.(n-1).(n+1)\ldots\nu}{\nu!} = \nu_q.\epsilon$$

and so, since $|\nu_q| < (\nu!)^2/2$, the number $q = m/n$ belongs to the set Π.

Given a (standard) point $r \in \mathbb{R}$ we denote by $mon_\Pi(r)$ the (external) set consisting of all the points belonging to Π which are infinitely close to r; that is,[1]

$$mon_\Pi(r) = \mathbf{st}_\Pi^{-1}(r) = \{x \in \Pi : \mathbf{st}(x) = r\} = mon(r) \cap \Pi.$$

[1]Note that, in general, $r \notin mon_\Pi(r)$.

This set will be called the **Π-monad** of the standard point $x \in \mathbb{R}$. The union of the Π-monads of all the real numbers

$$\Pi_b = \bigcup_{r \in \mathbb{R}} mon_\Pi(r) = \mathbf{st}_\Pi^{-1}(\mathbb{R})$$

forms the finite part of Π, while $\Pi_\infty \equiv \Pi \backslash \Pi_b$ is the remote part of the hyperfinite line.

Consider the set of all the internal functions defined on Π

$$\mathbf{F}_\Pi = \{F : \Pi \to {}^*\mathbf{C} | F \text{ is internal}\}$$

and suppose, wherever necessary, that each function $F \in \mathbf{F}_\pi$ is periodically extended to the set $\epsilon{}^*\mathbb{Z} \equiv \{\nu\epsilon \in {}^*\mathbb{R} : \nu \in {}^*\mathbb{Z}\}$. With componentwise definitions of addition, and multiplication by a hyperfinite number, $\mathbf{F}_\Pi$ is a ${}^*\mathbf{C}$-linear space of hyperfinite dimension κ^2. The system

$$\{\Xi\}_{j=-(\kappa^2/2),-(\kappa^2/2)+1,\ldots,0,\ldots,(\kappa^2/2)-1}$$

where, for each j, the internal function $\Xi_j(x)$, $x \in \Pi$ is defined by

$$\Xi_j(x) = \begin{cases} 1, & \text{if } x = j\epsilon \\ 0, & \text{if } x \neq j\epsilon \end{cases}$$

constitutes an algebraic basis for $\mathbf{F}_\Pi$. If, in addition, we define the product of two functions componentwise, then $\mathbf{F}_\Pi$ is an algebra.

Canonical Π-extension of a standard function Given a function $f : A \subset \mathbb{R} \to \mathbf{C}$, we shall always consider the extension of f to the whole of $\mathbb{R}$, setting $f(x) = 0$ for $x \notin A$, and we shall denote this extension by the same symbol f. If ${}^*f$ is the nonstandard extension of any such function f then we define ${}^*f_\Pi : \Pi \to {}^*\mathbf{C}$ by taking the restriction of ${}^*f$ to Π, periodically extended to $\epsilon{}^*\mathbb{Z}$. The function ${}^*f_\Pi$, which we call the Π-extension of f, belongs to $\mathbf{F}_\Pi$.

7.1.1 *S*Π-continuous functions

Definition 7.2 Let Ω be any (non-empty) subset of Π_b. An internal function $F \in \mathbf{F}_\Pi$ is said to be ***S*Π-continuous** on Ω if and only if

1. $F(x)$ is finite at every point $x \in \Omega$ and,
2. $\forall_{x,y}[x, y \in \Omega \Rightarrow [x \approx y \Rightarrow F(x) \approx F(y)]]$.

If $\Omega = \Pi_b$ then we simply say that the function F is *S*Π-continuous.

From the nonstandard definition of continuity (pointwise or uniform) given in Chapter 3 we obtain immediately the following results:

Theorem 7.3 Let $f : \mathbb{R} \to \mathbf{C}$ be a function continuous at the point $r \in \mathbb{R}$. Then the internal function ${}^*f_\Pi : \Pi \to {}^*\mathbf{C}$ is *S*Π-continuous on $mon_\Pi(r)$.

Corollary 7.1.1 Let $f : \mathbb{R} \to \mathbf{C}$ be a function continuous on the (standard) set $A \subset \mathbb{R}$. Then the internal function ${}^*f$ is *S*Π-continuous on $\mathbf{st}_\Pi^{-1}(A)$.

The converse of these results is not true: the function $^*f$ may have infinitesimal variation over the Π-monad of a point, but this fact by itself does not ensure that the variation is infinitesimal over the entire monad of the same point. Consider for example the Dirichlet function:

$$d(x) = \begin{cases} 1, & \text{if } \ x \in [0,1]\backslash\mathbf{Q} \\ 0, & \text{if } \ x \notin [0,1]\backslash\mathbf{Q} \end{cases} .$$

Since Π contains only hyperrational points, $^*d_\Pi(x)$ is zero for all $x \in \Pi$. Consequently $^*d_\Pi$ is a function $S\Pi$-continuous on Π_b whereas $^*d$ is not S-continuous on $^*[0,1] \subset {}^*\mathbb{R}$.

7.1.2 Π-derivatives and Π-integrals

.

The Π-**derivatives** (*descending* and *ascending* respectively) of any function F belonging to $\mathbf{F}_\Pi$ at a point $x \in \Pi$ are the quotients

$$D_+F(x) = \frac{F(x+\epsilon) - F(x)}{\epsilon}$$

and

$$D_-F(x) = \frac{F(x) - F(x-\epsilon)}{\epsilon} .$$

These are always well-defined hypercomplex numbers. Given any two functions $F, G \in \mathbf{F}_\Pi$ and a scalar $\alpha \in {}^*\mathbf{C}$, it is easy to confirm the following properties:

$$D_+(F+G) = D_+F + D_+G \qquad D_-(F+G) = D_-F + D_-G$$

$$D_+(\alpha F) = \alpha D_+F \qquad D_-(\alpha F) = \alpha D_-F.$$

This means that D_+ and D_- are linear operators on $\mathbf{F}_\Pi$. Iterating D_+ or D_- we obtain Π-derivatives of higher order than the first,

$$D_+^n F(x) = D_+^{n-1}(D_+F(x)) \qquad D_-^n F(x) = D_-^{n-1}(D_-F(x))$$

where $x \in \Pi, n = 1, 2, \ldots$, and we set $D_+^0 F(x) = D_-^0 F(x) = F(x)$.

The Π-**integral** operators (*descending* and *ascending* respectively)

$$S_+, S_- : \mathbf{F}_\Pi \to \mathbf{F}_\Pi$$

are defined, for each $F \in \mathbf{F}_\Pi$ and $x \in \Pi$, by

$$S_+F(x) = \sum_{-\kappa/2 \le y < x} \epsilon F(y)$$

and

$$S_-F(x) = \sum_{-\kappa/2 \le y \le x} \epsilon F(y) .$$

Once again, it is easy to verify that

$$D_+S_+F(x) = F(x),$$

and

$$D_-S_-F(x) = F(x) \quad .$$

This shows that S_+ and S_- are the respective inverse operators for D_+ and D_-.

Theorem 7.4 Let F be an internal function in $\mathbf{F}_\Pi$ such that
$F(0)$ is a finite number, and
D_+F is a function $S\Pi$-continuous on Π_b.
Then
1. F is $S\Pi$-continuous on Π_b
2. $\mathbf{st}F$ exists, is continuously differentiable on $\mathbb{R}$ and, in addition,

$$(\mathbf{st}F)'(t) = \mathbf{st} \circ D_+F(x)$$

where $x \in mon_\Pi(t)$ is arbitrary.

Proof (see [15]) Let $x, y \in \Pi_b$ and, without loss of generality, assume that $x < y$. Since the maximum of a hyperfinite set always exists we can define

$$M = \max_{x \leq z \leq y} |D_+f(z)|$$

and, by hypothesis, M must be a finite hyperreal number. Since

$$F(y) - F(x) = \sum_{x \leq z < y} [F(z+\epsilon) - F(z)] = \sum_{x \leq z < y} \epsilon D_+F(z)$$

then

$$|F(y) - F(x)| \leq \sum_{x \leq z < y} |D_+F(z)|$$

$$\leq M. \sum_{x \leq z < y} \epsilon = M(y - \epsilon - x)\epsilon \approx 0$$

whence we conclude that F is $S\Pi$-continuous.

To prove the second part of the theorem we have, by hypothesis, that D_+F is an $S\Pi$-continuous function and, therefore, it is enough to show that

$$\lim_{h \to 0} \frac{\mathbf{st}F(t+h) - \mathbf{st}F(t)}{h} = \mathbf{st}. \circ D_+F(x)$$

for any $t \in \mathbb{R}$ and $x \in \Pi$ such that $x \approx t$.

Let $r > 0$ be a given real number and suppose that $x = m\epsilon \approx t$. Since D_+F, by hypothesis, is $S\Pi$-continuous then for some real number $d > 0$ we have

$$\forall_s [s \in \Pi \Rightarrow [|s - x| < d \Rightarrow |D_+F(s) - D_+F(x)| < r/2]] \quad .$$

Suppose now that $u \in \mathbb{R}$ is such that $0 < |u - x| < d/2$ and choose a value for $s = n\epsilon \in \Pi$ such that $s \approx u$. Without any loss of generality we may suppose that $x < s$ and therefore

$$F(s) - F(x) = \sum_{j=m}^{n-1} [F((j+1)\epsilon) - F(j\epsilon)] = \sum_{j=m}^{n-1} \epsilon D_+F(j\epsilon)$$

whence

$$\frac{F(s) - F(x)}{(n-m)\epsilon} = \sum_{j=m}^{n-1} \frac{D_+F(j\epsilon)}{n-m} \quad . \tag{7.3}$$

Since $n\epsilon \in mon_\Pi(u)$ and $m\epsilon \in mon_\Pi(t)$ and $mon_\Pi(u) \cap mon_\Pi(t) = \emptyset$ then $(n-m)\epsilon \not\approx 0$ and therefore,the left-hand side of (7.3) is infinitely close to the fraction

$$\frac{F(u) - F(t)}{u - t} \quad .$$

From the $S\Pi$-continuity of D_+F it follows that

$$D_+F(x) - \frac{r}{2} < D_+F(j\epsilon) < D_+F(x) + \frac{r}{2}$$

for all $j = m, m+1, \ldots, n-1$, and in consequence the right-hand side of (7.3) satisfies the double inequality

$$D_+F(x) - \frac{r}{2} < \sum_{j=m}^{n-1} \frac{D_+F(j\epsilon)}{n-m} < D_+F(x) + \frac{r}{2} \quad .$$

Consequently we obtain

$$\left| \frac{F(u) - F(t)}{u - t} - \mathbf{st} \circ D_+F(x) \right| < r$$

which completes the proof. •

Corollary 7.4.1 Let F be an internal function in $\mathbf{F}_\Pi$ such that
$F(0), D_+F(0), \ldots, D_+^{r-1}(0)$ are finite numbers, and
D_+^r is an $S\Pi$-continuous function.
Then $\mathbf{st}F$ is a function of class C^r and, for $t \in \mathbb{R}$, and $j = 1, 2, \ldots, r$,

$$(\mathbf{st}F)^{(j)}(t) = \mathbf{st} \circ D_+^j F(x)$$

where $x \in mon_\Pi(t)$ is arbitrary.

A similar result can be obtained on replacing D_+ by D_-.]

Corollary 7.4.2 Let F be an internal function in $\mathbf{F}_\Pi$ such that
$F(0), D_+F(0), \ldots, D_+^{r-1}F(0), D_+F^r(0)$ are finite numbers, and
$D_+^{r+1}F$ is a function finite on Π_b.

Then $\mathbf{st}F$ is a function of C^r and, for $t \in \mathbb{R}$ and $j = 1, 2, \ldots, r$,

$$(\mathbf{st}F)^{(j)}(t) = \mathbf{st} \circ D_+^j F(x)$$

where $x \in mon_\Pi(t)$ is arbitrary.

A similar result can be obtained on replacing D_+ by D_-.]

Proof This corollary results from the fact that the second condition on F implies hat D_+F will be $S\Pi$-continuous on Π_b. •

7.2 The Loeb Counting Measure

First of all we give an introductory account of the more relevant aspects of the construction of measures devised by P.Loeb, following closely the exposition by T.Lindstrom in [27]. The theory of Loeb measure can be used to construct standard measures (such as, for example, Lebesgue measure), by means of nonstandard measures, which are generally much more simply defined. In particular this is the case with the so-called **counting measures**, with or without weights, in hyperfinite spaces. These spaces are founded on the set $\Pi \subset {}^*\mathbb{R}$, of hyperfinite internal cardinality $\natural(\Pi)$, which allows us to associate with each internal subset of Π the number of its elements. Thanks to the construction of P.Loeb every Radon measure over $\mathbb{R}$ can be represented (though not uniquely) as a function on Π into ${}^*\mathbf{C}$ (or equivalently, in the form of a vector of hyperfinite dimension in ${}^*\mathbf{C}^{\kappa^2}$). This idea can be further extended to the hyperfinite representation of generalised functions belonging to various well-known spaces. The material of this section depends essentially on the fundamental work of M.Kinoshita, in references [24] and [25].

7.2.1 The Loeb construction

Although the immediate objective of this chapter is the study of some topics of higher analysis in the hyperfinite set Π, there is little to be gained by just presenting the Loeb construction in such a restricted context. Accordingly instead of Π we shall consider a completely arbitrary internal subset X of ${}^*\mathbb{R}$. An **internal algebra** over X is any internal subset $\mathcal{A}$ of subsets of X which contains $\emptyset$ and X and which is closed under finite unions and complementation. A finitely additive internal measure defined on $\mathcal{A}$ is an internal function

$$\Xi : \mathcal{A} \to {}^*\mathcal{R}_0^+ \cup \{+\infty\}$$

which satisfies the following conditions:

(a) $\Xi(\emptyset) = 0$, and

(b) $A \cap B = \emptyset \Rightarrow \Xi(A \cup B) = \Xi(A) + \Xi(B)$

for any subsets $A, B \in \mathcal{A}$.

$(X, \mathcal{A}, \Xi)$ is an internal measure space. Since Ξ is a finitely additive function then, by hyperfinite induction, it is also *finitely additive: that is, if $(A_n)_{1 \le n \le \nu}$ is a hyperfinite family of ν disjoint subsets of $\mathcal{A}$, where $\nu \in {}^*\mathbb{N}$, then

$$\Xi\left(\bigcup_{n=1}^{\nu} A_n\right) = \sum_{n=1}^{\nu} \Xi(A_n) \quad .$$

However, in general, Ξ is not σ-finite. If $(A_n)_{n\in\mathbf{N}}$ is a sequence of internal sets of $\mathcal{A}$ the σ-union $\bigcup_{n\in\mathbf{N}} A_n$ is only internal (and therefore only belongs to $\mathcal{A}$) if there is a natural number p such that $\bigcup_{n\in\mathbf{N}} A_n = \bigcup_{n=1}^{p} A_n$: otherwise the σ-union $\bigcup_{n\in\mathbf{N}} A_n$ is an external object, not belonging to $\mathcal{A}$, and therefore does not have a meaningful Ξ-measure. However, as will be seen, if $(A_n)_{n\in\mathbf{N}}$ is a family of internal subsets of $\mathcal{A}$ then although the (generally external) set $\bigcup_{n\in\mathbf{N}} A_n$ may not belong to $\mathcal{A}$ it will be *"very close"* to a set of $\mathcal{A}$. That is, such a set can be *"approximated"* by elements of $\mathcal{A}$ in a sense to be made precise in what follows.

Given an internal (hyperreal) measure Ξ defined on $\mathcal{A}$ we can always construct from it a *real* measure

$$^0\Xi : \mathcal{A} \to \mathbb{R}_0^+ \cup \{+\infty\}$$

by means of the natural definition $^0\Xi(A) = \mathbf{st}\Xi(A)$, for all $A \in \mathcal{A}$ (where, by convention, we define $\mathbf{st}(r) = +\infty$ if $r \in {}^*\mathbb{R}_\infty^+ \cup \{+\infty\}$). The Loeb measure, Ξ_L, associated with the internal measure Ξ is basically an extension of $^0\Xi$ to a σ-additive measure. It turns out that such an extension requires in the first place that the algebra $\mathcal{A}$ of internal sets of X be extended to a σ-algebra containing sets of X which, although not necessarily internal, are in some sense *"sufficiently close"* to sets of $\mathcal{A}$. This notion will be made precise by means of the following concept of Ξ-approximation of an arbitrary subset of X by internal sets belonging to $\mathcal{A}$.

Definition 7.5 A subset B of X (internal or external) is said to be **Ξ-approximable** if for each real number $r > 0$ there exist internal sets of finite Ξ-measure, $A_r, C_r \in \mathcal{A}$ such that $A_r \subset B \subset C_r$ and

$$\Xi(C_r) - \Xi(A_r) < r \quad .$$

Since all the sets of $\mathcal{A}$ possess Ξ-measure then the required σ-extension of $\mathcal{A}$ should contain all sets *"sufficiently close"* to sets of $\mathcal{A}$. In particular, all the subsets $B \subset X$ which are Ξ-approximable, and for which

$$\Xi_L(B) = \inf\{^0\Xi(C) : C \in \mathcal{A} \wedge B \subset C\}$$

is a well-defined element of $\mathbb{R}_0^+$, satisfy this condition. If, in addition, B belongs to $\mathcal{A}$, then $\Xi_L(B) = {}^0\Xi(B)$.

The family of all Ξ-approximable subsets of X adjoined to $\mathcal{A}$ does not, in the general case, constitute a σ-algebra. In order to obtain a σ-algebra of subsets of X which includes $\mathcal{A}$ it is necessary to introduce other sets $B \subset X$ which, although not Ξ-approximable, are also *"close"* to elements of $\mathcal{A}$, in the sense that for all $F \in \mathcal{A}$ with finite Ξ-measure the set $B \cap F$ is Ξ-approximable. The purpose of the following definition is to deal with all such situations at once:

Definition 7.6 The **Loeb Algebra** generated by an internal algebra $\mathcal{A}$ is the family $\mathcal{A}_L$ which consists of all the subsets B of X (internal or external) for which $B \cap F$ is Ξ-approximable for every set $F \in \mathcal{A}$ of finite Ξ-measure.

If $\Xi(X) \in {}^*\mathbb{R}_b$ then $\mathcal{A}_L$ contains only those subsets of X which are Ξ-approximable: if X has unbounded Ξ-measure then $\mathcal{A}_L$ will contain other elements as well as the Ξ-approximable sets. It is necessary to prove, to begin with, that $\mathcal{A}_L$ really is a σ-algebra.

Theorem 7.7 $\mathcal{A}_L$ is a σ-algebra which extends $\mathcal{A}$.

Proof The inclusion $\mathcal{A} \subset \mathcal{A}_L$ follows immediately from the definition of $\mathcal{A}_L$. In particular we have $\emptyset, X \in \mathcal{A}_L$. To prove that $\mathcal{A}_L$ is a σ-algebra it is now enough to show that $\mathcal{A}_L$ is closed under complementation and countable unions.

(a) Let $B \in \mathcal{A}_L$, $r > 0$ and $F \in \mathcal{A}$ such that $\Xi(F) \in {}^*\mathbb{R}_b$. By the definition of $\mathcal{A}_L$ there exist internal sets $A, C \in \mathcal{A}$ such that

$$A \subset (B \cap F) \subset C \quad \text{and} \quad \Xi(C) - \Xi(A) < r \ .$$

Then $C^c \subset (B \cap F)^c \equiv B^c \cup F^c$ and $(B \cap F)^c \equiv B^c \cup F^c \subset A^c$. Therefore we have $C^c \cap F \subset [(B^c \cup F^c) \cap F] \subset A^c \cap F$ or, equivalently

$$C^c \cap F \subset (B^c \cap F) \subset A^c \cap F \ .$$

Since

$$\begin{aligned}(A^c \cap F) \backslash (C^c \cap F) &= (A^c \cap F) \cap (C^c \cap F)^c \\ &= (A^c \cap F) \cap (C \cup F^c) = (C \cap F) \backslash A \subset C \backslash A\end{aligned}$$

then

$$\Xi(A^c \cap F) - \Xi(C^c \cap F) < \Xi(C) - \Xi(A) < r$$

and therefore $B^c \in \mathcal{A}_L$.

(b) Let $(B_n)_{n\in\mathbb{N}}$ be a sequence of elements of $\mathcal{A}_L$. We now want to prove that the set $B = \bigcup_{n=1}^{\infty} B_n$ also belongs to $\mathcal{A}_L$. Let $F \in \mathcal{A}$ have finite Ξ-measure, and r be a given positive real number. Since, by hypothesis, $B_n \in \mathcal{A}_L, n = 1, 2, \ldots$, then for each $n \in \mathbb{N}$ there exist sets $C_n, A_n \in \mathcal{A}$ such that

$$A_n \subset (B_n \cap F) \subset C_n \quad \text{and} \quad \Xi(C_n) - \Xi(A_n) < r/2^{n+1} \ .$$

For each $n \in \mathbb{N}$ we have $A_n \subset B_n \cap F$, and since $\Xi(F)$ is a finite number it follows that $\bigcup_{k=1}^{n} A_k \subset F$ for all $n \in \mathbb{N}$ and the limit

$$\gamma = \lim_{n\to\infty} \left\{ {}^0\Xi\left(\bigcup_{k=1}^{n} A_k\right)\right\}$$

exists and is a finite number; consequently there exists $n_r \in \mathbb{N}$ for which

$${}^0\Xi\left(\bigcup_{k=1}^{n_r} A_k\right) > \gamma - (r/2).$$

Setting $A = \bigcup_{k=1}^{n_\epsilon} A_k$ we have $A \in \mathcal{A}$ and $A \subset B \cap F$.

To obtain the exterior approximation C we extend the sequence $(C_n)_{n\in\mathbb{N}}$ to the hypersequence $(C_n)_{n\in{}^*\mathbb{N}}$. Since $\mathcal{A}$, Ξ, and $(C_n)_{n\in\mathbb{N}}$ are all internal objects then, by the principle of internality, the set

$$\mathcal{M} = \{n \in {}^*\mathbb{N} : \bigcup_{k=1}^{n} C_k \in \mathcal{A} \ \text{ and } \ \Xi\left(\bigcup_{k=1}^{n} C_k\right) < \gamma + (r/2)\}$$

is internal. As $\bigcup_{k=1}^{n} A_k \subset \bigcup_{k=1}^{n} C_k$ and $\Xi(C_n) - \Xi(A_n) < r/2^{n+1}$, $n = 1, 2, \ldots$, then

$$\Xi\left(\bigcup_{k=1}^{n} C_k\right) < \Xi\left(\bigcup_{k=1}^{n} A_k\right) + \sum_{k=1}^{n} r/2^{k+1} < \gamma + (r/2)$$

from which it follows that $\mathbb{N} \subset \mathcal{M}$. By "overflow", $\mathcal{M}$ contains an infinite element $\nu \in {}^*\mathbb{N}_\infty$. Defining

$$C = \bigcup_{k=1}^{\nu} C_k \in \mathcal{A},$$

then since $B_n \subset C_n$ for all $n \in \mathbb{N}$ we get

$$A \subset (B \cap F) \subset C \quad \text{and} \quad \Xi(C) - \Xi(A) < r$$

and, therefore, $B \subset \mathcal{A}_L$, as required. •

Now consider the (external) function

$$\Xi_L : \mathcal{A}_L \to \mathbb{R}_0^+ \cup \{+\infty\}$$

defined by

$$\Xi_L(B) = \inf\{{}^0\Xi(C) : C \in \mathcal{A} \text{ and } B \subset C\} \tag{7.4}$$

for all $B \in \mathcal{A}_L$. We have

Theorem 7.8 Ξ_L is a complete measure defined on the algebra $\mathcal{A}_L$.

Proof In order to prove this theorem we must show that Ξ_L really is a (σ-additive) measure defined on $\mathcal{A}_L$ and further that Ξ_L is complete:[2] that is, we have to show that if $B, B' \in \mathcal{A}_L$ are two sets such that $\Xi_L(B) = 0$ and $B' \subset B$ then we also have $\Xi(B') = 0$. This last part follows at once from the relation

$$\{{}^0\Xi(C) : C \in \mathcal{A} \text{ and } C \supset B'\} \supset \{{}^0\Xi(C) : C \in \mathcal{A} \text{ and } C \supset B\}$$

which implies that

$$\begin{aligned} \Xi_L(B') &= \inf\{{}^0\Xi(C) : C \in \mathcal{A} \text{ and } C \supset B'\} \\ &\leq \inf\{{}^0\Xi(C) : C \in \mathcal{A} \text{ and } C \supset B\} = 0 \ . \end{aligned}$$

In particular, we have $\emptyset \subset B$ for all sets $B \in \mathcal{A}_L$ and therefore

$$\Xi_L(\emptyset) = 0 \ .$$

It remains to prove σ-addivity, that is that if $(B_n)_{n\in\mathbb{N}}$ is a sequence of disjoint elements of the Loeb algebra $\mathcal{A}_L$ then

$$\Xi_L\left(\bigcup_{n=1}^{\infty} B_n\right) = \sum_{n=1}^{\infty} \Xi_L(B_n) \ .$$

If for some $n \in \mathbb{N}$ we were to have $\Xi_L(B_n) = +\infty$ the equality would be trivially satisfied. Suppose now that $\Xi_L(B_n) < +\infty$ for every $n \in \mathbb{N}$. In this case B_n is Ξ-approximable and therefore given $r > 0$ there exist $C_n, A_n \in \mathcal{A}$ such that

$$A_n \subset B_n \subset C_n \ \text{ and } \ {}^0\Xi(C_n) - (r/2^{n+1}) < \Xi_L(B_n) < \ {}^0\Xi_L(A_n) + (r/2^{n+1}) \tag{7.5}$$

[2] Let $(\mathcal{M}, M, \nu)$ be an arbitrary measure space and suppose that for two subsets B and B' of X we have $B' \subset B$ and $\nu(B) = 0$. We cannot generally infer from this that $\nu(B') = 0$ since it may happen that B' does not belong to $\mathcal{M}$. When $\mathcal{M}$ contains all the subsets of all the sets of ν-measure zero, then $(\mathcal{M}, M, \nu)$ is said to be a complete measure space.

for all $n \in \mathbb{N}$.

There are two possibilities:

1. $\sum_{n=1}^{\infty} \Xi_L(B_n) = +\infty$.
Then it follows from (7.5) that the sum $\sum_{n=1}^{\infty} {}^0\Xi(A_n)$ is also infinite and therefore the set $\bigcup_{n\in\mathbf{N}} B_n$ may be internally Ξ-approximated by internal sets of the form $\bigcup_{k=1}^{n} A_k$ whose measures are finite for each n but are arbitrarily large for sufficiently large values of n. Hence,

$$\Xi_L\left(\bigcup_{n\in\mathcal{N}}\right) = +\infty$$

and so the equality is satisfied.

2. $\sum_{n=1}^{\infty} \Xi_L(B_n) = \gamma < +\infty$.
From (7.5) we have

$$\sum_{n=1}^{\infty} {}^0\Xi(C_n) - r/2 \leq \sum_{n=1}^{\infty} \Xi_L(B_n) \equiv \gamma \leq \sum_{n=1}^{\infty} {}^0\Xi(A_n) + r/2 \ .$$

If $m \in \mathbb{N}$ is sufficiently large then

$$\Xi\left(\bigcup_{n=1}^{m} A_n\right) = \sum_{n=1}^{m} \Xi(A_n) > \gamma - r/2 \ \text{ and } \ A \equiv \bigcup_{n=1}^{m} A_n \subset \bigcup_{n=1}^{m} B_n$$

where $A \in \mathcal{A}$. To obtain an exterior approximation we can argue as in the preceding theorem: $(C_n)_{n\in {}^*\mathbf{N}}$ being the internal extension of $(C_n)_{n\in\mathbf{N}}$, the set

$$\{n \in {}^*\mathbb{N} : \bigcup_{k=1}^{n} C_k \in \mathcal{A} \ \text{ and } \ \Xi\left(\bigcup_{k=1}^{n} C_k\right) < \gamma + r/2\}$$

is internal and contains all the natural numbers. By overflow this set contains an infinite hypernatural number $\nu \in {}^*\mathbb{N}_\infty$ for which therefore the relation

$$\bigcup_{n\in\mathbf{N}} B_n \subset \bigcup_{n=1}^{\nu} C_n \equiv C$$

where $C \in \mathcal{A}$, is satisfied. Consequently

$$\bigcup_{n=1}^{m} A_n \subset \bigcup_{n=1}^{\infty} B_n \subset \bigcup_{n=1}^{\nu} C_n \ \text{ and } \ \gamma - r/2 < \Xi_L\left(\bigcup_{n\in\mathcal{N}} B_n\right) < \gamma + r/2 \ .$$

Since $r > 0$ is arbitrary we obtain

$$\Xi_L\left(\bigcup_{n\in\mathbf{N}} B_n\right) = \gamma \equiv \sum_{n=1}^{\infty} \Xi_L(B_n)$$

as it was required to show.

Thus it has been proved that the (external) function Ξ_L is a complete measure defined on the Loeb algebra $\mathcal{A}_L$. •

This theorem ensures the consistency of the following definition:

Definition 7.9 We define the **Loeb measure** associated with the internal measure Ξ to be the (external) function $\Xi_L : \mathcal{A}_L \to \mathbb{R}_0^+ \cup \{+\infty\} \equiv \bar{\mathbb{R}}_0^+$ defined for every set $B \in \mathcal{A}_L$ by

$$\Xi_L(B) = \inf\{{}^0\Xi(C) : C \in \mathcal{A} \text{ and } B \subset C\} \ .$$

The triple $\mathcal{L}(\Xi) \equiv (X, \mathcal{A}_L, \Xi_L)$ is known as the **Loeb Measure Space** associated with the internal measure space $(X, \mathcal{A}, \Xi)$.

7.2.2 Hyperfinite representation of Lebesgue measure

As an example of an important application of the theory of Loeb measure we now study the hyperfinite representation of Lebesgue measure on $\mathbb{R}$ in terms of a counting measure. Let $X = \Pi$ and denote by $\mathcal{A}$ the algebra of all the internal subsets of Π. Each element A of $\mathcal{A}$, being an internal subset of Π, has internal hyperfinite cardinality $\leq \kappa^2$. On $\mathcal{A}$ we define an internal function $\Xi : \mathcal{A} \to {}^*\mathbb{R}_0^+$, setting for each $A \in \mathcal{A}$,

$$\Xi(A) = \epsilon.\natural(A) = \frac{\natural(A)}{\kappa} \ .$$

Given two sets $A, B \in \mathcal{A}$, if $A \cap B = \emptyset$ then $\natural(A \cup B) = \natural(A) + \natural(B)$ and therefore $(\Pi, \mathcal{A}, \Xi)$ is a hyperfinitely additive measure space. The measure Ξ is called the **Loeb counting measure** and the space $\mathbf{L}(\Xi) \equiv (\Pi, \mathcal{A}_L, \Xi_L)$ is known as the **Loeb counting measure space.**

Given any subset V of $\mathbb{R}$, denote by $\mathbf{st}_\Pi^{-1}(V)$ the subset of Π defined by

$$\mathbf{st}_\Pi^{-1}(V) = \{x \in \Pi \subset {}^*\mathbb{R} : \mathbf{st}(x) \in V\} \ .$$

Denote by $\mathcal{L}$ the family of subsets $V \subset \mathbb{R}$ for which $\mathbf{st}_\Pi^{-1}(V)$ belongs to $\mathcal{A}_L$, that is

$$\mathcal{L} = \{V \subset \mathbb{R} : \mathbf{st}_\Pi^{-1}(V) \in \mathcal{A}_L\}$$

and define the function $\lambda : \mathcal{L} \to \mathbb{R}_0^+ \cup \{+\infty\}$, by setting

$$\lambda(V) = \Xi_L(\mathbf{st}_\Pi^{-1}(V))$$

for each set $V \in \mathcal{L}$.

Theorem 7.10 $(\mathbb{R}, \mathcal{L}, \lambda)$ is the Lebesgue measure space, that is to say, the standard set $V \subset \mathbb{R}$ is Lebesgue measurable if and only if the (possibly external) set $\mathbf{st}_\Pi^{-1}(V) \subset \Pi$ is Ξ_L-measurable, and

$$\lambda(V) = \Xi_L(\mathbf{st}_\Pi^{-1}(V)) \ .$$

Proof To begin with note that the set

$$\mathbf{st}_\Pi^{-1}(\mathbb{R}) = \Pi \cap {}^*\mathbb{R}_b = \bigcup_{n \in \mathcal{N}} \{x \in \Pi : |x| \leq n\},$$

being a countable union of internal subsets of Π, is Ξ-measurable and therefore λ-measurable. The proof of the theorem now follows with the statement and proof of various auxiliary lemmas. We show first that λ coincides with Lebesgue measure on open intervals of $\mathbb{R}$.

Lemma 7.11 Given $a, b \in \mathbb{R}$ the open interval (a, b) is λ-measurable and $\lambda(a, b) = b - a$.

We have

$$\mathbf{st}_{\Pi}^{-1}(a, b) \equiv \{x \in \Pi : a < \mathbf{st}(x) < b\}$$
$$= \bigcup_{n=1}^{\infty} \{x \in \Pi : a + 1/n \leq x \leq b - 1/n\}$$

where for each $n \in \mathbb{N}$ the set

$$\{x \in \Pi : a + 1/n \leq x \leq b - 1/n\}$$

is an internal subset of Π and therefore belongs to $\mathcal{A}$. Hence, since $\mathcal{A}_L$ is a σ-algebra, the set $\mathbf{st}_{\Pi}^{-1}(a, b)$ belongs to $\mathcal{A}_L$, and is therefore Ξ-measurable and consequently λ-measurable.

If t is any real number, the set

$$\{\nu \in {}^*\mathcal{Z} \cap [-\kappa^2/2, (\kappa^2/2) - 1] : \nu\epsilon \leq t\}$$

is internal and hyperfinite, and therefore has a maximum. Let

$$\nu_t = \max\{\nu \in {}^*\mathcal{Z} \cap [-\kappa^2/2, +\kappa^2/2 - 1] : \nu\epsilon \leq t\} \quad .$$

Then $(0 \leq t - \nu_t\epsilon < \epsilon \approx 0)$ so that the quotient $\nu_t/\kappa \equiv \nu_t\epsilon$ is the largest element of Π less than or equal to t. Accordingly we can write

$$\natural\{x \in \Pi : a + 1/n \leq x \leq b - 1/n\} = \nu_{b-1/n} - \nu_{a+1/n}$$

whence

$$\Xi(\{x \in \Pi : a + 1/n \leq x \leq b - 1/n\}) = (\nu_{b-1/n} - \nu_{a+1/n})\epsilon = b - a - 2/n \quad .$$

Setting, for each $n \in \mathbb{N}$,

$$A_n = \{x \in \Pi : a + 1/n \leq x \leq b - 1/n\} \in \mathcal{A}$$

we have

$$\mathbf{st}._{\Pi}^{-1}(a, b) = A_1 \cup \left(\bigcup_{n \geq 2} (A_n \backslash A_{n-1}) \right) \in \mathcal{A}_L$$

and therefore

$$\Xi_L(\mathbf{st}._{\Pi}^{-1}(a, b)) = \ {}^0\Xi(A_1) + \sum_{n=2}^{\infty} \ {}^0\Xi(A_n \backslash A_{n-1})$$

$$= \lim_{n\to\infty}\{\ {}^0\Xi(A_n)\} = \lim_{n\to\infty}(b-a-2/n) = b-a$$

so that we then get

$$\lambda(a,b) = \Xi_L(\mathbf{st}._{\Pi}^{-1}(a,b)) = b-a,$$

as it was required to prove.

Lemma 7.12 $\mathcal{L}$ is a σ-algebra containing the algebra of all the Borel sets of $\mathbb{R}$.

Let $(V_n)_{n\in\mathbf{N}}$ be a sequence of subsets of $\mathcal{L}$. Then for each $n \in \mathbb{N}$, $\mathbf{st}_{\Pi}^{-1}(V_n)$ belongs to $\mathcal{A}_L$ and therefore $\bigcup_{n\in\mathbf{N}}\mathbf{st}_{\Pi}^{-1}(V_n)$ also belongs to $\mathcal{A}_L$. Since

$$\mathbf{st}_{\Pi}^{-1}\left(\bigcup_{n\in\mathbf{N}} V_n\right) = \{x \in \Pi : \mathbf{st}(x) \in \bigcup_{n\in\mathbf{N}} V_n\}$$

$$= \bigcup_{n\in\mathbf{N}}\{x \in \Pi : \mathbf{st}(x) \in V_n\} = \bigcup_{n\in\mathbf{N}} \mathbf{st}_{\Pi}^{-1}(V_n)$$

then $\bigcup_{n\in\mathbf{N}} V_n$ belongs to $\mathcal{L}$, proving that $\mathcal{L}$ is a σ-algebra. On the other hand from the lemma 7.11 it follows that $\mathcal{L}$ contains every open interval of $\mathbb{R}$ and therefore that $\mathcal{L}$ contains all the Borel subsets of $\mathbb{R}$.

Lemma 7.13 $(\mathbb{R}, \mathcal{L}, \lambda)$ is a complete measure space.

Since $\mathbf{st}_{\Pi}^{-1}(\emptyset) = \emptyset$ then

$$\lambda(\emptyset) = \Xi_L(\emptyset) = 0\ \ .$$

If $(V_n)_{n\in\mathbf{N}}$ is a sequence of mutually disjoint elements of $\mathcal{L}$, then $(\mathbf{st}_{\pi}^{-1}(V_n))_{n\in\mathbf{N}}$ is a sequence of mutually disjoint elements of $\mathcal{A}_L$ and,

$$\lambda\left(\bigcup_{n\in\mathbf{N}} V_n\right) = \Xi_L\left(\mathbf{st}_{\Pi}^{-1}\left(\bigcup_{n\in\mathbf{N}}(V_n)\right)\right) = \Xi_L\left(\bigcup_{n\in\mathbf{N}}(\mathbf{st}_{\Pi}^{-1}(V_n))\right)$$

$$= \sum_{n=1}^{\infty}\Xi_L(\mathbf{st}_{\Pi}^{-1}(V_n)) = \sum_{n=1}^{\infty}\lambda(V_n)$$

which proves that $(\mathbb{R}, \mathcal{L}, \lambda)$ is a measure space. It now remains to be shown that we are dealing with a complete measure space, for which it will be enough to verify that $\mathcal{L}$ contains all subsets of $\mathbb{R}$ of measure zero.

Since, for two subsets V and U of $\mathcal{E}$ we have

$$V \subset U \Rightarrow \mathbf{st}_{\Pi}^{-1}(V) \subset \mathbf{st}_{\Pi}^{-1}(U),$$

and since Ξ_L is a complete measure then from the conditions

$$\lambda(U) = 0 \ \text{ and } \ V \subset U$$

it follows that

$$\Xi_L(\mathbf{st}_{\Pi}^{-1}(V) \le \Xi_L(\mathbf{st}._{\Pi}^{-1}(U)) = \lambda(U) = 0$$

which means that we have $\lambda(V) = 0$ and therefore that $\mathcal{L}$ contains all subsets of sets of λ-measure zero.

It now follows from the three Lemmas 7.11, 7.12, 7.13, that λ is a complete measure over $\mathbb{R}$ which coincides with Lebesgue measure on open intervals (and therefore on all Borel sets) of $\mathbb{R}$. Thus λ is Lebesgue measure or some extension of Lebesgue measure to a σ-algebra which contains all Lebesgue-measurable sets. The following lemma is sufficient to show that the second alternative is not true.

Lemma 7.14 Let A be a λ-measurable subset of $\mathbb{R}$ such that $\lambda(A) = \alpha < +\infty$. Then A is Lebesgue-measurable with measure equal to α.

Note first that the general case of a set of arbitrary measure can always be reduced to this one since $V = \bigcup_{n\in\mathbf{N}}(V \cap I_n)$ where $(I_n)_{n\in\mathbf{N}}$ is a sequence of bounded intervals which covers $\mathbb{R}$.

From the definition of λ we have $\mathbf{st}_{\Pi}^{-1}(V) \in \mathcal{A}_L$ and $\Xi_L(\mathbf{st}_{\Pi}^{-1}(V)) = \alpha$. Given a real number $r > 0$ let $A, C \in \mathcal{A}$ be such that

$$A \subset \mathbf{st}_{\Pi}^{-1}(V) \subset C$$

with $\Xi_L(A) \equiv {}^0\Xi(A) > \alpha - r/2$ and $\Xi_L(C) \equiv {}^0\Xi(C) < \alpha + r/2$. Since $A \in \mathcal{A}$ it is internal and therefore (see theorem 5.14) $\mathbf{st}(A) \equiv V'$ is a closed subset of V. Further, since $A \subset \mathbf{st}_{\Pi}^{-1}(V')$, it is also true that

$$\lambda(V') = \Xi_L(\mathbf{st}_{\Pi}^{-1}(V') \geq \Xi_L(A) > \alpha - r/2 \ .$$

Similarly, since C is internal so also is $\Pi \backslash C$ and therefore $\mathbf{st}(\Pi \backslash C)$ is closed. Hence $U' = \mathbb{R} \backslash \mathbf{st}(\Pi \backslash C)$ is an open set containing V and such that

$$\lambda(U') = \Xi_L(\mathbf{st}_{\Pi}^{-1}(U')) < \alpha + r/2 \ .$$

Consequently there exists a closed set $V' \subset \mathbb{R}$ and an open set $U' \subset \mathbb{R}$ such that

$$V' \subset V \subset U' \quad \text{and} \quad \lambda(U') - \lambda(V') < r.$$

Since Lebesgue measure is complete and since $r > 0$ is arbitrary, it follows that V differs from U' by a set of measure zero or, equivalently, that

$$\lambda(V) = \lambda(U') \ .$$

But λ coincides with Lebesgue measure on open intervals and, therefore, since U' is an open set of $\mathbb{R}$, the Lebesgue measure of U' is equal to $\lambda(U')$. Hence the set V is measurable in the Lebesgue sense and has measure equal to $\lambda(U') = \alpha$. •

7.3 Integration

7.3.1 Generalities

In what follows the symbol $\mathcal{K}$ may be taken to stand for either the field $\mathbb{R}$ or the field ${}^*\mathbb{R}$. (In each case $\bar{\mathcal{K}}$ stands for the field $\mathcal{K}$ together with the elements $\pm\infty$ (with the usual algebraic conventions in measure theory). Let Γ be a measure (finitely additive or σ-additive) which takes values in $\bar{\mathcal{K}}_0^+$ and further let $(\mathbf{A}, \mathcal{A}, \Gamma)$ be a measure space

A function Θ, defined on $\mathbf{A}$, with values in $\bar{\mathcal{K}}$ is said to be **Γ-measurable** if sets of the form

$$\{x \in \mathbf{A} : a \leq \Theta(x) < b\} \quad \text{and} \quad \{x \in \mathbf{A} : \Theta(x) = \pm\infty\}$$

belong to $\mathcal{A}$ for any $a, b \in \mathcal{K}, (a < b)$. The first of these conditions can be replaced by other, equivalent, ones, such as that of the Γ-measurability of sets of the form $\{x \in \mathbf{A} : \Theta(x) \leq r\}$ or of the form $\{x \in \mathbf{A} : \Theta(x) < r\}$ etc., where r may be any element of $\mathcal{K}$.

Θ being a non-negative function, bounded on $\mathcal{K}$ and Γ-measurable, we define the integral (in the sense of Lebesgue) of Θ extended to $\mathbf{A}$ in the following way: Given a strictly increasing sequence (finite or hyperfinite respectively), $\alpha \equiv (\alpha_k)_{0 \leq k \leq p}$ of elements of $\mathcal{K}$ for which we have $\alpha_0 \leq \Theta(x) < \alpha_p$ for any $x \in \mathbf{A}$, we associate with it the sets

$$A_{\alpha,k} = \{x \in \mathbf{A} : \alpha_k \leq \Theta(x) < \alpha_{k+1}\}, \qquad k = 0, 1, \ldots, p-1 \ .$$

Since Θ is Γ-measurable then $A_{\alpha,k}$ belongs to $\mathcal{A}$ for any $k = 0, 1, \ldots, p-1$, and therefore the sum (respectively finite or hyperfinite)

$$\sum_{k=0}^{p-1} \alpha_k . \Gamma(A_{\alpha,k}) \tag{7.6}$$

is well-defined (possibly equal to $\pm\infty$). The Lebesgue integral of Θ extended to $\mathbf{A}$ relative to the measure Γ and denoted by

$$\Gamma(\Theta) \equiv \int_{\mathbf{A}} \Theta d\Gamma = \int_{\mathbf{A}} \Theta(x) d\Gamma(x)$$

is the supremum of the set of sums (7.6) formed under all possible conditions stated.

For an arbitrary bounded function Φ defined on $\mathbf{A}$, we always have the decomposition

$$\Phi = \Phi_+ - \Phi_-$$

where $\Phi_+ = \max\{0, \Phi\}$ and $\Phi_- = \max\{0, -\Phi\}$ are non-negative functions bounded on $\mathcal{K}$. We define the integral of Φ extended to $\mathbf{A}$ relative to Γ, by setting

$$\Gamma(\Phi) \equiv \int_{\mathbf{A}} \Phi d\Gamma = \int_{\mathbf{A}} \Phi_+ d\Gamma - \int_{\mathbf{A}} \Phi_- d\Gamma = \Gamma(\Phi_+) - \Gamma(\Phi_-) \ .$$

The function Φ is said to be **Γ-integrable** if both $\Gamma(\Phi_+)$ and $\Gamma(\Phi_-)$ are elements of $\mathcal{K}$ ($\mathbb{R}_0^+$ or ${}^*\mathbb{R}_0^+$ respectively).

Considering in particular $\mathbf{A} = \Pi$, $\mathcal{K} = \mathbb{R}$ and $\Gamma = \lambda$, if the function $\Phi = f : \mathbb{R} \to \mathbb{R}$ is λ-measurable (or, equivalently, measurable in the Lebesgue sense) we get

$$\lambda(f) \equiv \int_{\mathbb{R}} f d\lambda = \int_{\mathbb{R}} f(x) d\lambda(x)$$

which is the usual **Lebesgue integral** of a real function of a real variable. If we set $\mathbf{A} = \Pi \subset {}^*\mathbb{R}$, $\mathcal{K} = \mathbb{R}$ and $\Gamma = \Xi_L$ (the counting measure of Loeb), and if Φ is a Ξ-measurable function of the form $\tilde{f} : \Pi \to \mathbb{R}$, then the integral

$$\Xi_L(\tilde{f}) \equiv \int_{\Pi} \tilde{f} d\Xi_L = \int_{\Pi} \tilde{f}(x) d\Xi_L(x)$$

is called the **Loeb integral** of the (external) function $\tilde{f}$ extended to the hyperfinite line Π.

The Lebesgue *integral of internal functions on Π. We now consider the internal measure space $(\Pi, \mathcal{A}, \Xi)$ where $\mathcal{A}$ is the internal algebra of all internal subsets of Π, and Ξ is the counting measure. Given an internal function $F \in \mathbf{F}_\Pi$ we define similarly the Lebesgue *-integral of F extended to Π relative to the counting measure Ξ. With no loss of generality we may assume that F is a non-negative hyperreal function.[3]

Since for any pair of numbers $a, b \in {}^*\mathbb{R}, (a < b)$ the set

$$\{x \in \Pi : a \leq F(x) < b\}$$

is internal, then F is Ξ-measurable: moreover F, like the rest of all the functions in $\mathbf{F}_\Pi$, is *-bounded. The *-integral of F extended to Π

$${}^*\int_\Pi F(x) d\Xi(x)$$

is the supremum of the sums

$$\sum_{k=0}^{\nu-1} \alpha_k . \Xi(A_{\alpha,k})$$

when $\alpha \equiv (\alpha_k)_{0 \leq k \leq \nu}$ belongs to the set of all hyperfinite sequences such that $\alpha_0 \leq F(x) < \alpha_\nu$, $x \in \Pi$. As F assumes only a hyperfinite set of values with internal cardinality $\leq \kappa^2$, there exists a particular hyperfinite sequence

$$\alpha_F \equiv (\alpha_{F,k})_{k=0,1,\ldots,\nu_F}$$

such that for all $x \in \Pi$ there exists $k \in \{0, 1, \ldots, \nu_F\}$ for which $\alpha_{F,k_x} = F(x)$. For such a sequence the (generally hyperfinite) sum

$$\sum_{k=0}^{\nu_F - 1} \alpha_{F,k} . \Xi(A_{\alpha_F, k})$$

corresponds to the required supremum (maximum). Then

$${}^*\int_\Pi F(x) d\Xi(x) = \sum_{k=0}^{\nu_F - 1} \alpha_{F,k} . \Xi(A_{\alpha_{F,k}})$$

where each $\alpha_{F,k}$ can be expressed under the form $F(x)$ for an appropriate value of $x \in \Pi$; moreover,

$$A_{\alpha,k} = \{x \in \Pi : \alpha_k \leq F(x) < \alpha_{k+1}\} = \{x \in \Pi : \alpha_k = F(x)\}$$

[3] In fact, for $F = F_r + iF_i$ we have $F_r = F_{r+} - F_{r-}$ and $F_i = F_{i+} - F_{i-}$ where all the functions $F_{r+}, F_{r-}, F_{i+}, F_{i-}$ are internal non-negative hyperreal functions.

and therefore for $k = 0, 1, \ldots, \nu_F - 1$,

$$\alpha_{F,k}.\Xi(A_{\alpha_F,k}) = \epsilon F(x)\natural(A_{\alpha_F,k})$$

for any $x \in A_{\alpha_F,k}$. We may then write

$$\alpha_{F,k}.\Xi(A_{\alpha_F,k}) = \sum_{x\in A_{\alpha_F,k}} \epsilon F(x)$$

and therefore finally obtain

$$^*\int_\Pi F(x)d\Xi(x) = \sum_{k=0}^{\nu_F-1} \alpha_{F,k}.\Xi(A_{\alpha_F,k})$$

$$= \sum_{k-0}^{\nu_F-1}\left\{\sum_{x\in A_{\alpha_F,k}} \epsilon F(x)\right\} = \sum_{x\in\Pi} \epsilon F(x) \ .$$

That is, the Lebesgue *-integral of $F \in \mathbf{F}_\Pi$ extended to Π relative to the counting measure ξ is equal to the Π-integral of F extended to Π.

Example 7.15 Delta Function: canonical hyperfinite representation

Defining the internal function $\Delta_0 : \Pi \to \mathbb{R}$ by

$$\Delta_0(x) = \kappa.\Xi_0(x) = \begin{cases} \kappa & \text{if } \ x = 0 \\ 0 & \text{if } \ x \neq 0 \end{cases}$$

we obtain immediately

$$\sum_{x\in\Pi} \epsilon\Delta_0(x) = 1$$

and, additionally, if $g : \mathbb{R} \to \mathbf{C}$ denotes an arbitrary (standard) function, we have

$$\sum_{x\in\Pi} \epsilon\Delta_0(x)\,{}^*g_\Pi(x) = g(0) \ . \tag{7.7}$$

Note that the function Δ_0 samples the exact value of the function g at the origin without having to take standard parts. This fact makes it unnecessary to impose any continuity condition on the function g at the origin: it is enough that $g(0)$ be defined. If instead of Δ_0 we were to consider the internal function

$$\Delta_p(x) = \kappa\Xi_p(x) \quad , \qquad x \in \Pi$$

for some finite integer p different from zero then we would get

$$\sum_{x\in\Pi} \epsilon\Delta_p(x)\,{}^*g_\Pi(x) = {}^*g(p\epsilon)$$

and therefore to obtain the sampling property at the origin it would be necessary for the function g to be continuous at the origin. Then taking standard parts we would get

$$\mathbf{st}\left(\sum_{x\in\Pi} \Delta_p(x)\,{}^*g_\Pi(x)\right) = g(0) \quad .$$

If, however, p is an infinite hyperinteger such that $p\epsilon \in \Pi_b$ then we would obtain the sample of g at the point $\mathbf{st}(p\epsilon)$, provided that g were continuous in some neighbourhood of that point.

Suppose now that instead of the (nonstandard) Π-extension of a standard function we consider an arbitrary internal function $G(x), x \in \Pi$; we would then get

$$\sum_{x\in\Pi} \epsilon\Delta_0(x)G_\Pi(x) = G(0) \quad .$$

That is to say, the function G is sampled at the origin. Taking in particular $G = \Delta_0$ we have

$$\sum_{x\in\Pi} \epsilon\Delta_0^2(x) = \Delta_0(0) = \kappa$$

which agrees with results in the literature which *attribute to the integral of the square of the delta function the value* $+\infty$.

Now consider, as an example, the Π-derivative of Δ_0:

$$D_+\Delta_0(x) = \begin{cases} +\kappa^2 & \text{if } x = -\epsilon \\ -\kappa^2 & \text{if } x = 0 \\ 0 & \text{for other values of } x \end{cases} \quad .$$

Then

$$\sum_{x\in\Pi} \epsilon D_+\Delta_0(x) = 0$$

and for a (standard) function $g : \mathbb{R} \to \mathbf{C}$ differentiable at the origin

$$\sum_{x\in\Pi} \Delta_0(x)\,{}^*g_\pi(x) = \kappa\,{}^*g_\pi(-\epsilon) - \kappa g(0) \approx -g'(0) \quad .$$

7.3.2 Integration and Π-integration

Let $f : \mathbb{R} \to \mathbb{R}$ be a non-negative[4] function which is integrable over $\mathbb{R}$ (in the Lebesgue sense) and which is extended to $\bar{\mathbb{R}}$ by setting $f(\pm\infty) = 0$. Given any finite sequence $\alpha \equiv (\alpha_k)_{0\leq k\leq p}$ with $\alpha_0 \leq f(t) < \alpha_p, \quad t \in \mathbb{R}$, consider the sum

$$\sum_{k=0}^{p-1} \alpha_k.\lambda(A_{\alpha,k})$$

[4] Once again, the hypothesis of non-negativity does not present any loss of generality, since any real function may be decomposed into the difference of two non-negative functions.

where λ is Lebesgue measure on $\mathbb{R}$ and $A_{\alpha,k} = \{t \in \mathbb{R} : \alpha_k \leq f(t) < \alpha_{k-1}\}$, for $k = 0, 1, \ldots, p-1$. Taking into account the existing relation between Lebesgue measure and Loeb counting measure on the hyperfinite line we may write

$$\sum_{k=0}^{p-1} \alpha_k . \lambda(A_{\alpha,k}) = \sum_{k=0}^{p-1} \alpha_k . \Xi_L(\mathbf{st}_{\Pi}^{-1}(A_{\alpha,k})) \tag{7.8}$$

where, for each $k = 0, 1, \ldots, p-1$,

$$\begin{aligned} \mathbf{st}_{\Pi}^{-1}(A_{\alpha,k}) &= \mathbf{st}_{\Pi}^{-1}\{t \in \bar{\mathbb{R}} : \alpha_k \leq f(t) < \alpha_{k+1}\} \\ &= \{x \in \Pi : \mathbf{st}_{\infty} x \in \{t \in \bar{\mathbb{R}} : \alpha_k \leq f(t) < \alpha_{k+1}\}\} \\ &= \{x \in \Pi : \alpha_k \leq f \circ \mathbf{st}(x) < \alpha_{k+1}\} \end{aligned}$$

and $\mathbf{st}._{\infty} : \Pi \to \bar{\mathbb{R}}$ is the (external) function defined by

$$\mathbf{st}_{\infty} = \begin{cases} -\infty & \text{if } x \in \Pi_{\infty}^{-} \\ \mathbf{st}(x) & \text{if } x \in \Pi_b \\ +\infty & \text{if } x \in \Pi_{\infty}^{+} \end{cases} .$$

Denoting by $\tilde{f} : \Pi \to \mathbb{R}_0^+$ the (external) function defined by

$$\tilde{f}(x) = f \circ \mathbf{st}_{\infty}(x), \qquad x \in \Pi$$

we then have

$$\mathbf{st}_{\Pi}^{-1}(A_{\alpha,k}) = \{x \in \Pi : \alpha_k \tilde{f}(x) < \alpha_{k+1}\}$$

which, on the one hand, shows that f is Lebesgue-measurable if and only if $\tilde{f}$ is measurable in the Loeb sense. On the other hand, taking the supremum of all the sums in (7.8) when α ranges over all possible forms of the stated conditions, it shows that we have

$$\int_{\mathcal{R}} f d\lambda = \int_{\Pi} \tilde{f} d\Xi_L \quad . \tag{7.9}$$

Thus the Lebesgue integral of f extended to $\mathbb{R}$ is equal to the Loeb integral of $\tilde{f}$ extended to Π relative to the Loeb counting measure Ξ_L.

Now suppose in addition that f is a function which is defined and continuous on a compact $K \subset \mathbb{R}$ and which vanishes outside K. Then f is Lebesgue integrable and Riemann integrable and these integrals have the same value: in this case we obtain the double equality

$$\int_{\mathcal{R}} f d\lambda = \int_{\Pi} \tilde{f} d\Xi_L = \mathbf{st}\left(\sum_{x \in \Pi} \epsilon\, {}^*f_{\Pi}(x)\right) \tag{7.10}$$

which allows us to calculate the Loeb integral of $\tilde{f}$ (and therefore the Lebesgue integral of f) from the Π-integral of the (internal) function ${}^*f_{\Pi}$ which actually reduces to a simple hyperfinite sum.

Since the first equality in (7.10) is always valid for any Lebesgue-integrable function f, we shall now attempt to make an analysis (though not an exhaustive one) of other situations in which it is possible to express the Loeb integral of an external function $\tilde{f}$ in terms of the Π-integral of an appropriate internal function F. The Loeb integral occupies in this way a position between the (standard) Lebesgue integral and the (nonstandard) Π-integrals of internal functions. The interest of this study resides in the possibility of using representations of standard measures in terms of nonstandard measures which are generally conceptually much simpler. Although it is possible to study the Loeb integral in a more general context - that of an arbitrary Loeb measure space - the present study will be restricted to the case of Loeb counting measure on the hyperfinite line Π. To begin with we shall prove the following result:

Theorem 7.16 The projection of an arbitrary internal function $F \in \mathbf{F}_\Pi$, that is the external function $\mathbf{st}F : \Pi \to \bar{\mathbb{R}}$, is $\mathcal{A}_L$-measurable.

Proof It is enough to show that the sets $(\mathbf{st}F)^{-1}(\{\pm\infty\})$ and $(\mathbf{st}F)^{-1}((-\infty, r])$ belong to $\mathcal{A}_L$ for any $r \in \mathbb{R}$. As

$$\begin{aligned}(\mathbf{st}F)^{-1}((-\infty, r]) &= \{x \in \Pi : \mathbf{st}F(x) \leq r\} \\ &= \bigcap_{n \in \mathbf{N}} \{x \in \Pi : F(x) < r + 1/n\} \\ &= \bigcap_{n \in \mathbf{N}} F^{-1}((-\infty, r + 1/n])\end{aligned}$$

and F is $\mathcal{A}$-measurable, then $F^{-1}((-\infty, r + 1/n])$ belongs to $\mathcal{A}$ for every $n \in \mathbb{N}$, and therefore $(\mathbf{st}F)^{-1}((-\infty, r]) \in \mathcal{A}_L$. On the other hand

$$\begin{aligned}(\mathbf{st}F)^{-1}(\{\pm\infty\}) &= \{x \in \Pi : F(x) \in {}^*\mathbb{R}\} \\ &= \bigcap_{n \in N} \{x \in \Pi : |F(x)| > n\}\end{aligned}$$

and as, for each $n \in \mathbb{N}$, the set $\{x \in \Pi : |F(x)| > n\}$ is $\mathcal{A}$-measurable then $(\mathbf{st}F)^{-1}(\{\pm\infty\})$ is $\mathcal{A}_L$-measurable. •

In the general case the integrals (Π-integral and Loeb integral respectively)

$$\mathbf{st}\left(\sum_{x \in \Pi} \epsilon F(x)\right) \quad \text{and} \quad \int_\Pi \mathbf{st}F(x) d\Xi_L(x)$$

are distinct. In fact, if we consider for example the internal function defined by

$$\Delta_0(x) = \begin{cases} \kappa & \text{if } x = 0 \\ 0 & \text{if } x \neq 0 \end{cases}$$

we obtain

$$\mathbf{st}\left(\sum_{x \in \Pi} \epsilon \Delta_0(x)\right) = 1 \quad \text{while} \quad \int_\Pi \mathbf{st}\Delta_0(x) d\Xi_L(x) = 0$$

which confirms this assertion. But under certain conditions it is possible to guarantee the equality of the two integrals. The following definition is concerned with one such condition:

Definition 7.17 An internal function $F \in \mathbf{F}_\Pi$ is said to be **Ξ-finite** if it satisfies the following two conditions:
F is finite on the whole hyperfinite line, and
$\Xi\{x \in \Pi : F(x) \neq 0\}$ is a finite number.

Theorem 7.18 If $F \to {}^*\mathbb{R}$ is a Ξ-finite function then

$$\mathbf{st}\left(\sum_{x\in\Pi} \epsilon F(x)\right) = \int_\Pi \mathbf{st}F(x)d\Xi_L(x) \quad . \tag{7.11}$$

Proof Let $x > 0$ be an arbitrarily given real number, and define

$$A = \{x \in \Pi : F(x) \neq 0\}.$$

For each $k \in \mathbb{Z}$ consider the (internal) set

$$A_k = \{x \in A : kr \leq F(x) < (k+1)r\}.$$

Then the functions

$$F_r^+(x) = \sum_{k\in\mathbb{Z}} (k+1)r.1_{A_k}(x)$$

and

$$F_r^-(x) = \sum_{k\in\mathbb{Z}} (k-1)r.1_{A_k}(x)$$

(where 1_{A_k} denotes the characteristic function of the set A_k) are respectively the upper and lower approximants, both of the internal function F and of the external function $\mathbf{st}F$. Since $\sup|F(x)| \in {}^*\mathbb{R}_b$ there exists $k_0 \in \mathbb{N}$ such that $A_k = \emptyset$ for $|k| > k_0$ and, therefore, both F_r^+ and F_r^- are $\mathcal{A}$-measurable internal functions. Consequently,

$$\sum_{x\in\Pi} \epsilon F_t^-(x) \leq \sum_{x\in\Pi} \epsilon F(x) \leq \sum_{x\in\Pi} \epsilon F_t^+(x) \tag{7.12}$$

and

$$\int_\Pi \mathbf{st}F_r^-(x)d\Xi_L(x) \leq \int_\Pi \mathbf{st}F(x)d\Xi_L(x) \leq \int_\Pi \mathbf{st}F_r^+(x)d\Xi_L(x) \quad . \tag{7.13}$$

Since F_r^- and F_r^+ are internal functions (which assume standard values) and since

$$\int_\Pi \mathbf{st}F_r^-(x)d\Xi_L(x) = \sum_{|k|\leq k_0} (k-1)r.\Xi_L(A_k) = \sum_{|k|\leq k_0} (k-1)r\mathbf{st}\Xi(A_k)$$

$$= \mathbf{st}\left(\sum_{|k|\leq k_0} (k-1)r\Xi(A_k)\right) = \mathbf{st}\left(\sum_{x\in\Pi} \epsilon F_r^-(x)\right)$$

and similarly

$$\int_\Pi \mathbf{st}F_r^+(x)d\Xi_L(x) = \mathbf{st}\left(\sum_{x\in\Pi} \epsilon F_r^+(x)\right) \quad .$$

Then from (7.12) and (7.13) we get

$$\mathbf{st}\left(\sum_{x\in\Pi}\epsilon F_r^-(x)\right) \leq \mathbf{st}\left(\sum_{x\in\Pi}\epsilon F(x)\right) \leq \mathbf{st}\left(\sum_{x\in\Pi}\epsilon F_r^+(x)\right)$$

and

$$\mathbf{st}\left(\sum_{x\in\Pi}\epsilon F_r^-(x)\right) \leq \int_\Pi F(x)d\Xi_L(x) \leq \mathbf{st}\left(\sum_{x\in\Pi}\epsilon F_r^+(x)\right)$$

whence we obtain the inequality

$$\left|\mathbf{st}.\left(\sum_{x\in\Pi}\epsilon F(x)\right) - \int_\Pi \mathbf{st}F(x)d\Xi_L(x)\right| \leq \mathbf{st}\left(\sum_{x\in\Pi}\epsilon|F_r^+ - F_r^-|\right)$$

$$= \mathbf{st}\left(\sum_{x\in A}\epsilon(2r)\right) = 2r.\Xi(A) \quad .$$

The theorem now follows since $r \in \mathbb{R}^+$ is arbitrary. •

To obtain inequality (7.11) it is necessary to impose conditions on the internal function concerned: the function Δ_0, for example, shows that the equality is not necessarily valid when a function is not Ξ-finite. However the following is always true:

Theorem 7.19 $F \in \mathbf{F}_\Pi$ being a non-negative function, we always have the inequality

$$\int_\Pi F(x)d\Xi_L(x) \leq \mathbf{st}\left(\sum_{x\in\Pi}\epsilon F(x)\right) \tag{7.14}$$

where the value $+\infty$ may occur on either of the two sides.

Proof If the right-hand side is equal to $+\infty$ the result is trivial and there is nothing to prove. Suppose then that

$$\mathbf{st}\left(\sum_{x\in\Pi}\epsilon F(x)\right) < +\infty$$

and for each hypernatural $n \in {}^*\mathbb{N}$ let $F_n : \Pi \to {}^*\mathbb{R}$ be the function defined by

$$F_n(x) = \begin{cases} F(x) & \text{if } \frac{1}{n} \leq F(x) \leq n \\ 0 & \text{if } F(x) < \frac{1}{n} \text{ or } F(x) > n \end{cases} \quad .$$

If n is a finite number then F_n is an internal Ξ-finite function: in fact

$$\frac{1}{n}\Xi(\{x \in \Pi : F_n(x) \neq 0\}) = \frac{1}{n}(\{x \in \Pi : \frac{1}{n} \leq F(x) \leq n\})$$

$$= \frac{1}{n}\left(\sum_{x\in\{1/n\leq F\leq n\}}\epsilon\right) \leq \sum_{x\in\{1/n\leq F\leq n\}}\epsilon F(x) \leq \sum_{x\in\Pi}\epsilon F(x)$$

and therefore we have

$$\Xi(\{x \in \Pi : F_n(x) \neq 0\}) \leq n \sum_{x \in \Pi} \epsilon F(x)$$

which proves the assertion under the given hypothesis. Then, in accordance with Theorem 7.18, we obtain the following:

$$\int_{\Pi} \mathbf{st} F_{\Pi}(x) d\Xi_L(x) = \mathbf{st}\left(\sum_{x \in \Pi} \epsilon F_n(x)\right) \leq \mathbf{st}\left(\sum_{x \in \Pi} \epsilon F(x)\right) .$$

For each $n \in \mathbb{N}$ the function $\mathbf{st} F_n$ is Loeb integrable and

$$0 \leq \mathbf{st} F_1 \leq \mathbf{st} F_2 \leq \ldots \leq \mathbf{st} F_n \leq \ldots \quad \text{and} \quad \lim_{n \to \infty} \mathbf{st} F_n = \mathbf{st} F .$$

Hence, since $\mathbf{st}.F$ is $\mathcal{A}_L$-measurable, we can use the **monotone convergence theorem** to obtain the result

$$\int_{\Pi} \mathbf{st} F(x) d\Xi_L(x) = \lim_{n \to \infty} \int_{\Pi} \mathbf{st} F_n(x) d\Xi_L(x)$$

$$= \lim_{n \to \infty} \mathbf{st}\left(\sum_{x \in \Pi} \epsilon F_n(x)\right) \leq \mathbf{st}\left(\sum_{x \in \Pi} \epsilon F(x)\right)$$

(where the limits must be understood in the usual sense when $n \in \mathbb{N}$). •

There are functions which are not Ξ-finite but for which, nevertheless, (7.11) still holds. In order to obtain other conditions, more general than Ξ-finiteness, for which this equality is valid, we introduce now the notion of $S\Pi$-integrability.

Definition 7.20 An internal function $F \in \mathbf{F}_{\Pi}$ is said to be $S\Pi$**-integrable** if and only if it satisfies the following conditions:

1. the sum $\sum_{x \in \Pi} \epsilon |F(x)|$ is finite;
2. $\forall_A [A \in \mathcal{A} \Rightarrow [\Xi(A) \approx 0 \Rightarrow \sum_{x \in A} |F(x)| \approx 0]]$;
3. $\forall_A [A \in \mathcal{A} \Rightarrow [F(x) \approx 0, \forall_{x \in A} \Rightarrow \sum_{x \in A} |F(x)| \approx 0]]$.

We denote by $\mathbf{L}^1_{\Pi}$ the subspace of functions in $\mathbf{F}_{\Pi}$ which satisfy condition 1, and by $S\mathbf{L}^1_{\Pi}$ the subspace of $\mathbf{F}_{\Pi}$ formed by all functions which are $S\Pi$-integrable (that is, those functions of $\mathbf{L}^1_{\Pi}$ which satisfy in addition the conditions 2 and 3).

Theorem 7.21 If $\mathbf{F} \in \mathcal{F}_{\Pi}$ is a Ξ-finite function then it belongs to $\mathbf{SL}^1_{\Pi}$.

Proof If F is a Ξ-finite function there exists a real number $M > 0$ such that $|F(x)| \leq M$ for all $x \in \Pi$, and because of this $\Xi(\{x \in \Pi : F(x) \neq 0\})$ is finite. Then

$$\sum_{x \in \Pi} \epsilon |F(x)| = \sum_{x \in \{F \neq 0\}} \epsilon |F(x)|$$

$$\leq M. \sum_{x \in \{F \neq 0\}} \epsilon = M.\Xi(\{x \in \Pi : F(x) \neq 0\})$$

and therefore $\sum_{x\in\Pi} \epsilon|F(x)|$ is finite, or equivalently, $F \in \mathbf{L}^1_\Pi$. Taking $A \in \mathcal{A}$ such that $\Xi(A) \approx 0$ we have

$$\sum_{x\in A} \epsilon|F(x)| \leq M.\Xi(A) \approx 0 \ .$$

Finally, if $A \in \mathcal{A}$ is such that $F(x) \approx 0$ for all $x \in A$ then, taking into account that $\Xi(\{F \neq 0\})$ is finite, by hypothesis,

$$\begin{aligned}\sum_{x\in A} \epsilon|F(x)| &= \sum_{x\in A\cap\{F\neq 0\}} \epsilon|F(x)| \\ &\leq \max_{x\in A} |F(x)|.\Xi(\{x \in \Pi : F(x) \neq 0\}) \approx 0 \ .\end{aligned}$$

It follows then that $F \in \mathbf{SL}^1_\Pi$. •

Note that F is $S\Pi$-integrable if and only if both the functions

$$F_+ \equiv \max\{0, F\} \quad \text{and} \quad F_- \equiv \max\{0, -F\}$$

are $S\Pi$-integrable. Then we have

Theorem 7.22 The internal function $F \in \mathbf{F}_\Pi$ is $S\Pi$-integrable if and only if $\mathbf{st}F$ is Loeb integrable; in this case the equality

$$\mathbf{st}\left(\sum_{x\in A} \epsilon F(x)\right) = \int_A \mathbf{st}F(x)d\Xi_L(x)$$

is true for any internal subset A of Π.

Proof It is enough to prove the theorem when F is a non-negative hyperreal function since the general case can be reduced to this. We consider in the first place the case in which $A = \Pi$. From Theorem 7.19 we have

$$\int_\Pi \mathbf{st}F(x)d\Xi_L(x) \leq \mathbf{st}\left(\sum_{x\in\Pi} \epsilon F(x)\right) .$$

Suppose that this inequality is strict. That is to say, suppose that we have

$$\mathbf{st}\left(\sum_{x\in\Pi} \epsilon F(x)\right) = \alpha \quad \text{and} \quad \int_\Pi \mathbf{st}F(x)d\Xi_L(x) = \beta$$

with $\beta < \alpha$. If for each $n \in \mathbb{N}$ we define

$$F_n(x) = \begin{cases} F(x) & \text{if } \frac{1}{n} \leq F(x) \leq n \\ 0 & \text{if } F(x) < \frac{1}{n} \text{ or } F(x) > n \end{cases}$$

then, from the monotone convergence theorem,

$$\lim_{n\to\infty} \mathbf{st}\left(\sum_{x\in\Pi} \epsilon F_n(x)\right) = \lim_{n\to} \int_\Pi \mathbf{st}F_n(x)d\Xi_L(x) = \int_\Pi \mathbf{st}F(x)d\Xi_L(x) = \beta < \frac{\alpha+\beta}{2} \ .$$

The set

$$\left\{n \in {}^*\mathbb{N} : \sum_{x\in\Pi} \epsilon F_n(x) < \frac{\alpha+\beta}{2}\right\}$$

is internal and contains all the (standard) natural numbers; by overflow it will contain an infinite number $\nu \in {}^*\mathbb{N}_\infty$.

Seeing that, for any hyperreal number $n \in {}^*\mathbb{N}$, we have

$$n\Xi(\{x \in \Pi : F(x) > n\} = n. \sum_{x\in\{F>n\}} \epsilon < \sum_{x\in\{F>n\}} \epsilon F(x) \leq \sum_{x\in\Pi} \epsilon F(x)$$

and that the last member, by hypothesis, is a finite number, then for the infinite hypernatural number $\nu \in {}^*\mathbb{N}_\infty$ chosen above, we have

$$\Xi(\{x \in \Pi : F(x) > \nu\}) \leq \frac{1}{\nu} \sum_{x\in\Pi} \epsilon F(x) \approx 0.$$

Taking into account the $S\Pi$-integrability of the function F we have,

$$\mathbf{st}\left(\sum_{x\in\{F>\nu\}} \epsilon F(x)\right) = 0 \text{ and } \mathbf{st}\left(\sum_{x\in\{F<1/\nu\}} \epsilon F(x)\right) = 0 \ .$$

Therefore

$$\alpha = \mathbf{st}\left(\sum_{x\in\Pi} \epsilon F(x)\right) = \mathbf{st}\left(\sum_{x\in\{1/\nu<F<\nu\}} \epsilon F_\nu(x)\right) +$$

$$+\mathbf{st}\left(\sum_{x\in\{F>\nu\}} \epsilon F(x)\right) + \mathbf{st}\left(\sum_{x\in\{F<1/\nu\}} \epsilon F(x)\right)$$

$$= \mathbf{st}\left(\sum_{x\in\Pi} \epsilon F_\nu(x)\right) \leq \frac{\alpha+\beta}{2} < \alpha \ ,$$

and so the assumption (that the inequality is strict) results in a contradiction.

We now consider the case of an arbitrary internal subset $A \subset \Pi$. Denoting by 1_A the characteristic function of the set A, the function $F1_A$ is $S\Pi$-integrable. We then obtain the result

$$\int_A \mathbf{st}F d\Xi_L = \int_R (\mathbf{st}F)1_A d\Xi_L$$

$$= \int_\Pi (\mathbf{st}F1_A) d\Xi_L = \sum_{x\in\Pi} \epsilon F(x)1_A(x) = \sum_{x\in A} \epsilon F(x) \ .$$

To prove the converse suppose that $\mathbf{st}F$ is Loeb integrable and that the equality is valid. Then F has a finite Π-integral, since the two integrals are infinitely close. Moreover, for each set $A \in \mathcal{A}$ such that $\Xi(A) \approx 0$,

$$\sum_{x\in A} \epsilon F(x) \approx \int_A \mathbf{st}F d\Xi_L = 0 \ .$$

On the other hand, if $A \in \mathcal{A}$ is such that $F(x) \approx 0$ for $x \in A$, then

$$\sum_{x \in A} \epsilon F(x) \approx \int_A \mathbf{st} F d\Xi_L = 0 \quad .$$

Consequently, F is $S\Pi$-integrable. •

According to Theorem 7.16, the projection of any internal function $F \in \mathbf{F}_\Pi$ is $\mathcal{A}_L$-measurable. That is to say, if $\tilde{f} : \Pi \to \mathbb{R}$ is given by an expression of the form $\tilde{f} = \mathbf{st} F$ for some internal function $F \in \mathbf{F}_\Pi$, then $\tilde{f}$ is $\mathcal{A}_L$-measurable. If, in addition, F is $S\Pi$-integrable then, from Theorem 7.22, $\tilde{f}$ is Loeb integrable and,

$$\int_\Pi \tilde{f}(x) d\Xi_L(x) = \mathbf{st}\left(\sum_{x \in \Pi} \epsilon F(x)\right)$$

or equivalently the Loeb integral of the (external) function $\tilde{f}$ can be expressed in the form of a hyperfinite sum which is much easier to work with.

From the point of view of the integral calculus it is not actually necessary to put $\tilde{f} = \mathbf{st} F$: in fact it is enough in the general case that this equality is satisfied Ξ_L-*almost everywhere*. Accordingly we have the following definition:

Definition 7.23 An internal function $F(x)$, $x \in \Pi$, is said to be a **lifting** of the external $\mathcal{A}_L$-measurable function $\tilde{f}(x)$, $x \in \Pi$, if and only if the equality $\tilde{f}(x) = \mathbf{st} F(x)$ is valid in Ξ_L-measure. If, in addition, F belongs to SL^1_Π then it is said to be an **$S\Pi$-integrable lifting** for the function f.

When an external function $\tilde{f}$ possesses a lifting we shall say, for the moment, that we are dealing with a **Ξ-relative internal function.**

Although not all the functions $\tilde{f} : \Pi \to \mathbb{R}$ are Ξ_L-relative internal, these form a very important class for which it is permissible to transfer to the domain of internal functions calculations involving external functions with which it is always much more difficult to work. The following theorem gives an example of this transfer.

Theorem 7.24 Let $f : \mathbb{R} \to \mathbf{C}$ be a (standard) continuous function which is Lebesgue integrable and which tends monotonely to zero at infinity. Then

$$\int_{\mathbb{R}} f d\lambda = \mathbf{st}\left(\sum_{x \in \Pi} \epsilon \, {}^* f_\Pi(x)\right) \quad .$$

Proof Without any loss of generality we need only consider functions of the type $f : \mathbb{R}_0^+ \to \mathbb{R}_0^+$ which are extended to $\bar{\mathbb{R}}_0^+$ by setting $f(+\infty) = 0$. We then have

$$\int_0^{+\infty} f d\lambda = \int_{\Pi_0^+} \tilde{f}(x) d\Xi_L(x)$$

where the integral on the right-hand side is the Loeb integral relative to Loeb counting measure. It is now enough to show that, under the conditions of the theorem, the internal function ${}^*f_\Pi : \Pi_0^+ \to {}^*\mathbb{R}_0^+$ is an $S\Pi$-integrable lifting of the external function $\tilde{f} \equiv f \circ \mathbf{st}_\infty : \Pi_0^+ \to \bar{\mathbb{R}}_0^+$. In the first place we show that

$$\tilde{f}(x) = \mathbf{st} \circ {}^*f_\Pi(x)$$

for all $x \in \Pi_0^+$. In fact,

1. if $x \in \Pi_0^+$ is infinite then, on the one hand, $\tilde{f}(x) = f \circ \mathbf{st}_\infty(x) = 0$; on the other hand, from the conditions imposed on the function f we have ${}^*f_\Pi(x) \approx 0$ for x infinite, and so $\mathbf{st} \circ {}^*f_\Pi(x) = \tilde{f}(x) = 0$.

2. if $x \in \Pi_0^+$ is finite then $\mathbf{st}x \in \mathbb{R}_0^+$; taking into account the continuity of f we would have ${}^*f_\Pi(x) \approx f(\mathbf{st}x)$ and, therefore, $f(x) = f(\mathbf{st}x) = \mathbf{st} \circ {}^*f_\Pi(x)$.

It now remains to be shown that ${}^*f_\Pi$ belongs to SL^1_Π, that is that it satisfies the three following conditions:

(a) $\sum_{\Pi_0^+} \epsilon\, {}^*f_\Pi(x)$ is finite,

(b) for all $A \in \mathcal{A}$, if $\Xi(A) \approx 0$ then $\sum_A \epsilon\, {}^*f_\Pi(x) \approx 0$.

(c) for all $A \in \mathcal{A}$, if ${}^*f_\Pi(x) \approx 0$ then for every $x \in A$, $\sum_A \epsilon\, {}^*f_\Pi(x) \approx 0$.

Seeing that ${}^*f_\Pi(x)$, $x \in \Pi$, is finitely bounded, then taking into account that

$$\sum_A \epsilon\, {}^*f_\Pi(x) \leq \{\max_{x\in A} |{}^*f_\Pi(x)|\}.\Xi(A)$$

condition (b) follows at once.

The proof now proceeds with the following Lemma:

Lemma 7.25 The hyperfinite sum

$$\sum_{x=\gamma_1}^{\gamma_2} \epsilon\, {}^*f_\Pi(x)$$

s infinitesimal for any points $\gamma_1, \gamma_2 \in \Pi_\infty^+$ (with $\gamma_1 \leq \gamma_2 < \frac{\kappa}{2}$).

Proof of Lemma 7.25 Without loss of generality we may assume that γ_1 and γ_2 re infinite hyperintegers: in fact if this were not so, that is, if we were to have a sum f the form

$$\sum_{x=\gamma_1'}^{\gamma_2'} \epsilon\, {}^*f_\Pi(x)$$

hen defining

$$\gamma_1 = \max\{\gamma \in {}^*\mathbb{N}_\infty : \gamma \leq \gamma_1'\} \quad \text{and} \quad \gamma_2 = \min\{\gamma \in {}^*\mathbb{N}_\infty : \gamma \geq \gamma_2'\}$$

e would obtain the inequality

$$\sum_{x=\gamma_1'}^{\gamma_2'} \epsilon\, {}^*f_\Pi(x) \leq \sum_{x=\gamma_1}^{\gamma_2} \epsilon\, {}^*f_\Pi(x)$$

which shows that if the right-hand member is infinitesimal then so also is the left-hand one. Since

$$\sum_{x=\gamma_1}^{\gamma_2} \epsilon\, {}^*f_\Pi(x) = \sum_{j=\gamma_1\kappa}^{\gamma_2\kappa} \epsilon\, {}^*f_\Pi(j\epsilon) = \sum_{n=\gamma_1}^{\gamma_2} \left\{ \sum_{m=0}^{\kappa-1} \epsilon\, {}^*f_\Pi(n+m\epsilon) \right\}$$

and since f is monotone, we have

$$\sum_{x=\gamma_1}^{\gamma_2} \epsilon\, {}^*f_\Pi(x) \leq \sum_{n=\gamma_1}^{\gamma_2} \epsilon\, {}^*f_\Pi(n) \left\{ \sum_{m=0}^{\kappa-1} \epsilon \right\} = \sum_{n=\gamma_1}^{\gamma_2} \epsilon\, {}^*f_\Pi(n) \quad .$$

From the integral test for convergence we know that the series and the integral

$$\sum_{n=1}^{\infty} f(n) \quad \text{and} \quad \int_0^{+\infty} f(x)dx$$

have the same behaviour. Since the function f is integrable then the series is convergent. From the nonstandard treatment of the convergence of series we can then conclude that the hyperfinite sum

$$\sum_{n=\gamma_1}^{\gamma_2} \epsilon\, {}^*f_\Pi(n)$$

is infinitesimal for any choice of the infinite hypernatural numbers γ_1 and γ_2 (with $\gamma_1 \leq \gamma_2 < \kappa/2$).

Let $r > 0$ now be an arbitrary given real number and define

$$M_r = \left\{ n \in {}^*\mathbb{N} : \sum_{j=n\kappa}^{\frac{\kappa^2}{2}-1} \epsilon\, {}^*f_\Pi(j\epsilon) < r \right\} \quad .$$

From Lemma 7.25 it follows that M_r contains arbitrarily small positive infinite numbers; consequently, by underflow, M_r contains a finite number $n_r \in \mathbb{N}$ such that

$$\forall_n [n \in {}^*\mathbb{N} : [n_r \leq n \leq \kappa/2 \Rightarrow \sum_{j=n\kappa}^{\kappa^2/2-1} \epsilon\, {}^*f_\Pi(j\epsilon) < r]].$$

Since

$$\sum_{x\in\Pi_0^+} \epsilon\, {}^*f_\Pi(x) = \sum_{j=0}^{n_r\kappa-1} \epsilon\, {}^*f_\Pi(j\epsilon) + \sum_{j=n_r\kappa}^{\kappa^2/2-1} \epsilon\, {}^*f_\Pi(j\epsilon) < \sum_{j=0}^{n_r\kappa-1} \epsilon\, {}^*f_\Pi(j\epsilon) + r$$

and in addition we have

$$\sum_{j=0}^{n_r\kappa-1} \epsilon\, {}^*f_\Pi(j\epsilon) \leq n_r \{ \max_{0\leq x\leq n_r} ({}^*f_\Pi(x)) \} < +\infty$$

hen (a) follows immediately.

In order to prove (c) we proceed in the following way: (i) if $\Xi(A)$ is finite it follows that

$$\sum_{x\in A} \epsilon\, {}^*f_\Pi(j\epsilon) = \{\max_{x\in A}({}^*f_\Pi(x))\}.\Xi(A) \approx 0.$$

(ii) if $\Xi(A)$is not finite then A certainly contains a remote point in Π^+_∞.

From Lemma 7.25 it follows that for each real number $r > 0$ there exists a (standard) number $n_r \in \mathbb{N}$ such that

$$\sum_{x\in A\cap[n_r\kappa,\kappa/2-\epsilon]} \epsilon\, {}^*f_\Pi(x) < r$$

whence

$$\sum_{x\in A\cap[0,n_r\kappa-\epsilon]} \epsilon\, {}^*f_\Pi(x) \approx 0 \quad .$$

Then, finally,

$$\sum_{x\in A} \epsilon\, {}^*f_\Pi(x) < r,$$

and the result follows since the real number $r > 0$ is arbitrary. •

Chapter 8

Hyperfinite Representation of Distributions

In Chapter 5, section 5.2, we gave an elementary presentation of the way in which NSA could be used to represent certain types of linear functionals (defined over suitable function spaces) in terms of simple internal functions. The so-called *"functions"* of Heaviside and Dirac are two examples of a much larger class of objects, the so-called **distributions** of Laurent Schwartz. These include, in particular, all functions locally integrable over $\mathbb{R}$. Returning once again to this subject we consider it now in a more general form framed in the context of hyperfinite analysis. Every Schwartz distribution over $\mathbb{R}$ can be represented, although not uniquely, by an internal function of $\mathbf{F}_{\Pi}$ with appropriate characteristics.

8.1 Sketch of the Theory of Distributions

The standard theory of Schwartz distributions is based fundamentally on the notion of topological vector space: we therefore begin with a resumé of the development of the concept of distribution as an element of the topological dual of a suitable space of test functions. For deeper study of the theory of distributions reference may be made to any of the standard books dealing with this special subject, in particular to references [53], [10] and [17].

8.1.1 Test function spaces

A non-empty set $\mathcal{V}$ is a **topological vector space** if and only if the following conditions are satisfied:

1. $\mathcal{V}$ is a vector space.
2. $\mathcal{V}$ is a topological space.
3. The algebraic operations of $\mathcal{V}$ - addition and multiplication by a scalar - are continuous relative to the topology of $\mathcal{V}$.

Since $\mathcal{V}$ is a vector space its topology may be completely specified if we are given a base of neighbourhoods of the origin: that is to say, a family $N(0)$ of neighbourhoods of $0 \in \mathcal{V}$ such that every neighbourhood of 0 contains an element of $N(0)$. For any other point $x \in \mathcal{V}$ the family of sets $x + N(0)$ constitutes a base of neighbourhoods of x. When $N(0)$ is a countable system each point $x \in \mathcal{V}$ possesses a countable neighbourhood base: it is then said that $\mathcal{V}$ satisfies the **first countability axiom.** This property is particularly important since, in such spaces, topological relations can be established in terms of the language of sequences.

A subset A of $\mathcal{V}$ is said to be **convex** if $x \in A$ and $y \in A$ always imply that $\alpha x + (1-\alpha)y \in A$ for every real number α such that $0 < \alpha < 1$. The vector space $\mathcal{V}$ is said to be **locally convex, (l.c.)**, if every neighbourhood of the origin contains a convex neighbourhood of the origin.

A **space of test functions** is a topological vector space whose elements are functions ϕ defined on some subset of $\mathbb{R}$ (or, more generally, of $\mathbb{R}^n$ for some $n \in \mathbb{N}$) with values in $\mathbb{C}$. The most important test function spaces in this context are those which have the structure of a so-called **Fréchet space**, or which are a (countable) union of such spaces. We recall first the concept of **Banach space**.

Banach spaces Topological vector spaces constitute, in a certain sense, a natural generalisation of the relatively familiar concept of a normed vector space.

A vector space $\mathcal{V}$ is said to be normed if there exists a function

$$\|.\| : \mathcal{V} \to \mathbb{R}$$

which to each $x \in \mathcal{V}$ assigns a real number $\|x\|$ (called the **norm** of x) with the following properties:

1. $\|x\| = 0 \quad \Rightarrow x = 0$ for any $x \in \mathcal{V}$,
2. $\|x + y\| \leq \|x\| + \|y\|$ for any $x, y \in \mathcal{V}$.
3. $\|\lambda x\| = |\lambda|.\|x\|$ for each $x \in \mathcal{V}$ and every scalar $\lambda \in \mathbb{C}$.

The norm $\|.\|$ generates a topology on $\mathcal{V}$: for each real number $r > 0$ the set defined by

$$N_r = \{x \in \mathcal{V} : \|x\| < r\}$$

is a neighbourhood of the origin. This topology is locally convex and satisfies the first countability axiom (each point has a countable base). The topological relations can therefore be described in terms of convergent sequences.

A sequence $(x_n)_{n\in\mathbb{N}}$ of elements of $\mathcal{V}$ is said to be a **Cauchy sequence** if and only if to each real number $r > 0$ we can assign an integer $n_r \in \mathbb{N}$ such that

$$\forall_{m,n}[m, n \in \mathbb{N} \Rightarrow [m, n \geq n_r \Rightarrow \|x_m - x_n\| < r]] \quad .$$

The normed vector space $(\mathcal{V}, \|.\|)$ is said to be a **Banach space** if it is complete relative to the metric induced by the norm: that is, if every Cauchy sequence of elements of $\mathcal{V}$ converges to an element of $\mathcal{V}$.

Particularly important in this context is the space of functions infinitely differentiable on $\mathbb{R}$ and with compact support. If K is any compact in $\mathbb{R}$ then $\mathcal{C}_c^\infty(K)$ denotes the set of all functions of class $\mathcal{C}^\infty$ on $\mathbb{R}$ with support contained in the compact K. $\mathcal{C}_c^\infty(K)$ is a vector space and, for each (fixed) $p \in \mathbb{N}_0$, the function

$$\|\phi\|_p = \max_{0\leq r\leq p}\{\sup_{x\in K}|\phi^r(x)|\} \quad , \quad \phi \in \mathcal{C}_c^\infty(K) \tag{8.1}$$

defines a norm on $\mathcal{C}_c^\infty(K)$ which induces on this space a topology of sets of the form

$$N_{p,r}(0) = \{\phi \in \mathcal{C}_c^\infty(K) : \|\phi\|_p < r\}$$

defined for all real numbers $r > 0$. These constitute a neighbourhood base at the origin.

The normed vector space $(\mathcal{C}_c^\infty(K), \|\phi\|_p)$ is not a Banach space; it is not complete. Completing it, however, we obtain the space $(\mathcal{C}_c^p(K), \|.\|_p)$ formed of all the functions of class $\mathcal{C}^p$ (p-times continuously differentiable on $\mathbb{R}$) with compact support contained in K. The topological vector space thus obtained is usually denoted more simply by $\mathcal{D}_{K,p}$.

Fréchet Spaces Suppose that $(\|.\|_p)_{p\in\mathbf{N}}$ is a sequence of norms defined on the vector space $\mathcal{V}$. This sequence determines on $\mathcal{V}$ an infinity of topological structures - one for each $p \in \mathbb{N}_0$. We now attempt to define on $\mathcal{V}$ a topology which will be, in a sense, the synthesis of all those topologies: such a topology can be defined by choosing for a base of neighbourhoods of the origin the sets of the form

$$N_{p,r} = \{x \in V : \|x\|_0 < r, \|x\|_1 < r, \ldots, \|x\|_p < r\}$$

where p is a non-negative integer and $r > 0$ is a real number. $\mathcal{V}$ is thereby equipped with the structure of a locally convex topological vector space.

A sequence of vectors $(x_n)_{n\in\mathbf{N}}$ converges to the null vector in $\mathcal{V}$ if and only if $\lim_{n\to\infty} \|x_n\|_p = 0$ for every $p \in \mathbb{N}_0$; in a similar way, $(x_n)_{n\in\mathbf{N}}$ is a Cauchy sequence if and only if $\|x_m - x_n\|_p$ tends to zero when m and n tend to infinity, for every $p \in \mathbb{N}_0$.

The topological vector space $\mathcal{V}$ with the **multinorm** $(\|.\|_p)_{p\in\mathbf{N}_0}$ can be given the structure of a metric space by defining

$$d(x,y) = \sum_{p=0}^{\infty} 2^{-p} \frac{\|x-y\|_p}{1 + \|x-y\|_p}$$

and it can be verified that the topology referred to above is equivalent to the topology induced by this metric. If, in addition, the topological vector space $\mathcal{V}$ is complete, then we say that it is a **Fréchet space.**

Considering again the case of the space $\mathcal{C}_c^\infty(K)$ and the family of norms $(\|.\|)_{p\in\mathbf{N}_0}$ defined in (8.1) we obtain, for each compact $K \subset \mathbb{R}$, a Fréchet space: we denote this space more simply by $\mathcal{D}_K$. The space $\mathcal{D}_K$ can be obtained as the countable intersection

$$\mathcal{D}_K = \bigcap_{p=0}^{\infty} \mathcal{D}_{K,p}$$

where $(\mathcal{D}_{K,p})_{p\in\mathcal{N}_0}$ is a decreasing sequence of Banach spaces.

Strict inductive limits of topological spaces Suppose we are given a sequence $(\mathcal{V}_m)_{m\in\mathbf{N}}$ of topological vector spaces satisfying the following conditions,

1. $\mathcal{V}_1 \subset \mathcal{V}_2 \subset \ldots \mathcal{V}_m \subset \ldots$
2. each $\mathcal{V}_m$ is locally convex, and

3. the topology of $\mathcal{V}_m$ is equal to the topology which is induced on $\mathcal{V}_m$ by that of $\mathcal{V}_{m+1}$.

The vector space defined for the countable union

$$\mathcal{V} = \bigcup_{m=1}^{\infty} \mathcal{V}_m$$

can be equipped with the so-called **inductive limit topology** by defining a neighbourhood base of the origin in $\mathcal{V}$ as the family of convex sets $N \subset \mathcal{V}$ such that, for each $m = 1, 2, \ldots$, the sets $N \cap \mathcal{V}_m$ constitute a neighbourhood base of the origin in the locally convex topological vector space $\mathcal{V}_m$.

An important example of an inductive limit of topological spaces in the immediate context is given by the space of test functions $\mathcal{D} \equiv \mathcal{D}(\mathbb{R})$ formed by all the functions infinitely differentiable on $\mathbb{R}$ with compact support. Let $(K_m)_{m\in\mathbb{N}}$ be a sequence of compacts in $\mathbb{R}$ such that

1. $K_1 \subset K_2 \subset \cdots \subset K_m \subset \ldots$,
2. $\mathbb{R} = \bigcup_{m\in\mathbb{N}} K_m$.

The space of test functions

$$\mathcal{D} = \bigcup_{m=1}^{\infty} \mathcal{D}_{K_m}$$

is equipped with the inductive limit topology defined for the sequence of Fréchet spaces $(\mathcal{D}_{K_m})_{m\in\mathcal{N}}$: a subset N of $\mathcal{D}$ is a neighbourhood of the origin in $\mathcal{D}$ if and only if, for each compact $K \subset \mathbb{R}$, the subset $N \cap \mathcal{D}_K$ is a neighbourhood of the origin in $\mathcal{D}_K$. Then a sequence $(\phi_n)_{n\in\mathcal{N}}$ of test functions converges in $\mathcal{D}$ to zero if and only if there exists a fixed compact $K \subset \mathbb{R}$ such that

1. $\phi_n \in \mathcal{D}_K$ for all $n \in \mathbb{N}$,
2. $\phi_n \to 0$ (when $n \to \infty$) in the space $\mathcal{D}_K$, that is to say,
3. $\phi_n^{(j)} \to 0$ (when $n \to \infty$) uniformly for each $j \in \mathbb{N}_0$.

8.1.2 Duality and Schwartz distributions

Given a topological vector space $\mathcal{V}$ (over $\mathbb{C}$), a **functional** defined over $\mathcal{V}$ is any mapping $\nu : \mathcal{V} \to \mathbb{C}$ which to each $x \in \mathcal{V}$ assigns a complex number, generally denoted by $< \nu, x >$ or $\nu[x]$. The functional is said to be **linear** if it satisfies the equality

$$< \nu, x + \alpha y > = < \nu, x > + \alpha < \nu, y >$$

for all $x, y \in \mathcal{V}$ and $\alpha \in \mathbb{C}$ and is said to be **continuous** at a point $x_0 \in \mathcal{V}$ if for each real number $r > 0$ there exists a neighbourhood $N_r(x_0)$ of $x_0 \in \mathcal{V}$ such that

$$| < \nu, x > - < \nu, x_0 > | < r$$

for any $x \in N_r(x_0)$. The functional ν is said simply to be continuous if and only if it is continuous at all the points of $\mathcal{V}$. If the functional ν is linear then it is continuous if and only if it is continuous at the origin.

The set of all linear continuous functionals over $\mathcal{V}$, generally denoted by $\mathcal{V}'$, is called the **topological dual** of the space $\mathcal{V}$.

In the particular case when $\mathcal{V}$ is the space $\mathcal{D}$ of infinitely differentiable functions of compact support in $\mathbb{R}$, the elements of the dual topology $\mathcal{D}'$ are called **Schwartz distributions** over $\mathbb{R}$. It can be shown that a linear functional over $\mathcal{D}$ is continuous if and only if it is locally bounded. This is the content of the following theorem whose proof can be found in any of the specialised texts on distributions.

Theorem 8.1 A linear functional $\nu : \mathcal{D} \to \mathbb{C}$ is a Schwartz distribution if and only if it satisfies one of following equivalent conditions:

1. If $(\phi_n)_{n\in\mathbb{N}}$ is a sequence of functions of $\mathcal{D}$ converging to zero (relative to the inductive limit topology) then the sequence of numbers $(<\nu,\phi_n>)_{n\in\mathbb{N}}$ converges to zero in $\mathbb{C}$.

2. For each compact $K \subset \mathbb{R}$ there exists an integer $p \geq 0$ and a constant $M > 0$ such that the inequality

$$|<\nu,\phi>| \leq M.\|\phi\|_p \tag{8.2}$$

is satisfied for every function $\phi \in \mathcal{D}$ whose support is contained in K.

For each compact $K \subset \mathbb{R}$ let $p_K \in \mathbb{N}_0$ be the smallest integer for which the inequality (8.2) is satisfied, and let $\{p_k\}_{K\subset\mathbb{R}}$ be the family corresponding to all the compacts of $\mathbb{R}$. If this family has a maximum (in $\mathbb{N}_0$) then ν is said to be a distribution of **finite order** equal to $\max_{K\subset\mathbb{R}} p_K$; otherwise, ν is said to be a distribution of **infinite order**.

Examples of distributions A locally integrable function f defines a distribution ν_f over $\mathbb{R}$, if we set

$$\nu_f[\phi] \equiv <\nu_f,\phi> = \int_{-\infty}^{+\infty} f(x)\phi(x)dx$$

for every function $\phi \in \mathcal{D}$. In fact, ν_f is clearly a linear functional and if ϕ has support in the compact $K = [a,b]$ then

$$|<\nu_f,\phi>| \leq \left\{\sup_{a\leq x\leq b} |\phi(x)|\right\} \int_a^b |f(x)|dx = M.\|\phi\|_0$$

where $M \equiv M([a,b])$ is a positive constant which does not depend on the function ϕ. $\nu_f \in \mathcal{D}'$ is thus a distribution of order zero.

A distribution generated by a locally integrable function in the way described above is said to be **regular**. Any distribution which cannot be generated by this process in terms of a function locally integrable over $\mathbb{R}$ is said to be **singular**. The linear functional over $\mathcal{D}$ defined by

$$\phi \mapsto \delta[\phi] \equiv <\delta,\phi> = \phi(0)$$

is a singular distribution which will be recognised as what is usually described as the delta "function". Strictly, it should be called the delta *distribution*. Since, for any $\phi \in \mathcal{D}$,

$$|<\delta,\phi>| = |\phi(0)| \leq \|\phi\|_0$$

it follows that δ is a singular distribution of order zero.

Another example of a singular distribution is associated with the function x^{-1}. Since x^{-1} is not locally integrable it cannot generate a regular distribution; for an arbitrary $\phi \in \mathcal{D}$, the integral

$$\int_{-\infty}^{+\infty} \frac{\phi(x)}{x} dx$$

is not generally convergent. We can, however, use this integral to generate a singular distribution simply by taking its *"Cauchy Principal Value"*.

Denoting the distribution generated by this process by $\mathbf{x}^{-1} \equiv Pv.[x^{-1}]$ we would have, for some $b > 0$ (dependent on ϕ),

$$\mathbf{x}^{-1}[\phi] \equiv < \mathbf{x}^{-1}, \phi > = Pv. \int_{-\infty}^{+\infty} \frac{\phi(x)}{x} dx$$

$$= \lim_{r \downarrow 0} \left\{ \int_{+r}^{+\infty} \frac{\phi(x)}{x} + \int_{-\infty}^{-r} \frac{\phi(x)}{x} dx \right\}$$

$$= \lim_{r \downarrow 0} \int_{r}^{b} \frac{\phi(x) - \phi(-x)}{x} dx,$$

for any $x \in \mathbb{R}$. From the Mean Value Theorem we have

$$\phi(x) = \phi(0) + x.\phi'(\xi_+) \quad \text{and} \quad \phi(-x) = \phi(0) - x.\phi'(\xi_-)$$

where $\min\{0, x\} < \xi_+ < \max\{0, x\}$ and $\min\{0, -x\} < \xi_- < \max\{0, -x\}$. Then,

$$< \mathbf{x}^{-1}, \phi > = \lim_{r \downarrow 0} \int_{r}^{b} [\phi'(\xi_+) + \phi'(\xi_-)] dx$$

$$= \int_{0}^{b} [\phi'(\xi_+) + \phi'(\xi_-)] dx,$$

and therefore

$$| < \mathbf{x}^{-1}, \phi > | \leq 2b.\|\phi\|_1$$

which means that $\mathbf{x}^{-1}$ is a singular distribution of order 1.

Algebraic operations with distributions Two distributions $\nu_1, \nu_2 \in \mathcal{D}'$ are said to be equal if and only if the equality

$$< \nu_1, \phi > = < \nu_2, \phi >$$

is satisfied for every function $\phi \in \mathcal{D}$.

The **sum** of ν_1 and ν_2, denoted by $\nu_1 + \nu_2$, is the distribution defined by

$$< \nu_1 + \nu_2, \phi > = < \nu_1, \phi > + < \nu_2, \phi >$$

for every $\phi \in \mathcal{D}$. The **product** of a scalar $a \in \mathbf{C}$ with a distribution $\nu \in \mathcal{D}'$, denoted by $a\nu$, is the distribution defined by

$$< a\nu, \phi >= a < \nu, \phi >$$

for every $\phi \in \mathcal{D}$. With these operations $\mathcal{D}$ is a vector space over $\mathbf{C}$ where the null distribution $\mathbf{0}$ is defined by

$$< \mathbf{0}, \phi >= 0, \quad \text{for every} \quad \phi \in \mathcal{D}.$$

A distribution $\nu \in \mathcal{D}'$ is said to be null over an arbitrary open set $A \subset \mathbb{R}$ if and only if we have

$$< \nu, \phi >= 0$$

for every function $\phi \in \mathcal{D}$ with support contained in the open set A. We call the **support** of $\nu \in \mathcal{D}'$ the complement of the largest open set $A \subseteq \mathbb{R}$ on which ν is null. Thus, the support of a distribution is always a closed set in $\mathbb{R}$. If, in addition, the support of ν is bounded then ν is said to be a **distribution of compact support**. The distribution δ, for example, is a distribution of compact support: it is null in every open set which does not contain the origin (which is therefore the support of δ).

Given a distribution $\nu \in \mathcal{D}'$ and a number $a \in \mathbb{R}$ we define the **translate** of ν to be the distribution denoted by $\tau_a\nu$ which is defined by

$$< \tau_a\nu, \phi >=< \nu, \tau_{-a}\phi >=< \nu(x), \phi(x+a) >$$

for all $\phi \in \mathcal{D}$.

Generalised differentiation ν being a distribution in $\mathcal{D}'$ we define the derivative of ν in the distributional (or generalised) sense to be the distribution $D\nu$ given by

$$< D\nu, \phi >=< \nu, (-\phi') >$$

for all $\phi \in \mathcal{D}$ (where ϕ' denotes the derivative of ϕ in the usual sense). Taking into account the regular properties of the functions ϕ it is immediately clear that, in this sense, *every distribution is infinitely differentiable.*

Let f be a function continuously differentiable (in the usual sense). Then f generates a regular distribution ν_f defined by

$$< \nu_f, \phi >= \int_{-\infty}^{+\infty} f(x)\phi(x)dx$$

for every $\phi \in \mathcal{D}$. The generalised derivative $D\nu_f$ is given by

$$\begin{aligned} < D\nu_f, \phi > &=< \nu_f, (-\phi') > \\ &= \int_{-\infty}^{+\infty} f(x)[-\phi'(x)]dx \\ &= \int_{-\infty}^{+\infty} f'(x)\phi(x)dx =< \nu_{f'}, \phi > \end{aligned} \qquad (8.3)$$

for every $\phi \in \mathcal{D}$, where $\nu_{f'}$ is the regular distribution generated by the continuous function f'. From (8.3) it is easy to understand that the generalised derivative of the regular distribution generated by a continuously differentiable function f is equal to the regular distribution generated by the usual derivative f' of the function f. In this case we can say that differentiation in $\mathcal{D}'$ forms a generalisation of the (usual) notion of derived function.

Every continuous function f defined on $\mathbb{R}$ is locally integrable and therefore defines a regular distribution ν_f which, without any danger of confusion, we may identify with the continuous function which itself generates it. In this sense we may consider $\mathcal{C} \equiv \mathcal{C}(\mathbb{R})$, the set of all functions continuous on $\mathbb{R}$, as a vector subspace of $\mathcal{D}'$. Then every function $f \in \mathcal{C} \subset \mathcal{D}'$ is infinitely differentiable in the distributional sense.. Conversely we have the following result. [1]

Theorem 8.2 (Structure Theorem) Locally, every distribution is a derivative of finite order of a continuous function.

This theorem means that for any distribution $\nu \in \mathcal{D}'$ whatsoever, the following is always the case: given a compact $K \subset \mathbb{R}$, there always exists an integer $p_K \in \mathbb{N}_0$ and a function f_K defined and continuous on an open neighbourhood of K, such that

$$< \nu, \phi >= (-1)^{p_K} \int_{-\infty}^{+\infty} f_K(x)\phi^{(p_K)} dx$$

for every $\phi \in \mathcal{D}_K$. Or, equivalently, on $K \subset \mathbb{R}$ the distribution is the derivative of order $p_K \in \mathbb{N}_0$ of the continuous function f.

If ν is a distribution of finite order then there exists a number $p \in \mathbb{N}_0$ and a function $f \in \mathcal{C}$ which does not depend on K, such that $\nu = D^p f$, satisfying 8.2 not merely locally, but also globally. We generally denote by $\mathcal{D}'_{fin}$ the subspace of $\mathcal{D}'$ formed by all distributions over $\mathbb{R}$ of finite order.

Example 8.3 $(a_n)_{n \in \mathbf{N}}$ being a sequence of real numbers which tends in absolute value to infinity (as $n \to \infty$), the distribution given by

$$\nu \equiv \sum_{n=1}^{\infty} D^{n-1}(\tau_{a_n}\delta)$$

is, in each compact, a derivative of finite order of a continuous function, but it is not globally a derivative (of finite order) of a continuous function defined on $\mathbb{R}$. We treat it as an example of a distribution of *infinite* order.

8.1.3 Axiomatic definition of finite order distributions

The fact that, locally, every distribution is a derivative of finite order of a continuous function led the Portuguese mathematician Jose Sebastiao e Silva (1914-1972) to propose a very simple axiomatic definition for finite order distributions over the real line. This method of constructing the theory of distributions, which we outline briefly

[1] A proof of this theorem can be found in any standard text of the theory of distributions, for example in [52].

below, forms the basis of the treatment adopted in the book by J.Campos Ferreira [10].

Every function $f \in \mathcal{C}$ has a primitive on $\mathbb{R}$: in fact, for any arbitrarily fixed $a \in \mathbb{R}$, the function defined by

$$g(x) = \int_a^x f(t)dt, \qquad x \in \mathbb{R}$$

is a primitive of f belonging to $\mathcal{C}$. If $\mathcal{I}_a$ denotes the indefinite integration operator with origin at the point a, then $g = \mathcal{I}_a f$ and

$$\mathcal{C}^1 = \{\mathcal{I}_a f + c : f \in \mathcal{C}\} \ .$$

More generally, if we define successive powers of the operator $\mathcal{I}_a$ by

$$\mathcal{I}_a^0 f = f, \ \ \mathcal{I}_a^{n+1} f = \mathcal{I}_a(\mathcal{I}_a^n f), \ \ n = 0, 1, 2, \ldots$$

and denote by P_n the set of polynomial functions with degree less than n (setting $P_0 = \{0\}$) then

$$\mathcal{C}^n = \{\mathcal{I}_a^n f + p : f \in \mathcal{C}, p \in P_n\} \ .$$

Further, setting

$$\mathcal{C}^\infty = \bigcap_{n=1}^{\infty} \mathcal{C}^p$$

we get the following chain of (strict) inclusions

$$\mathcal{C}^\infty \subset \ldots \subset \mathcal{C}^{n+1} \subset \mathcal{C}^n \subset \ldots \subset \mathcal{C}^1 \subset \mathcal{C} \ . \tag{8.4}$$

The definition of an operator of generalised differentiation (which coincides with the usual differentiation in $\mathcal{C}^1$) allows us to extend this chain of inclusions in the opposite sense, creating superspaces of the space $\mathcal{C}$ of continuous functions whose elements are the successive generalised derivatives of those functions. We obtain in this way the chain

$$\mathcal{C}^\infty \subset \ldots \subset \mathcal{C}^n \subset \ldots \subset \mathcal{C}^1 \subset \mathcal{C} \subset \mathcal{C}_1 \subset \ldots \subset \mathcal{C}_m \subset \ldots \subset \mathcal{C}_\infty \ . \tag{8.5}$$

Thus in the axiomatic theory of J.Sebastiao e Silva the **finite order distributions** over $\mathbb{R}$ are the elements of a space, denoted by $\mathcal{C}_\infty$, in which there is defined a mapping

$$D : \mathcal{C}_\infty \to \mathcal{C}_\infty$$

and which satisfies the following axioms:

D1. Every function in $\mathcal{C}$ is a distribution over $\mathbb{R}$.

D2. To each distribution ν over $\mathbb{R}$ there corresponds a distribution $D\nu$, called the **derivative** of ν, such that if $\nu = f$ with $f \in \mathcal{C}^1$ then $D\nu = f'$.

D3. If ν is a distribution over $\mathbb{R}$ then there exists a function $g \in \mathcal{C}$ and a number $r \in \mathbb{N}_0$ such that $\nu = D^r g$.

D4. Given two functions $f, g \in \mathcal{C}$ and a natural number $r \in \mathbb{N}_0$, the equality $D^r f = D^r g$ is satisfied if and only if $f - g$ is a polynomial of degree less than n.

The model of J.Sebastiao e Silva. To show that the set of axioms **D1, D2, D3** and **D4** is consistent, J.Sebastiao e Silva constructed the following simple, algebraic model:

Consider the Cartesian product $\mathbb{N}_0 \times \mathcal{C}$ together with a relation $[\equiv]$ defined for ordered pairs $(r, f), (s, g) \in \mathbb{N}_0 \times \mathcal{C}$ by

$$(r, f)[\equiv](s, g) \Leftrightarrow \mathcal{I}_a^s f - \mathcal{I}_a^r g \in P_{r+s} \tag{8.6}$$

where $a \in \mathbb{R}$ is an arbitrary point (fixed once for all),

Exercise 8.4 *Show that the relation $[\equiv]$ defined in (8.6) is an equivalence relation.*

Taking into account the result of exercise 8.4 we can decompose the set $\mathbb{N}_0 \times \mathcal{C}$ into equivalence classes

$$[r, f] = \{(s, g) \in \mathbb{N}_0 \times \mathcal{C} : (s, g)[\equiv](r, f)\}$$

and denote by $\mathcal{C}_\infty \equiv \mathcal{C}(\mathbb{R})$ the quotient set

$$\mathcal{C}_\infty = (\mathbb{N}_0 \times \mathcal{C})/[\equiv] \quad .$$

Now consider the mapping

$$\iota : \mathcal{C} \to \mathcal{C}_\infty$$

defined by $f \mapsto \iota(f) = [0, f]$: for any two functions $f, g \in \mathcal{C}$ the equality $\iota(f) = \iota(g)$ implies that $(0, f)[\equiv](0, g)$ which, in its turn, means that $f = g$. Thus ι is an injective mapping which embeds $\mathcal{C}$ into $\mathcal{C}_\infty$. Identifying $\mathcal{C}$ with $\iota(\mathcal{C}) \subset \mathcal{C}_\infty$ gives a verification of Axiom **D1**.

Now define the **generalised differentiation** operator

$$D : \mathcal{C}_\infty \to \mathcal{C}_\infty$$

by setting for each $r \in \mathbb{N}_0$ and each $f \in \mathcal{C}$

$$D[r, f] = [r + 1, f] \quad .$$

This mapping does not depend on the particular representative chosen for the class. In fact if we were to have $(r, f)[\equiv](s, g)$ then

$$\mathcal{I}_a^s - \mathcal{I}_a^r g \in \mathcal{P}_{r+s}$$

and therefore

$$\mathcal{I}_a^{s+1} f - \mathcal{I}_a^{r+1} g \in \mathcal{P}_{r+s+1} \subset \mathcal{P}_{r+s+2}$$

whence we would get the following chain of implications

$$[r, f] = [s, g] \Rightarrow [r + 1, f] = [s + 1, g] \Rightarrow D[r, f] = D[s, g] \quad .$$

The class $[r + 1, f] \in \mathcal{C}_\infty$ is the generalised derivative of the class $[r, f]$. If f belongs to $\mathcal{C}^1 \subset \mathcal{C}$ then, since $f - \mathcal{I}_a f'$ is a polynomial constant function, we would have

$$D[0, f] = [1, f] = [0, f'] = \iota(f') \quad (\equiv f')$$

thus verifying the axiom **D2**.

Finally, for each $[r, f] \in \mathcal{C}_\infty$ we have

$$[r, f] = D[r-1, f] = D^2[r-2, f] = \ldots = D^r[0, f] = D^r \iota(f) \equiv D^r f$$

which shows that axiom **D3** is also satisfied.

Thus $\mathcal{C}_\infty$, together with the operator D and with the identification of $\mathcal{C}$ with $\iota(\mathcal{C}) \subset \mathcal{C}_\infty$, constitutes a model for the axiomatic theory of J.Sebastiao e Silva, and the following result may be established (see [10]):

Theorem 8.5 Every model for the axioms **D1, D2, D3** and **D4** of J.Sebastiao e Silva for finite order distributions is isomorphic with $\mathcal{C}_\infty$.

From this theorem it follows that the set $\mathcal{D}'_{fin}$ of all Schwartz distributions of finite order is isomorphic with $\mathcal{C}_\infty$.

8.2 *Finite Representation of Distributions

Since classical Mathematical Analysis does not allow the use of infinite or infinitesimal numbers phenomena which involve the actual infinite, whether directly or indirectly, do not admit simple direct representation in terms of ordinary functions. This makes it necessary to introduce distributions and other generalised functions either as functionals in the theory of Schwartz, or as formal derivatives in the theory of J.Sebastiao e Silva, for example. In NSA which has a much richer numerical universe, it is possible to present these theories in a more direct form: distributions (and other generalised functions) together with the various operations which we carry out on them, can be represented as functions in the usual sense of the term and with the corresponding operations upon them. In what follows we develop such a nonstandard approach to distributions in the context of the hyperfinite line rather than that of the hyperreal continuum. This has the additional advantage of treating these objects in discrete terms.

8.2.1 Introduction

Recall that a function $F \in \mathbf{F}_\Pi$ is said to be $S\Pi$-continuous if

1. $F(x)$ is a finite number for every finite $x \in \Pi$, and
2. $\forall_{x,y}[x, y \in \Pi_b \Rightarrow [x \approx y \Rightarrow F(x) \approx F(y)]]$.

We denote by $\mathbf{SC}_\Pi \equiv \mathbf{SC}_\Pi(\mathrm{I\!R})$ the set comprising all the functions $F \in \mathbf{F}_\Pi$ which are **S**Π-continuous. Given a function $F \in \mathbf{F}_\Pi$ we denote by J_+F the function of $\mathbf{F}_\Pi$ defined by

$$J_+F(x) = \sum_{t \in J_0(x)} \epsilon F(t)$$

where $J_0(x) = [0 \ldots x)$ if $x > 0$, $J_0(x) = (x \ldots 0]$ if $x < 0$ and $J_+F(0) = 0$.[2]

[2] Similarly we define J_-F by suitably modifying the definition of $J_0(x)$. Note that the choice of the origin for one of the extremities of the Π-interval $J_0(x)$ is made for reasons of convenience of exposition. There is no substantial alteration if we choose another finite point of Π.

Theorem 8.6 If F is a function belonging to $\mathbf{SC}_\Pi$ the function J_+F (or, respectively, the function J_-F) also belongs to $\mathbf{SC}_\Pi$.

Proof For any $x \in \Pi_b$, considering the operator J_+, for example, we have

$$|J_+F(x)| \le \sum_{t \in J_0(x)} \epsilon |F(t)| \le \left\{ \max_{t \in J_0(x)} |F(t)| \right\} |x|$$

which shows that J_+F is a function finite on Π_b. On the other hand, for any $x \in \Pi_b$ we have

$$|J_+F(y) - J_+F(x)| \le \sum_{t=\min\{x,y\}}^{\max\{x,y\}-\epsilon} \epsilon |F(t)| \le C_F |y-x|$$

where $C_F = \max\{|F(t)| : \min\{x,y\} \le t < \max\{x,y\}\} \in {}^*\mathbf{C}_b$.
Hence, if $x \approx y$ then $J_+F(x) \approx J_+F(y)$ and therefore J_+F is a function belonging to $\mathbf{SC}_\Pi$, as was to be shown. •

Successive application of the operator J_+ (or J_-) to functions of $\mathbf{SC}_\Pi$ always leads to functions of $\mathbf{SC}_\Pi$. The same is not true for the operator D_+ (or D_-): the fact that F belongs to $\mathbf{SC}_\Pi$ does not guarantee that D_+F (or D_-F) again belongs to $\mathbf{SC}_\Pi$. However, for any $F \in \mathbf{SC}_\Pi$ we do always have $D_+F \in \mathbf{F}_\Pi$ (and $D_-F \in \mathbf{F}_\Pi$).

8.2.2 The module $\mathbf{D}_\infty$ (over ${}^*\mathbf{C}_b$)

Let F be an internal function in $\mathbf{SC}_\Pi$. Then $\mathbf{st}F$ is a function continuous on $\mathbb{R}$ which generates a (regular) distribution belonging to $\mathcal{D}'$. Denoting by ν_F either the standard function $\mathbf{st}(F)$ or the distribution which it generates, according to which is the more convenient, we have

$$< \nu_F, \phi >= \int_F \nu_F(t)\phi(t)dt = \int_{\mathbf{st}_\Pi^{-1}(K)} (\mathbf{st} \circ F)\phi_\Pi d\Xi_L \tag{8.7}$$

where $\mathbf{st}\circ F$ and $\phi_\Pi \equiv \mathbf{st} \circ {}^*\phi_\Pi$ are external functions defined on Π with values in $\bar{\mathbf{C}}$ and K is a compact of $\mathbb{R}$ containing the support of ϕ. Since $F\,{}^*\phi_\Pi$ is an $\mathbf{S}\Pi$-integrable lifting of the function $(\mathbf{st}\circ F)\phi_\Pi$ we can replace the Loeb integral in (8.7) by a hyperfinite sum obtained as follows,

$$< \nu_F, \phi >= \mathbf{st}\left(\sum_{x \in {}^*K_\Pi} \epsilon F(x)\, {}^*\phi_\Pi(x) \right) \tag{8.8}$$

where ${}^*K_\Pi \equiv {}^*K \cap \Pi$, (note that $\mathbf{st}_\Pi^{-1}(K) \setminus {}^*K_\Pi$ has Ξ_L-measure zero).

The mapping $\phi \mapsto {}^*\phi_\Pi$ is linear and continuous, in the sense that if ϕ_n tends to zero in $\mathcal{D}$ then ${}^*\phi_{n,\Pi}(x)$ tends to zero for each $x \in \Pi$. Therefore, every internal function $F \in \mathbf{SC}_\Pi$ generates a distribution in this way. Since the mapping $f \mapsto {}^*f_\Pi$ embeds $\mathcal{C}$, (the space of functions continuous on $\mathbb{R}$), in $\mathbf{SC}_\Pi$ and the distribution generated by the (standard) function f coincides with ν_f (with $F = {}^*f_\Pi$), then the mapping

$$\mathbf{st}_{\mathcal{D}} : \mathbf{SC}_{\Pi} \to \mathcal{D}'$$
$$F \mapsto \mathbf{st}_{\mathcal{D}}(F) = \nu_F$$

where $\nu_F : \mathcal{D} \to \mathbf{C}$ is defined by (8.8), establishes a surjective correspondence between $\mathbf{SC}_{\Pi}$ and the subspace of $\mathcal{D}'$ comprising all the regular distributions generated by functions continuous on $\mathbb{R}$.

Let F be an internal function in $\mathbf{SC}_{\Pi}$ and let $\phi \in \mathcal{D}$ be a (standard) function with support contained in the compact $K \subset \mathbb{R}$. Then, taking theorem 7.4 into account, we have

$$\begin{aligned}\sum_{x\in\Pi} \epsilon D_+ F(x)\, {}^*\phi_{\Pi}(x) &= \sum_{x\in\Pi} [F(x+\epsilon) - F(x)]\, {}^*\phi_{\Pi}(x) \\ &= \sum_{x\in\Pi} \epsilon F(x)\, (D_- {}^*\phi_{\Pi}(x)) \\ &\approx \int_{st_{\Pi}^{-1}(K)} (\mathbf{st} \circ F)(-\phi')_{\Pi} d\Xi_L \\ &= < \nu_F, (-\phi') > = < D\nu_F, \phi >\end{aligned}$$

where $D\nu_F$ is the generalised derivative of the distribution $\nu_F \in \mathcal{D}'$.

We denote by $\mathbf{D}_+(\mathbf{SC}_{\Pi})$ the set of the first order Π-derivatives of all the internal functions in $\mathbf{SC}_{\Pi}$. Since for each $F \in \mathbf{SC}_{\Pi}$ we have $F = D_+(J_+F)$ then $\mathbf{SC}_{\Pi} \subset \mathbf{D}_+(\mathbf{SC}_{\Pi})$. We can then extend the mapping $\mathbf{st}_{\mathcal{D}}$ to $\mathbf{D}_+(\mathbf{SC}_{\Pi})$ by setting

$$\mathbf{st}_D(D_+F) = D\nu_F = D(\mathbf{st}_D F) \quad .$$

The same idea can be generalised to Π-derivatives of arbitrary **finite** order of the functions in $\mathbf{SC}_{\Pi}$. Thus if $F \in \mathbf{SC}_{\Pi}$ and $\phi \in \mathcal{D}_K$, with K an arbitrary compact in $\mathbb{R}$, we would obtain for any $j \in \mathbb{N}_0$

$$\begin{aligned}\sum_{x\in\Pi} \epsilon D_+^j F(x)\, {}^*\phi_{\Pi}(x) &= \sum_{x\in\Pi} \epsilon F(x) \left((-1)^j D_-^j\, {}^*\phi_{\Pi}(x)\right) \\ &\approx \int_{\mathbf{st}_{\Pi}^{-1}(K)} (\mathbf{st} \circ F)\left((-1)^j (\phi^j)_{\Pi}\right) d\Xi_L \\ &= < \nu_F, (-1)^j \phi^j > = < D^j \nu_F, \phi > \quad .\end{aligned}$$

That is to say, $\mathbf{st}_{\mathcal{D}}(D_+^j F) = D^j(\mathbf{st}_{\mathcal{D}} F)$. If, for each $j \in \mathbb{N}_0$, we denote by $\mathbf{D}_+^j(\mathbf{SC}_{\Pi})$ the set of the Π-derivatives of order j of all the internal functions in SC_{Π} then we have the inclusions $\mathbf{D}_+^j(\mathbf{SC}_{\Pi}) \subset \mathbf{D}_+^{j+1}(\mathbf{SC}_{\Pi})$ and

$$\mathbf{D}_{\infty} \equiv \mathbf{D}_{\infty}(\mathbb{R}) = \bigcup_{j=0}^{\infty} \mathbf{D}_+^j(\mathbf{SC}_{\Pi})$$

is an (external) set which comprises the Π-derivatives of finite order of all the internal functions in $\mathbf{SC}_{\Pi}$. Since the translate $\tau_{\epsilon}F$ of $F \in \mathbf{SC}_{\Pi}$ also belongs to $\mathbf{SC}_{\Pi}$ and, moreover, $D_- = D_+ \circ \tau_{\epsilon}$, then $\mathbf{D}_{\infty}$ can be obtained using indifferently either of the two operators D_+ or D_- of Π-differentiation on $\mathbf{SC}_{\Pi}$. More generally, since for each

$\in \mathbf{SC}_\Pi$ and $q \in \mathbb{N}_0$ we have $F = D_-^q(J_-^q F)$ where $J_-^q F \in \mathbf{SC}_\Pi$, then every element f $\mathbf{D}_\infty$ is a finite sum of internal functions of the form,

$$D_+^p D_-^q F$$

ith $F \in \mathbf{SC}_\Pi$ and $p, q \in \mathbb{N}_0$.

We can now extend the mapping $\mathbf{st}_\mathcal{D}$ to the whole set $\mathbf{D}_\infty$ in the following way: r each $\Phi \in \mathbf{D}_\infty$ there exists $F \in \mathbf{SC}_\Pi$ and $j \in \mathbb{N}_0$ such that $\Phi = D_+^j F$; then $\mathbf{t}_\mathcal{D}\Phi = D^j \nu_F$. Note that $\mathbf{st}_\mathcal{D}$ does not depend on the representation of Φ as the nite order derivative of a particular internal function in $\mathbf{SC}_\Pi$. In fact suppose that e also have $\Phi = D_+^m G$ with $G \in \mathbf{SC}_\Pi$ and $m \in \mathbb{N}_0$ (which, without any loss of enerality, we may suppose $\geq j$). From the equality $D_+^j F = D_+^m G$ it follows that $^{m-j}_{+} F + P_m = G$ where P_m is a polynomial of degree $< m$ (and with coefficients in C). For $\phi \in \mathcal{C}$ we obtain

$$< D^m \nu_G, \phi > = < \nu_G, (-1)^m \phi^{(m)} > \approx \sum_{x \in \Pi} \epsilon G(x) \left((-1)^m D_+^{m\,*} \phi_\Pi(x)\right)$$

$$= \sum_{x \in \Pi} \left(J_+^{m-j} F(x) + P_m(x)\right) \left((-1)^m D_-^{m\,*} \phi_\Pi(x)\right)$$

$$= \sum_{x \in \Pi} \epsilon F(x) \left((-1)^j D_-^{j\,*} \phi_\Pi(x)\right) + \sum_{x \in \Pi} \epsilon D_+^m P_m(x)\, {}^*\phi_\Pi(x)$$

$$\approx < \nu_F, (-1)^j \phi^{(j)} > = < D^j \nu_F, \phi >$$

nd, therefore, $D^m \nu_G = D^j \nu_F$, as we wished to show.

The mapping $\mathbf{st}_\mathcal{D} : \mathbf{D}_\infty \to \mathcal{D}'$ is clearly linear; the **kernel** of this mapping, $\mathcal{M}_\infty \equiv \mathcal{M}(\mathbf{st}_\mathcal{D}) \subset \mathbf{D}_\infty$ is formed of all the internal functions which generate the null istribution by this process: that is, all the internal functions which are Π-derivatives f finite order of the functions infinitesimal in $\mathbf{SC}_\Pi$.[3] The quotient space $\mathbf{D}_\infty / \mathcal{M}_\infty$ a vector space which is isomorphic to $\mathcal{C}_\infty$, the space of distributions of finite order f J.Sebastiao e Silva which, in turn, is isomorphic to $\mathcal{D}_{fin}$, the space of Schwartz istributions of finite order.

efinition 8.7 We call the internal functions in $\mathbf{D}_\infty$ **pre-distributions of finite rder**, and we call the quotient classes $[\Phi]$ belonging to $\mathbf{D}_\infty / \mathcal{M}_\infty$ **Π-distributions f finite order**.

n $\mathbf{D}_\infty / \mathcal{M}_\infty$ we define a differentiation operator D, by setting for any class $[\Phi]$

$$D[\Phi] = [D_+ \Phi].$$

dentifying $\mathbf{SC}_\Pi / \mathcal{M}_\infty$ with $\mathcal{C}$, the quotient space $\mathbf{D}_\infty / \mathcal{M}_\infty$ equipped with the operator D gives a new model of the Silva axiomatic theory of finite order distributions. he pre-distributions are internal functions in $\mathbf{F}_\Pi$ which do not assume excessively igh infinite values in Π_b, as shown by the following theorem:

[3]Note that the functions of $\mathcal{M}_\infty$ are not, in themselves, necessarily infinitesimal.

Theorem 8.8 For every pre-distribution $\Phi \in \mathbf{D}_\infty$ there exists a non-negative (finite) integer $m \equiv m_\Phi$ such that for each compact $K \subset \mathbb{R}$,

$$|\Phi(x)| \leq C_{\Phi,K}.\kappa^m \in {}^*K_\Pi \tag{8.9}$$

where $C_{\Phi,K}$ is a positive finite constant (which depends, in general, on the function Φ and the compact K).

Proof The inequality (8.9) is clearly satisfied for every internal function $\Phi \in \mathbf{SC}_\Pi$ with $m = 0$. If $\Phi = D_+F$ with $F \in \mathbf{SC}_\Pi$ then

$$\Phi(x) = D_+F(x) = \kappa[F(x+\epsilon) - F(x)]$$

and therefore for each compact $K \subset \mathbb{R}$ we have

$$\max_{x \in {}^*K_\Pi} |\Phi(x)| \leq 2\{\max_{t \in {}^*K_\Pi} |F(t)|\}.\kappa$$

which shows that (8.9) is satisfied with $C_{\Phi,K} = 2\max_{t\in {}^*K_\Pi}|F(t)| \in {}^*\mathbb{R}_b$ and $m = 1$.

Suppose now that the inequality is satisfied for all internal functions of the form D_+^jF with $F \in \mathbf{SC}_\Pi$; then for $\Phi = D_+^{j+1}F$ we obtain

$$\max_{x \in {}^*K_\Pi} |\Phi(x)| \leq 2\{\max_{x \in {}^*K_\Pi} |D_+^jF(t)|\}.\kappa \leq C_{\Phi,K}^{[j+1]}.\kappa^{j+1}$$

where $C_{\Phi,K}^{[j+1]}$ is, for each given number $j \in \mathbb{N}_0$ a bounded positive constant. Note that if the function $F \in \mathbf{SC}_\Pi$ is such that some of its earlier Π-derivatives belong to $\mathbf{SC}_\Pi$ then the exponent of κ may be lower than the order of Π-differentiation. •

Now let G_∞ be the subset of $\mathbf{F}_\Pi$ comprising all internal functions Φ for which (8.9) is satisfied for some $m \in \mathbb{N}_0$ and every compact $K \subset \mathbb{R}$, with a finite constant $C_{\Phi,K} > 0$. G_∞ is a Π-differential algebra which contains $\mathbf{D}_\infty$ as a submodule. In G_∞ the product of two pre-distributions always makes sense, although the result of this operation will not in general be a pre-distribution. However, imposing appropriate restrictions on the factors, the product of two elements of $\mathbf{D}_\infty$ may again be an element of $\mathbf{D}_\infty$: suppose, for example, that $\Theta, \Phi \in \mathbf{D}_\infty$ are two internal functions such that

$$D_+\Theta \in \mathbf{SC}_\Pi \quad \text{and} \quad \Phi \in \mathbf{D}_+(\mathbf{SC}_\Pi) \ .$$

Then there exists $F \in \mathbf{SC}_\Pi$ such that $\Phi = D_+F$ and therefore

$$\begin{aligned}\Theta\Phi = \Theta D_+F &= D_+(\Theta F) - (D_+\Theta)F - \epsilon(D_+\Theta)(D_+F) \\ &= D_+[\Theta F - J_+((D_+\Theta)F)] - \epsilon(D_+\Theta)(D_+F) \\ &= D_+[\Theta F - J_+((D_+\Theta)F)] - (D_+\Theta)[F(x+\epsilon) - F(x)] \ .\end{aligned}$$

Consequently

$$\Theta\Phi = D_+[G - J_+((D_+\Theta)(F(x+\epsilon) - F(x))]$$

where $G = \Theta F - J_+((D_+\Theta)F)$ is an internal function belonging to $\mathbf{SC}_\Pi$. Since $F \in \mathbf{SC}_\Pi$ we must have

$$F(x+\epsilon) \approx F(x)$$

for all $x \in \Pi_b$ and therefore the internal function

$$(D_+\Theta)F$$

belongs to $\mathbf{SC}_\Pi$ and is infinitesimal in Π_b, the same being true for the function

$$J_+((D_+\Theta)F) \quad .$$

For any $\phi \in \mathcal{D}$ we have

$$\left|\sum_{x\in\Pi} \epsilon(D_+\Theta(x))[F(x+\epsilon)-F(x)]\, {}^*\phi_\Pi(x)\right| \leq$$

$$\left\{\sup_{x\in {}^*K_\Pi} |F(x+\epsilon)-F(x)|\right\}.\sum_{x\in\Pi} \epsilon|D_+\Theta(x)|.|\, {}^*\phi_\Pi(x)| \approx 0$$

where K denotes a compact of $\mathbb{R}$ containing the support of ϕ. Consequently

$$\sum_{x\in\Pi} \epsilon\Theta(x)\Phi(x)\, {}^*\phi_\Pi(x) = \sum_{x\in\Pi} \epsilon(D_+G)\, {}^*\phi_\Pi(x)$$

$$- \sum_{x\in\Pi} \epsilon(D_+\Theta(x))[F(x+\epsilon)-F(x)]\, {}^*\phi_\Pi(x)$$

$$\approx \sum_{x\in\Pi} \epsilon(D_+G)\, {}^*\phi_\Pi(x)$$

and therefore

$$\mathbf{st}\left(\sum_{x\in\Pi} \epsilon\Theta(x)\Phi(x)\, {}^*\phi_\Pi(x)\right) = \mathbf{st}\left(\sum_{x\in\Pi} \epsilon(D_+G(x))\, {}^*\phi_\Pi(x)\right)$$

or equivalently

$$\mathbf{st}_D(\Theta\Phi) = \mathbf{st}_D(D_+G) = D(\mathbf{st}_D G)$$

which shows that $\Theta\Phi$ and D_+G belong to the same Π-distribution. The product $\Theta\Phi$ is equal to the first order Π-derivative of an $\mathbf{S}\Pi$-continuous function, being moreover a pre-distribution.

More generally, if $\Theta, \Phi \in \mathbf{D}_\infty$ are two internal functions such that

$$D_+^m\Theta \in \mathbf{SC}_\Pi \quad \text{and} \quad \Phi \in \mathbf{D}_+^m(\mathbf{SC}_\Pi)$$

for some $m \in \mathbb{N}_0$, then we will have $\Phi = D_+^m F$ with $F \in \mathbf{SC}_\Pi$ and we can prove by induction on m that

$$\Theta\Phi - D_+^m\left(\sum_{j=0}^{m}(-1)^{j+1}.{}^mC_j J_+^{m-j}[(D_+^{m-j}\Theta)F]\right)$$

is an internal function such that

$$\mathbf{st}_D\left(\Theta\Phi - \sum_{j=0}^{m}(-1)^{j+1}.{}^mC_j J_+^{m-j}[(D_+^{m-j}\Theta)F]\right) = 0 \quad .$$

We can then say that the product of the pre-distributions $\Theta, \Phi \in \mathbf{D}_\infty$, under the indicated conditions, is a pre-distribution. (Note that

$$G \equiv \sum_{j=0}^{m} (-1)^{j+1}.{}^mC_j J_+^{m-j}[(D_+^{m-j}\Theta)F]$$

belongs to $\mathbf{SC}_\Pi$).

8.2.3 The module $\mathbf{D}_\Pi$ (over ${}^*\mathbf{C}_b$)

Let $\mathbf{D}_\Pi$ be the set consisting of all internal functions $F \in \mathbf{F}_\Pi$ which have the following property: for each compact $K \subset \mathbb{R}$ there exists $\Phi_K \in \mathbf{D}_\infty$ such that $\Phi = \Phi_K$ for all $x \in {}^*K_\Pi$. Each function in $\mathbf{D}_\Pi$ determines a family

$$\{\Phi_K\}_{K \subset \mathbb{R}}$$

which satisfies the following condition:

if $K, L \subset \mathbb{R}$ are two compact sets with $K \subset L$ then $\Phi_K = \Phi_L$ on ${}^*K_\Pi$.

Such a family of functions of $\mathbf{D}_\infty$ is said to be **compatible**. Conversely, given a compatible family $\{\Phi_K\}_{K \subset \mathcal{R}}$ of functions in $\mathbf{D}_\infty$, let Φ be the function such that

$$\Phi_{|K} = \Phi_K \quad \text{on} \quad {}^*K_\Pi$$

for every compact $K \subset \mathbb{R}$. Then $\Phi \in \mathbf{D}_\Pi$.

If Φ is a function in $\mathbf{D}_\infty$, the "constant" $\{\Phi\}_{K \subset \mathbb{R}}$ is certainly a compatible family and therefore defines an element of $\mathbf{D}_\Pi$; hence $\mathbf{D}_\infty \subset \mathbf{D}_\Pi$. Every internal function in $\mathbf{D}_\Pi$ is said to be a **global pre-distribution.** The pre-distributions of finite order are global pre-distributions but the converse is not true, as is shown by the following example.

Example 8.9 Given the internal function

$$\Delta_0(x) = \begin{cases} \kappa, & \text{if } x = 0 \\ 0, & \text{otherwise} \end{cases}$$

it is easy to verify that

$$D_+^m \Delta_0(x) = \begin{cases} (-1)^j.{}^mC_j \kappa^{m+1} & \text{if } x = -(m-j)\epsilon, \quad j = 0, 1, \ldots, m \\ 0, & \text{otherwise} \end{cases}$$

For any $\phi \in \mathcal{D}$ we have

$$\mathbf{st}\left(\sum_{x \in \Pi} \epsilon D_+^m \Delta_0(x)\, {}^*\phi_\Pi(x)\right) = (-1)^m \phi^m(0) = < D^m\delta, \phi > \quad .$$

$D_+^m \Delta_0$ is, for each $m \in \mathbb{N}_0$, a function in $\mathbf{D}_\infty$, the same being true for any finite linear combination (over ${}^*\mathbf{C}_b$) of finite-order Π-derivatives of translates of Δ_0. However, the internal function

$$\Phi(x) = \sum_{n=-\kappa/2}^{\kappa/2-1} D_+^{|n|} \Delta_0(x-n) \equiv \sum_{n=-\kappa/2}^{\kappa/2-1} D_+^{|n|} \tau_n \Delta_0(x)$$

does not belong to $\mathbf{D}_\infty$, although as we shall see it does belong to $\mathbf{D}_\Pi$. Note that for finite n the function $D_+^{|n|}\tau_n\Delta_0$ is null outside the Π-monad of n, while for n infinite it is null outside the interval $[n-1/2, n+1/2]$ which is entirely contained in the remote part of the hyperfinite line. Thus for every compact set $K \subset \mathbb{R}$ the intersection of ${}^*K_\Pi$ with the support of $D_+^{|n|}\tau_n\Delta_0$ is empty for $|n| \in {}^*\mathbb{N}_\infty$. Each ${}^*K_\Pi$ intersects the support of only a finite number of Π-derivatives of finite order of translates of Δ_0. Consequently the restriction of Φ to ${}^*K_\Pi$ is equal to a Π-derivative of finite order of a function in $\mathbf{SC}_\Pi$.

The mapping $\mathbf{st}_D$, defined on $\mathbf{D}_\infty$, can now be extended to the whole module $\mathbf{D}_\Pi$ by setting for each $\Phi \in \mathbf{D}_\infty$,

$$\mathbf{st}_D\Phi = \{\mathbf{st}_{D_K}\Phi_K\}_{K\subset\mathbb{R}}$$

where $\mathbf{st}_{D_K}\Phi_K$ denotes the restriction of $\mathbf{st}_\mathcal{D}\Phi_K$ to $\mathcal{D}_K$ for each $K \subset \mathbb{R}$. That is,

$$< \mathbf{st}_D\Phi, \phi >= \mathbf{st}\left(\sum_{x\in {}^*K_\Pi} \epsilon\Phi_K(x) \, * \phi_\Pi(x)\right)$$

for any test function $\phi \in \mathcal{D}_K$.

The mapping $\mathbf{st}_\mathcal{D}$ is linear. Its kernel $\mathcal{M} \equiv \mathcal{M}_\Pi(\mathbf{st}_\mathcal{D})$ comprises all the internal functions of $\mathbf{D}_\Pi$ giving rise to the null distribution. Then $\mathbf{D}_\Pi/\mathcal{M}_\Pi$ is a vector space whose elements, the classes $[\Phi]$, are called **global Π-distributions**. Note that for each compact $K \subset \mathbb{R}$ there exists a number $m_K \in \mathcal{N}_0$ and a continuous function $F_K \in \mathbf{SC}_\Pi$ such that $\Phi_K = D_+^{m_K}F_K$ on ${}^*K_\Pi$.

From theorem 8.8 it follows that if Φ is a function in $\mathbf{D}_\Pi$ then for each compact $K \subset \mathbb{R}$ there exists a bounded constant $C_{\Phi,K} > 0$ and a number $m_K \in \mathbb{N}_0$ such that

$$|\Phi(x)| \leq C_{\Phi,K}\kappa^{m_K} \tag{8.10}$$

for all $x \in {}^*K_\Pi$.

Let G_Π be the set formed by all internal functions $\Phi \in \mathbf{F}_\Pi$ which satisfy the following property: for each compact $K \subset \mathbb{R}$ there exists a bounded constant $C_{\Phi,K} > 0$ and a number $m_K \in \mathbb{N}_0$ for which the inequality (8.10) is satisfied for all $x \in {}^*K_\Pi$ The set G_Π constitutes a Π-differential algebra containing the module $\mathbf{D}_\Pi$ as a submodule and G_∞ as a subalgebra. Then the global pre-distributions can be multiplied in G_Π although in general the product of two pre-distributions will not be a pre-distribution. However, if $\Theta \in \mathbf{D}_\Pi$ is an internal function such that $D_+^j\Theta \in \mathbf{SC}_\Pi$ for all (finite) $j \in \mathbb{N}_0$, then for any $\Phi \in \mathbf{D}_\Pi$ the function $\Theta\Phi$ is again a pre-distribution.

Chapter 9

Hyperfinite Fourier Analysis

9.1 Standard Fourier Transformation

The (classical) Fourier Transform is defined on L^1, the space of all (standard) functions integrable over $\mathbb{R}$ in the sense of Lebesgue, as follows

$$\mathcal{F}[f](\xi) \equiv \hat{f}(\xi) = \int_{-\infty}^{+\infty} f(t)\exp(-2\pi i\xi t)dt, \qquad \xi \in \mathbb{R}. \tag{9.1}$$

From the integrability of f it follows immediately that $\hat{f} = \mathcal{F}[f]$ is a well defined function, bounded and continuous on $\mathbb{R}$.

Now let f, g be two functions in L^1 with Fourier transforms $\hat{f}$ and $\hat{g}$ respectively. Then the functions $f\hat{g}$ and $\hat{f}g$ belong to L^1 and we obtain successively

$$\begin{aligned}\int_{-\infty}^{+\infty} f(t)\hat{g}(t)dt &= \int_{-\infty}^{+\infty} f(t)\left\{\int_{-\infty}^{+\infty} g(t)\exp(-2\pi i\xi t)d\xi\right\} dt \\ &= \int_{-\infty}^{+\infty} g(\xi)\left\{\int_{-\infty}^{+\infty} f(t)\exp(-2\pi i\xi t)dt\right\} d\xi \\ &= \int_{-\infty}^{+\infty} \hat{f}(\xi)g(\xi)d\xi\end{aligned}$$

where the change of order of integration can be justified by appeal to the Fubini theorem. We get, therefore, the result

$$\int_{-\infty}^{+\infty} f(t)\hat{g}(t)dt = \int_{-\infty}^{+\infty} \hat{f}(\xi)g(\xi)d\xi \tag{9.2}$$

which is known as the **Parseval equation.**

The Fourier transform can be extended, in classical terms, to other classes of functions, in particular to the space L^2 of functions square integrable over $\mathbb{R}$. However since definition (9.1) imposes very restrictive integrability conditions, any sufficiently ample generalisation of this transform requires us to work in the context of generalised functions. The first generalisation considered here involves the so-called tempered distributions of Schwartz, whose definition is given briefly as follows.

9.1.1 Tempered distributions

A function $\phi \in \mathcal{C}^\infty$ is said to be **rapidly decreasing** if it satisfies the conditions

$$\gamma_{p,q} \equiv \sup_{x\in\mathbb{R}} |x^p \phi^q(x)| < \infty$$

for all numbers $p, q \in \mathbb{N}_0$. The vector space of all such functions is denoted by $\mathcal{S} = \mathcal{S}(\mathbb{R})$.

Exercise 9.1 *Show that the function defined by*

$$\phi(x) = \exp(-x^2), \qquad x \in \mathbb{R}$$

is an element of $\mathcal{S}$.

For each $p, q \in \mathbb{N}_0$ the mapping $\gamma_{p,q} : \mathcal{S} \to \mathbb{R}_0^+$ defines a seminorm. The space $\mathcal{S}$ with the topology generated by the family of seminorms

$$\{\gamma_{p,q}\}_{p,q\in\mathbb{N}_0}$$

is a Fréchet space. A sequence $(\phi_n)_{n\in\mathbb{N}}$ is convergent relative to this topology if and only if

1. for every pair of numbers $p, q \in \mathbb{N}_0$ there exists a real constant $C_{p,q} > 0$ such that
$$|x^p \phi_n^{(q)}(x)| \leq C_{p,q}$$
for all $x \in \mathbb{R}$ and for all $n \in \mathbb{N}$; and

2. for each $q \in \mathbb{N}_0$ the sequence $(\phi_n^{(q)})_{n\in\mathbb{N}}$ converges uniformly on every compact in $\mathbb{R}$.

Exercise 9.2 *Show that* $\mathcal{D}$ *is a dense vector subspace of* $\mathcal{S}$. *HINT: Consider the function defined by*

$$\theta(x) = \begin{cases} \exp[x^2/(x^2-1)], & \text{if } |x| < 1, \\ 0, & \text{if } |x| \geq 1 \end{cases} .$$

and, setting $\theta_n(x) = \theta(x/n)$ *for* $n = 1, 2, \ldots,$ *show that for any* $\phi \in \mathcal{S}$ *the sequence of functions in* $\mathcal{D}$, *given by* $\phi_n = \phi\theta_n$, $n = 1, 2, \ldots,$ *converges in* $\mathcal{S}$ *to* ϕ. *[See [17], p.102, for example.]*

Let ϕ be any function of $\mathcal{S}$. Then $\phi \in L^1$ and therefore $\hat{\phi}$ is a well defined function of class $\mathcal{C}^\infty$ and we have

$$\hat{\phi}(\xi) = \int_{-\infty}^{+\infty} \phi(t)\exp(-2\pi i\xi t)dt, \qquad \xi \in \mathbb{R}$$

$$= \frac{1}{2\pi i\xi}\int_{-\infty}^{+\infty} \phi'(t)\exp(-2\pi i\xi y)dt$$

$$= (\frac{1}{2\pi i\xi})^2 \int_{-\infty}^{+\infty} \phi''(t)\exp(-2\pi i\xi y)dt$$

and so, in general,

$$\hat{\phi}(\xi) = (\frac{1}{2\pi i\xi})^p \int_{-\infty}^{+\infty} \phi^{(p)}(t)\exp(-2\pi i\xi y)dt$$

from which it follows that for any $p \in \mathbb{N}_0$ we have

$$|(2\pi i\xi)^p\hat{\phi}(\xi)| \leq \int_{-\infty}^{+\infty} |\phi^{(p)}(t)|dt < +\infty \quad .$$

More generally we can show in a similar way that for any $p, q \in \mathbb{N}_0$ we have

$$|(2\pi i\xi)^p\hat{\phi}^{(q)}(\xi)| \leq \int_{-\infty}^{+\infty} |[(-2\pi it)^q\phi(t)]^{(q)}|dt < +\infty,$$

which allows us to conclude that $\hat{\phi}$ also belongs to $\mathcal{S}$. Conversely, using the inverse Fourier transform, we can prove the implication $\hat{\phi} \in \mathcal{S} \Rightarrow \phi \in \mathcal{S}$. Hence we have

$$\forall_\phi[\phi \in \mathcal{S} \Leftrightarrow \hat{\phi} \in \mathcal{S}] \quad .$$

It is also possible to show that if $(\phi_n)_{n\in\mathbb{N}}$ is a sequence of functions converging to zero in $\mathcal{S}$, then the sequence $(\hat{\phi}_n)_{n\in\mathbb{N}}$ itself converges to zero in $\mathcal{S}$. That is to say, $\mathcal{F}: \mathcal{S} \to \mathcal{S}$ is a topological vector automorphism.

Exercise 9.3 *Determine the (classical) Fourier transform of the function referred to in Exercise 9.1.*

$\mathcal{S}' \equiv \mathcal{S}'(\mathbb{R})$ is the topological dual of $\mathcal{S}$, that is the vector space of all bounded (continuous) linear functionals defined over the topological vector space $\mathcal{S}$. Since $\mathcal{D}$ is a dense vector subspace of $\mathcal{S}$ (Exercise 9.2) it follows from well known results of Functional Analysis that we have the (actually proper) inclusion

$$\mathcal{S}' \subset \mathcal{D}'$$

which means that each element of $\mathcal{S}$ is (at least up to isomorphism) a Schwartz distribution. The elements of $\mathcal{S}'$ are called **tempered distributions.**

The tempered distributions can be given an equivalent description in the theory of Sebastiao e Silva: *the tempered distributions are all those elements of $\mathcal{C}_\infty$ which can be obtained by finite order differentiation of continuous polynomially increasing functions.*[1]

Example 9.4 The function $f(x) = e^x \cos(e^x)$, $x \in \mathbb{R}$, is not a function of slow growth in $\mathbb{R}$. However it does generate a tempered distribution since we have

$$f(x) = D\{\sin(e^x)\}$$

and $\sin(e^x)$, $x \in \mathbb{R}$, is a function of slow growth in $\mathbb{R}$.

[1]Functions whose increase in modulus is bounded by a polynomial in $\mathbb{R}$. Such functions are also said to be of **slow growth**.

9.1.2 Fourier transforms in $\mathcal{S}'$

Definition 9.5 The generalised Fourier transform of a tempered distribution $\nu \in \mathcal{S}'$ is defined to be that (tempered) distribution $\hat{\nu} \equiv \mathcal{F}[\nu]$ which satisfies the Parseval type equation

$$< \hat{\nu}, \phi >=< \nu, \hat{\phi} >$$

for every function $\phi \in \mathcal{S}$.

From (9.3) it is immediately clear that the transform of a tempered distribution is again a tempered distribution, and it is easily confirmed that $\mathcal{F} : \mathcal{S}' \to \mathcal{S}'$ is a linear continuous transformation actually constituting a topological vector automorphism on $\mathcal{S}'$.

Example 9.6 Since for any $a \in \mathbb{R}$ and any $\phi \in \mathcal{S}$ we have

$$\begin{aligned} < \tau_a \delta(t), \phi(t) > &= \hat{\phi}(a) \\ &= \int_{-\infty}^{+\infty} \phi(\xi) \exp(-2\pi i a \xi) d\xi =< \exp(-2\pi i a \xi, \phi(\xi) > \end{aligned}$$

we can therefore write

$$\mathcal{F}[\tau_a \delta] \equiv \mathcal{F}[\delta_a] = \exp(-2\pi i a)$$

where the equality is satisfied in the sense of tempered distributions.

9.2 The Π-Fourier Transforms

9.2.1 Preliminary definitions

If F is any function in $\mathbf{F}_\Pi$ then for each (fixed) $y \in \Pi$ the hyperfinite sum

$$\hat{F}(y) = \sum_{x \in \Pi} \epsilon \, {}^*\exp_\Pi(-2\pi i x y) F(x) \tag{9.4}$$

is a well defined hypercomplex number. Thus, (9.4) defines for each $F \in \mathbf{F}_\Pi$ a new internal function $\hat{F} : \Pi \to {}^*\mathbf{C}$ which also belongs to $\mathbf{F}_\Pi$. In order to be able to recover the original function F from this new function $\hat{F}$ we must first establish the following result:

Lemma 9.7 The following relations are valid in $\mathbf{F}_\Pi$.

$$\Delta_0(x) = \sum_{y \in \Pi} \epsilon \, {}^*\exp_\Pi(2\pi i x y) = \sum_{y \in \Pi} \epsilon \, {}^*\exp_\Pi(-2\pi i x y)$$

for any $x \in \Pi$.

Proof Since for any $x, y \in \Pi$ we may write $x = -\frac{\kappa}{2} + m\epsilon$ and $y = -\frac{\kappa}{2} + j\epsilon$, with $m, j = 0, 1, 2, \ldots, \kappa^2 - 1$, then using the formula for the sum of a geometric series we get

$$\sum_{y\in\Pi} \epsilon\, {}^*\exp_\Pi(2\pi i x y) = \sum_{j=0}^{\kappa^2-1} \epsilon e^{2\pi i x(-\kappa/2+j\epsilon)} = \epsilon e^{-\pi i\kappa x}\frac{1-e^{2\pi i\epsilon\kappa^2 x}}{1-e^{2\pi i\epsilon x}}$$

$$= \epsilon e^{-\pi i\kappa x}\frac{1-e^{2\pi i\kappa x}}{1-e^{2\pi i\epsilon x}} = \begin{cases}\kappa & \text{if } x=0\\ 0 & \text{if } x\neq 0\end{cases}$$

$$= \Delta_0(x)\ .$$

We can verify the other equality similarly.

Now multiplying both sides of (9.4) by $\epsilon\, {}^*\exp_\Pi(2\pi i t y)$ and summing over $y \in \Pi$ for each fixed $t \in \Pi$ we get

$$\sum_{y\in\Pi}\epsilon\hat{F}(y)\, {}^*\exp_\Pi(2\pi i t y) = \sum_{y\in\Pi}\epsilon\left\{\sum_{x\in\Pi}\epsilon\, {}^*\exp_\Pi(-2\pi i x y)F(x)\right\}.\, {}^*\exp_\Pi(2\pi i t y)$$

$$= \sum_{x\in\Pi}\epsilon F(x)\left\{\sum_{y\in\Pi}\epsilon\, {}^*\exp_\Pi[2\pi i y(t-x)]\right\}$$

$$= \sum_{x\in\Pi}\epsilon F(x)\Delta_0(t-x) = F(t) \tag{9.5}$$

thus allowing recovery of the function F from the function $\hat{F}$.

Definition 9.8 The **Π-Fourier Transform** of a function $F \in \mathbf{F}_\Pi$ is the function $\hat{F} \in \mathbf{F}_\Pi$ defined by (9.4). Conversely the function F defined by the first member of (9.5) is described as the **inverse Π-Fourier transform** of the function $\hat{F} \in \mathbf{F}_\Pi$.

Denoting the direct and inverse Π-Fourier transforms by $\mathcal{F}_\Pi$ and $\bar{\mathcal{F}}_\Pi$ respectively, we write $\hat{F} = \mathcal{F}_\Pi[F]$ and $F = \bar{\mathcal{F}}_\Pi[\hat{F}]$, noting that $\mathcal{F}_\Pi \circ \bar{\mathcal{F}}_\Pi = \bar{\mathcal{F}}_\Pi \circ \mathcal{F}_\Pi =$**id**. The transforms $\mathcal{F}_\Pi$ and $\bar{\mathcal{F}}_\Pi$ form linear automorphisms over $\mathbf{F}_\Pi$: in fact, if for any $F \in \mathbf{F}_\Pi$ we were to have

$$\hat{F}(y) = \mathcal{F}_\Pi[F](y) = 0, \qquad y \in \Pi,$$

then multiplying both sides by $\epsilon\, {}^*\exp(2\pi i t y)$ and summing over $y \in \Pi$ we would get

$$F(x) = \sum_{y\in\Pi}\epsilon\hat{F}(y)\, {}^*\exp_\Pi(2\pi i x y) = 0, \qquad x \in \Pi\ .$$

Similarly it follows from $F = \bar{\mathcal{F}}_\Pi[\hat{F}] = 0$ that $\hat{F} = 0$, giving the following equivalence

$$\forall_F[F \in \mathbf{F}_\Pi \Rightarrow [\hat{F} \equiv \mathcal{F}_\Pi[F] = 0 \Leftrightarrow F = \bar{\mathcal{F}}_\Pi[\hat{F}] = 0]]\ .$$

Π-Fourier transform of Π-derivative Let $F \in \mathbf{F}_\Pi$ be an internal function whose Π-Fourier transform is a function $\hat{F} \in \mathbf{F}_\Pi$. Then

$$\begin{aligned}\mathcal{F}_\Pi[D_+F](y) &= \sum_{x\in\Pi} \epsilon D_+F(x)\, {}^*\exp_\Pi(-2\pi ixy)\\ &= \sum_{x\in\Pi} [F(x+\epsilon) - F(x)]\, {}^*\exp_\Pi(-2\pi ixy)\\ &= \sum_{x\in\Pi} F(x)[{}^*\exp_\Pi[-2\pi i(x-\epsilon)y] - {}^*\exp_\Pi(-2\pi ixy)]\\ &= \sum_{x\in\Pi} \epsilon F(x)\, {}^*\exp_\Pi(-2\pi ixy)\frac{{}^*\exp_\Pi(-2\pi i\epsilon y) - 1}{\epsilon}\\ &= \lambda(y)\sum_{x\in\Pi} \epsilon F(x)\, {}^*\exp_\Pi(-2\pi ixy) = \lambda\hat{F}(y),\end{aligned}$$

where $\lambda : \Pi \to {}^*\mathbf{C}$ is the internal function defined by

$$\begin{aligned}\lambda(y) = \frac{{}^*\exp_\Pi(-2\pi i\epsilon y) - 1}{\epsilon} &= 2\pi iy\, {}^*\exp_\Pi(i\pi\epsilon y)\frac{{}^*\sin_\Pi(\pi\epsilon y)}{\pi\epsilon y}\\ &= 2\pi iy\, {}^*\exp_\Pi(i\pi\epsilon y)\, {}^*\mathrm{sinc}_\Pi(xy)\end{aligned}$$

and ${}^*\mathrm{sinc}_\Pi(y) = (\pi y)^{-1}\, {}^*\sin_\Pi(\pi y)$ for all $y \in \Pi$. Taking into account that for any $\alpha \in \mathbb{R}$ such that $|\alpha| \leq \pi/2$ we have $(2/\pi)|\alpha| \leq |\sin\alpha| \leq |\alpha|$, we obtain

$$4|x| \leq |\lambda(x)| = |\bar{\lambda}(x)| \leq 2\pi|x|, \qquad x \in \Pi, \tag{9.6}$$

the importance of which will be seen in what follows. Denoting by Λ the operator of multiplication of a function of $\mathbf{F}_\Pi$ by $\lambda(.)$ we then have $\mathcal{F}_\Pi \circ D_+ = \Lambda \circ \mathcal{F}_\Pi$ from which there follows immediately

$$D_+ = \bar{\mathcal{F}}_\Pi \circ \Lambda \circ \mathcal{F}_\Pi \quad \text{and} \quad \Lambda = \mathcal{F}_\Pi \circ D_+ \circ \bar{\mathcal{F}}_\Pi \ .$$

Similarly we can obtain other relations between these operators. In the general case,

$$\mathcal{F}_\Pi[D_+^p D_-^q \Lambda^r \bar{\Lambda}^s F] = \Lambda^p(-\bar{\Lambda})^q(-D_-)^r(D_+)^s\mathcal{F}_\Pi[F],$$

$$\text{or} \quad D_+^p D_-^q(-\Lambda)^r\bar{\Lambda}^s F = \bar{\mathcal{F}}_\Pi[\Lambda^p(-\bar{\Lambda})^q D_-^r D_+^s \hat{F}]$$

for $p, q, r, s \in {}^*\mathbb{N}_0$ (where simple juxtaposition of the operators here indicates the composition of those operators).

Π-convolution Given two internal functions $F, G \in \mathbf{F}_\Pi$ we define the Π-convolution of F and G to be the function $F \star G$ given by

$$F \star G(x) = \sum_{y\in\Pi} \epsilon F(x-y)G(y) = \sum_{y\in\Pi} \epsilon F(y)G(x-y),$$

and by simple application of the given definition we can verify that

$$\mathcal{F}_\Pi[F \star G] = \hat{F}.\hat{G} \ \text{ and } \ \bar{\mathcal{F}}_\Pi[\hat{F}.\hat{G}] = F \star G \ .$$

9.2.1.1 The Π-Fourier transform as generalisation of the classical transform

We denote by $\mathcal{F}_0$ the restriction of the operator $\mathcal{F}$ to the subspace $\mathcal{C}_0 \cap L^1$ consisting of all continuous functions in L^1 which *"vanish at infinity"*.[2] We now show that $\mathcal{F}_\Pi$ forms an extension of $\mathcal{F}_0$ in the following sense:

Theorem 9.9 If $f \in \mathcal{C}_0 \cap L^1$ then the following equality is valid for all $y \in \Pi_b$,

$$\mathcal{F}_0[f](\mathbf{st}y) = \mathbf{st}\left(\mathcal{F}_\Pi[{}^*f_\Pi](y)\right) \quad .$$

Proof For any (fixed) $\xi \in \mathbb{R}$ let y be any point chosen arbitrarily in $\mathbf{st}_\Pi^{-1}(\xi)$. Defining, for each $t \in \mathbb{R}$,

$$f_y(t) = f(t)\exp[-2\pi i(\mathbf{st}y)t]$$

and extending this function to the closed line by putting $f_y(\pm\infty) = 0$, we consider the (external) function $f_y \circ \mathbf{st}_\infty(x)$, $x \in \Pi$. Then,

$$\mathcal{F}_0[f](y) = \int_{\mathbb{R}} f_y(t)dt = \int_{\mathbf{st}_\Pi^{-1}(\mathbb{R})} f_y \circ \mathbf{st}_\infty(x)d\Xi_L(x) = \int_\Pi f_y \circ \mathbf{st}_\infty(x)d\Xi_L(x)$$

where the last two integrals relate to Loeb counting measure on the hyperfinite line. The proof will be completed if it can be shown that the equality

$$\int_\Pi f_y \circ \mathbf{st}_\infty(x)d\Xi_L(x) = \mathbf{st}\left(\sum_{x\in\Pi} \epsilon\, {}^*f_\Pi(x)\, {}^*\exp_\Pi(-2\pi i x y)\right) \tag{9.7}$$

is valid for all $y \in \Pi_b$. Furthermore for this it is necessary to show that the internal function

$${}^*f_\Pi(x)\, {}^*\exp_\Pi(-2\pi i x y), \qquad x \in \Pi$$

constitutes, for each $y \in \Pi_b$, an $\mathbf{S}\Pi$-integrable lifting of the (external) function $f_y \circ \mathbf{st}_\infty(x)$, $x \in \Pi$. In the first place we have

$$\mathbf{st}({}^*f_\Pi(x)\, {}^*\exp(-2\pi i x y)) = \mathbf{st}({}^*f_\Pi(x)).\mathbf{st}({}^*\exp_\Pi(-2\pi i x y))$$

and therefore

- if $x \in \Pi_b$ then since f and exp are continuous,

$$\begin{aligned}\mathbf{st}\left({}^*f_\Pi(x)\, {}^*\exp_\Pi(-2\pi i x y)\right) &= f(\mathbf{st}x).\exp[-2\pi i(\mathbf{st}x)(\mathbf{st}y)] \\ &= f_y \circ \mathbf{st}_\infty(x);\end{aligned}$$

- if $x \in \Pi_\infty$ then, since $f \in \mathcal{C}_0 \cap L^1$, we will have ${}^*f_\Pi(x) \approx 0$; moreover, the function ${}^*\exp_\Pi(-2\pi i x y)$ is finitely bounded, and so

$$\begin{aligned}\mathbf{st}\left({}^*f_\Pi(x)\, {}^*\exp_\Pi(-2\pi i x y)\right) &= f(\mathbf{st}x).\exp[-2\pi i(\mathbf{st}x)(\mathbf{st}y)] \\ &= 0 = f_y \circ \mathbf{st}_\infty(x) \quad .\end{aligned}$$

[2] We say that a function $f : \Pi \to C$ "vanishes at infinity" if to every real number $r > 0$ there corresponds a compact $K_r \subset \mathbb{R}$ such that $|F(x)| < r$ for all $x \notin K_r$.

It now remains to be shown that the internal function $^*f_\Pi(x)\,^*\exp_\Pi(-2\pi ixy)$ is an SΠ-integrable function for each fixed $y \in \Pi_b$; that is, that it satisfies the conditions

1. $\sum_{x\in\Pi} \epsilon[^*f_\Pi(x)]$ is finite;
2. if $A \subset \Pi$ is internal and $\Xi(A) \approx 0$ then $\sum_{x\in A} \epsilon|^*f_\Pi(x)| \approx 0$;
3. if $A \subset \Pi$ is internal and $^*f_\Pi(x) \approx 0$ is satisfied for all $x \in A$ then

$$\sum_{x\in A} \epsilon|^*f_\Pi(x)| \approx 0 \quad .$$

In order to show that 1, 2 and 3 are satisfied we proceed exactly as in the case of the proof of theorem 7.24. It then only remains to verify that we have the equality

$$\mathbf{st}\left(\sum_{x\in\Pi} \epsilon\,^*f_\Pi(x)\,^*\exp_\Pi(-2\pi ixy)\right) = \mathbf{st}\left(\sum_{x\in\Pi} \epsilon\,^*f_\Pi(x)\,^*\exp_\Pi[-2\pi ix(\mathbf{st}y)]\right)$$

which is equivalent to showing that the internal function

$$\hat{F}(y) = \sum_{x\in\Pi} \epsilon\,^*f_\Pi(x)\,^*\exp_\Pi(-2\pi ixy)$$

is SΠ-continuous on Π_b. For $y, z \in \Pi_b$ we have that

$$|\hat{F}(y) - \hat{F}(z)| \leq \sum_{x\in\Pi} \epsilon|^*f_\Pi(x)|.|1 - \,^*\exp_\Pi[-2\pi ix(y-z)]| \quad .$$

Seeing that $f \in \mathcal{C}_0 \cap L^1$, and again taking into account Lemma 7.25, then given a real number $r > 0$ we can affirm that the set

$$\left\{x \in \Pi : \sum_{|t|\geq x} \epsilon|^*f_\Pi(t)| < \frac{r}{3}\right\}$$

contains arbitrarily small positive infinite points: since this set is internal then, by underflow, there exists $x_r \in \Pi_b^+$ such that $\sum_{|x|>x_r} \epsilon|^*f_\Pi(t)| < \frac{r}{3}$, and so

$$|\hat{F}(y) - \hat{F}(z)| \leq \left\{\sum_{|x|\leq x_r} + \sum_{|x|>x_r}\right\} \epsilon|^*f_\Pi(x)|.|1 - \,^*\exp_\Pi[-2\pi ix(y-z)]|$$

$$\leq \sum_{|x|\leq x_r} \epsilon|^*f_\Pi(x)|.|1 - \,^*\exp_\Pi[-2\pi ix(y-z)]| + 2\sum_{|x|>r} \epsilon|^*f_\Pi(x)|$$

$$< \frac{2r}{3} + \sum_{|x|\leq x_r} \epsilon|^*f_\Pi(x)|.|1 - \,^*\exp_\Pi[-2\pi ix(y-z)]| \quad .$$

If $y \approx z$ and x is finite then $2\pi ix(y-z) \approx 0$ and therefore

$$\sum_{|x|\le x_r} \{\epsilon|{}^*f_\Pi(x)|.|1 - {}^*\exp_\Pi[-2\pi ix(y-z)]| \le$$

$$\left\{\max_{|x|\le x_r} |1 - {}^*\exp_\Pi[-2\pi ix(y-z)]|\right\} \sum_{|x|\le x_r} \epsilon|{}^*f_\Pi(x)| \approx 0 < \frac{r}{3} \ .$$

Then,

$$|\hat{F}(y) - \hat{F}(z)| < r$$

and as this inequality is satisfied for every real number $r > 0$ it follows that

$$\forall_{y,z}[y, z \in \Pi_b \Rightarrow [y \approx z \Rightarrow \hat{F}(y) \approx \hat{F}(z)]] \ .$$

That is to say, $\hat{F}$ is $S\Pi$-continuous on Π_b and therefore

$$\hat{f}(\mathbf{st}y) = \mathbf{st}(\hat{F}(y)), \qquad y \in \Pi_b \tag{9.8}$$

This completes the proof of the theorem. •

Given a function $f \in C_0 \cap L^1$ we can obtain its classical Fourier transform by means of (9.8), where $\hat{F}$ is the $S\Pi$-continuous internal function defined by

$$\hat{F}(y) = \sum_{x\in\Pi} \epsilon\, {}^*f_\Pi(x)\, {}^*\exp_\Pi(-2\pi ixy) \ .$$

From this expression we can conclude that, for each $y \in \Pi$, we have

$$|\hat{F}(y)| \le \sum_{x\in\Pi} \epsilon|{}^*f_\Pi(x)|$$

which confirms the fact that, in this case, $\hat{f}$ is a function bounded and continuous on $\mathbb{R}$.

In the case of an arbitrary function continuous on $\mathbb{R}$ it is not possible to obtain a result as simple as that established in (9.8). However if the function is of slow growth over $\mathbb{R}$ then that result can be recovered, in the sense of tempered distributions.

9.2.2 Hyperfinite representation of tempered distributions

Let $\mathbf{SC}_{\Pi,t}$ be the submodule (over ${}^*\mathbf{C}_b$) of $\mathbf{SC}_\Pi$ formed by all functions which are $S\Pi$-continuous on Π_b and which satisfy a relation of the type

$$|F(x)| \le C_{F},(1 + |\lambda(x)|^2)^r \tag{9.9}$$

for all $x \in \Pi$, some (finite) integer $r \in \mathbb{N}_0$ and a finite hyperreal constant $C_p > 0$. The shadow of a function of this type is a function of slow growth over $\mathbb{R}$. For an test function $\hat{\phi} \in \mathcal{S}$ the function $(\mathbf{st}F)\hat{\phi}$ belongs to $\mathcal{C}_0 \cap L^1$ and therefore F generates a tempered distribution $\nu_F \in \mathcal{S}'$ in accordance with the following

$$< \nu_F, \hat{\phi} >= \mathbf{st}\left(\sum_{x\in\Pi} \epsilon F(x)\, {}^*\phi_\Pi(x)\right) \quad . \tag{9.10}$$

The internal function $F \in \mathbf{SC}_{\Pi,t}$ thus constitutes a hyperfinite representation for the tempered regular distribution generated by the continuous function of slow growth $\mathbf{st}F$. In this way we can define the mapping

$$\mathbf{st}_{\mathcal{S}} : \mathbf{SC}_\Pi \rightarrow \mathcal{S}'$$

$$F \mapsto \mathbf{st}_{\mathcal{S}}F = \nu_F \quad .$$

From the equality

$$\sum_{x\in\Pi} \left(D_+^p F(x)\right)\, {}^*\hat{\phi}_\Pi(x) = (-1)^p \sum_{x\in\Pi} \epsilon F(x) D_-^p \left({}^*\hat{\phi}_\Pi(x)\right),$$

it follows that, in accordance with theorem 7.4 (and corollaries), for $p \in \mathbb{N}_0$ we have

$$\mathbf{st}\left(D_-^p\, {}^*\hat{\phi}_\Pi(x)\right) = \hat{\phi}^{(p)} \circ \mathbf{st}(x)$$

for all $x \in \Pi_b$. Thus $(\mathbf{st}F)\mathbf{st}\left(D_-^p\, {}^*\hat{\phi}_\Pi(x)\right)$ belongs to $\mathcal{C}_0 \cap L^1$, and we can extend the mapping $\mathbf{st}_{\mathcal{S}}$ to all internal functions which are obtained by applying the operator D_+ (or D_-) a finite number of times to functions of $\mathbf{SC}_{\Pi,t}$. This gives

$$< \mathbf{st}_{\mathcal{S}}(D_+^p F), \hat{\phi} >= \mathbf{st}\left(\sum_{x\in\Pi} \epsilon(D_+^p F(x))\, {}^*\hat{\phi}_\Pi(x)\right)$$

$$= \mathbf{st}\left((-1)^p \sum_{x\in\Pi} \epsilon F(x) D_-^p ({}^*\hat{\phi}_\Pi(x))\right)$$

$$=< \mathbf{st}_{\mathcal{S}}F, (-1)^p \hat{\phi}^{(p)} >=< D^p \mathbf{st}_{\mathcal{S}}F, \hat{\phi} >$$

or equivalently,

$$\mathbf{st}(D_+^p F) = D^p(\mathbf{st}_{\mathcal{S}}F)$$

where equality must be interpreted in the sense of tempered distributions. Similarly we get

$$\mathbf{st}_{\mathcal{S}}(D_-^q F) = D^q(\mathbf{st}_{\mathcal{S}}F)$$

for $q \in \mathbb{N}_0$ and any function F in $\mathbf{SC}_{\Pi,t}$. More generally, for $p, q \in \mathbb{N}_0$ and $F \in \mathbf{SC}_{\Pi,t}$ we would get

$$\mathbf{st}_{\mathcal{S}}(D_+^p D_-^q F) = D^{p+q}(\mathbf{st}_{\mathcal{S}}F) \quad .$$

Definition 9.10 A **tempered pre-distribution** is any internal function that is equal to a finite sum of functions of the form

$$D_+^p D_-^q F \tag{9.11}$$

where $F \in \mathbf{SC}_{\Pi,t}$ and $p, q \in \mathbb{N}_0$. The (external) set of all tempered pre-distributions

$$\mathbf{T}_\Pi \equiv \mathbf{T}_\Pi(\mathcal{R}) = \bigcup_{p=0}^{\infty} \bigcup_{q=0}^{\infty} \mathbf{D}_+^p \mathbf{D}_-^q (\mathbf{SC}_{\Pi,t}) \tag{9.12}$$

constitutes a submodule (over ${}^*\mathbf{C}_b$) of $\mathcal{D}_\infty$.

9.2.3 The Π-Fourier Transform on $\mathbf{T}_\Pi$

Let F be any tempered pre-distribution, Then there exist numbers $p, q \in \mathbb{N}_0$ such that

$$F = D_+^p D_-^q \Phi$$

where Φ is an internal function in $\mathbf{SC}_{\Pi,t}$; that is, F is a function SΠ-continuous on Π_b which satisfies an inequality of the form (9.9). In order to show that $\hat{F} = \mathcal{F}_\Pi[F]$ is also a tempered pre-distribution we need to establish some preliminary results which will be used in the sequel.

Lemma 9.11 For any $n \in \mathbb{N}_0$ the following equalities hold:

$$D_+^n \lambda(x) = \frac{\lambda(\epsilon)^n}{\epsilon} {}^*\exp_\Pi(2\pi i \epsilon x) = \frac{\lambda(\epsilon)^n}{\epsilon} + \lambda(\epsilon)^n \lambda(x)$$

$$D_-^n \lambda(x) = \frac{-\bar{\lambda}(\epsilon)^n}{\epsilon} {}^*\exp_\Pi(2\pi i \epsilon x) = -\frac{\bar{\lambda}(\epsilon)^n}{\epsilon} + \bar{\lambda}(\epsilon)^n \lambda(x)$$

$$D_+^n \bar{\lambda}(x) = \frac{\bar{\lambda}(\epsilon)^n}{\epsilon} + \lambda(\epsilon)^n \bar{\lambda}(x)$$

$$D_-^n \bar{\lambda}(x) = \frac{-\lambda(\epsilon)^n}{\epsilon} + \lambda(\epsilon)^n \bar{\lambda}(x)$$

where $|\lambda(\epsilon)^n|/\epsilon \approx (2\pi)^n \epsilon^{n-1}$.

Proof We prove the first identity by induction over $n \in \mathbb{N}$; the remainder are proved similarly. For $n = 1$ we have

$$\begin{aligned} D_+\lambda(x) &= \frac{{}^*\exp_\Pi[2\pi i \epsilon(x+\epsilon)] - {}^*\exp_\Pi(2\pi i \epsilon x)}{\epsilon^2} \\ &= {}^*\exp_\Pi(2\pi i \epsilon x) \frac{{}^*\exp_\Pi(2\pi i \epsilon^2) - 1}{\epsilon^2} = \frac{\lambda(\epsilon)}{\epsilon} {}^*\exp_\Pi(2\pi i \epsilon x) \\ &= \frac{\lambda(\epsilon)}{\epsilon}(\epsilon\lambda(x) + 1) = \frac{\lambda(\epsilon)}{\epsilon} + \lambda(\epsilon)\lambda(x) \quad . \end{aligned}$$

Suppose now that the identity is valid for some given natural number n. Then we will have

$$\begin{aligned} D_+^{n+1}\lambda(x) &= D_+\left(D_+^n \lambda(x)\right) = D_+\left(\frac{\lambda(\epsilon)^n}{\epsilon} + \lambda(\epsilon)^n \lambda(x)\right) = \lambda(\epsilon)^n D_+ \lambda(x) \\ &= \lambda(\epsilon)^n \left(\frac{\lambda(\epsilon)}{\epsilon} + \lambda(\epsilon)\lambda(x)\right) = \frac{\lambda(\epsilon)}{\epsilon}^{n+1} + \lambda(\epsilon)^{n+1}\lambda(x) \end{aligned}$$

which means that the identity is valid for any $n \in \mathbb{N}$.

To verify the last part of the Lemma it is enough to take account of the fact that

$$\frac{\lambda(\epsilon)}{\epsilon} = \frac{{}^*\exp_\Pi(2\pi i\epsilon^2) - 1}{\epsilon^2} \approx 2\pi i$$

and the result follows immediately by induction. •

Definition 9.12 Let $\mathbf{T}_{\Pi,0}$ be the subset formed by all finite sums of the form

$$\sum_{j=1}^{n} \alpha_j F_j$$

with $n \in \mathbb{N}$, $F_j \in \mathbf{T}_\Pi$ and $\alpha_j \approx 0$ for $j = 1, 2, \ldots, n$. Two functions $F, G \in \mathbf{T}_\Pi$ are said to be $[S]$**-equivalent** , and we write $F \ [S] \ G$, if and only if $F - G \in \mathbf{T}_{\Pi,0}$.

Lemma 9.13 Let G be any function in $\mathbf{SC}_{\Pi,t}$. Then given any numbers $p, q, r, s \in \mathbb{N}_0$, there exists a finite number, say $v \in \mathbb{N}$, of functions G_j (with $j = 1, 2, \ldots, v$) belonging to $\mathbf{SC}_{\Pi,t}$ such that

$$\lambda^r \bar{\lambda}^s D_+^p D_-^q G \ \ [S] \ \ \sum_{j=1}^{v} D_+^{p_j} D_-^{q_j} G_j \tag{9.14}$$

where $p_j, q_j, \in \mathbb{N}_0$ for $j = 1, 2, \ldots, v$.

Proof In order to simplify the exposition we shall denote the formula (9.14) with the variables $p, q, r, s \in \mathbb{N}_0$ by $\phi(p, q, r, s)$.

The formula $\phi(p, q, 0, 0)$ is trivially satisfied for $p, q \in \mathbb{N}_0$. For given $r \in \mathbb{N}_0$ suppose that the formula $\phi(p, q, r, 0)$ is valid; that is to say,

$$\lambda^r D_+^p D_-^q G \ \ [S] \ \ \sum_{j=1}^{v} D_+^{p_j} D_-^{q_j} G_j$$

with $p_j, q_j, j = 1, 2, \ldots, v$ belonging to $\mathbb{N}_0$ and G_j, $j = 1, 2, \ldots, v$, belonging to $\mathbf{SC}_{\Pi,t}$. Then

$$\lambda^{r+1} D_+^p D_-^q G = \lambda(\lambda^r D_+^p D_-^q G) \ \ [S] \ \ \lambda\left(\sum_{j=1}^{r} D_+^{p_j} D_-^{q_j} G_j\right) \tag{9.15}$$

and therefore, to prove $\phi(p, q, r+1, 0)$ it is enough to show that

$$\lambda D_+^p D_-^q G \ \ [S] \ \ D_+^p D_-^q(\lambda G) - 2\pi i(p D_+^{p-1} D_-^q G + q D_+^p D_-^{q-1} G) \ . \tag{9.16}$$

We prove (9.16) in the first place for $p \in \mathbb{N}_0$ and $q = 0$. That is, we show the validity of the relation

$$\lambda D_+^p G \ \ [S] \ \ D_+^p(\lambda G) - 2\pi i p D_+^{p-1} G$$

for any $p \in \mathbb{N}$. Let $p = 1$. Seeing that

$$\begin{aligned} D_+(\lambda G) = \lambda(x+\epsilon) D_+ G + (D_+\lambda) G &= \{\epsilon(D_+\lambda) + \lambda\} D_+ G + (D_+\lambda) G \\ &= \epsilon(D_+\lambda)(D_+ G) + \lambda(D_+ G) + (D_+\lambda) G, \end{aligned}$$

then, taking Lemma 9.11 into account, we get

$$D_+(\lambda G) = \epsilon\left(\frac{\lambda(\epsilon)}{\epsilon} + \lambda(\epsilon)\lambda\right)(D_+G) + \lambda(D_+G) + \left(\frac{\lambda(\epsilon)}{\epsilon} + \lambda(\epsilon)\lambda\right)G$$

$$= \lambda(D_+G) + \lambda(\epsilon)(D_+G) + \epsilon\lambda(\epsilon)\lambda(D_+G) + \frac{\lambda(\epsilon)}{\epsilon}G + \lambda(\epsilon)\lambda G$$

and therefore

$$\lambda D_+G = D_+(\lambda G) - \lambda(\epsilon)(D_+G) - \epsilon\lambda(\epsilon)(D_+G) - \frac{\lambda(\epsilon)}{\epsilon}G - \lambda(\epsilon)\lambda G$$

$$= D_+(\lambda G) - \lambda(\epsilon)\{(D_+G) + \epsilon(D_+G) + G\} - \frac{\lambda(\epsilon)}{\epsilon}G \ .$$

Since $\lambda(\epsilon)/\epsilon = 2\pi i + \eta$, with $\eta \approx 0$, then $\lambda(\epsilon) = 2\pi i\epsilon + \epsilon\eta$ and $\epsilon\lambda(\epsilon) = 2\pi i\epsilon^2 + \epsilon^2\eta$ and so

$$\lambda D_+G = D_+(\lambda G) - 2\pi iG - \{\eta G + \epsilon(2\pi i + \eta)[(1+\epsilon\lambda)(D_+G) + \lambda G]\}$$

from which there follows

$$\lambda D_+G \ \ [S] \ \ D_+(\lambda G) - 2\pi iG$$

thus proving the proposition for $p = 1$.

Suppose now that the proposition is valid for a given $p \in \mathbb{N}$, that is that

$$\lambda D_+^pG \ \ [S] \ \ D_+^p(\lambda G) - 2\pi ipD_+^{p-1}G \ .$$

Then

$$\begin{aligned}\lambda D_+^{p+1}G &= \lambda D_+(D_+^pG)\\ &[S] \ \ D_+(\lambda D_+^pG) - 2\pi iD_+^pG\\ &[S] \ \ D_+(D_+^p(\lambda G) - 2\pi ipD_+^{p-1}G) - 2\pi iD_+^pG\\ &= D_+^{p+1}(\lambda G) - 2\pi ipD_+^pG - 2\pi iD_+^pG\\ &= D_+^{p+1}(\lambda G) - 2\pi i(p+1)D_+^pG\end{aligned}$$

which shows that (9.16) is true for $p+1$ if it is true for p. Similarly we can prove (9.16) for any $p, q \in \mathbb{N}_0$ whatsoever.

Returning to the general proof of the Lemma, from (9.16) we have

$$\lambda^{p+1}D_+^pD_-^qG = \lambda(\lambda^rD_+^pD_-^qG)$$

$$[S] \ \ \lambda\left(\sum_{j=1}^{v} D_+^{p_j}D_-^{q_j}G_j\right) = \sum_{j=1}^{v}\lambda\left(D_+^{p_j}D_-^{q_j}G_j\right)$$

$$[S] \ \ \sum_{j=1}^{v}\left\{D_+^{p_j}D_-^{q_j}(\lambda G_j) - 2\pi i(p_jD_+^{p_j-1}D_-^{q_j}G_j + q_jD_+^{p}D_-^{q_j-1}G_j\right\}$$

which proves $\phi(p, q, r, 0)$ for any $p, q, r \in \mathbb{N}_0$.
Finally by a procedure similar to that used to prove (9.16), after the relation

$$\bar{\lambda} D_+^p D_-^q G \quad [S] \quad D_+^p D_-^q G + 2\pi i \left(p D_+^{p-1} D^q G + q D_+^p D_-^{q-1} D G \right), \tag{9.17}$$

is established, we can show that $\phi(p, q, r, s)$ implies $\phi(p, q, r, s+1)$, thus completing the overall proof of the lemma. •

Theorem 9.14 The Π-Fourier transform of a tempered pre-distribution $F \in \mathbf{T}_\Pi$ is again a tempered pre-distribution $\hat{F} \in \mathbf{T}_\Pi$.

Proof From (9.13) we obtain, for each $y \in \Pi$,

$$\hat{F}(y) = \sum_{x \in \Pi} \epsilon F(x) {}^* \exp_\Pi(-2\pi i x y) = \sum_{x \in \Pi} \left(D_+^p D_-^q \Phi(x) \right) {}^* \exp_\Pi(-2\pi i x y)$$

$$= \lambda(y)^p (-\bar{\lambda}(y))^q \sum_{x \in \Pi} \epsilon \Phi(x) {}^* \exp_\Pi(-2\pi i x y) \quad .$$

Considering (9.9), let $m \in \mathbb{N}_0$ be such that

$$G_m(x) = \frac{\Phi(x)}{(1 + |\lambda(x)|^2)^m}, \qquad x \in \Pi$$

is an **S**Π-integrable function. Then

$$\hat{F}(y) = \lambda(y)^p (-\bar{\lambda}(y))^q \sum_{x \in \Pi} \epsilon (1 + |\lambda(x)|^2)^m G_m(x) {}^* \exp_\Pi(-2\pi i x y)$$

$$= \lambda(y)^p (-\bar{\lambda}(y))^q (1 - D_+ D_-)^m \hat{G}_m(y),$$

where

$$\hat{G}_m(y) = \sum_{x \in \Pi} \epsilon G_m(x) {}^* \exp(-2\pi i x y), \qquad y \in \Pi,$$

is a function belonging to $\mathbf{SC}_{\Pi,t}$. From Lemmas 9.11 and 9.13 it now follows that

$$F(y) \quad [S] \quad \sum_{j=1}^{v} D_+^{p_j} D_-^{q_j} \hat{G}_{m,j}(y) \tag{9.18}$$

where $\hat{G}_{m,j}$, $j = 1, 2, \ldots, v$ are functions of $\mathbf{SC}_{\Pi,t}$. Since the two sides only differ by an infinitesimal function (belonging to $\mathbf{T}_{\Pi,0}$), we obtain

$$\mathbf{st}_S F = \mathbf{st}_S \left(\sum_{j=1}^{v} D_+^{p_j} D_-^{q_j} \hat{G}_{m,j}(y) \right)$$

which proves the theorem. •

Let $F \equiv \mathcal{F}_\Pi[F]$, where F is a tempered pre-distribution. Then $\hat{F}$ generates a (standard) tempered distribution $\mathbf{st}_S \hat{F}$. It remains to show that $\mathbf{st}_S \hat{F}$ is the (standard) generalised Fourier transform of $\mathbf{st}_S F$.

Theorem 9.15 For any $F \in \mathbf{T}_\Pi$ we have

$$\mathcal{F}[\mathbf{st}_\mathcal{S} F] = \mathbf{st}_\mathcal{S} \hat{F} \quad .$$

Proof For any function $\phi \in \mathcal{S}$ we have

$$< \mathbf{st}_\mathcal{S} \hat{F}, \phi > \approx \sum_{y \in \Pi} \epsilon \hat{F}(y) \, {}^*\phi_\Pi(y)$$

$$= \sum_{y \in \Pi} \epsilon \left\{ \sum_{x \in \Pi} \epsilon F(x) \, {}^*\exp_\Pi(-2\pi i x y) \right\} {}^*\phi_\Pi(y)$$

$$= \sum_{x \in \Pi} \epsilon F(x) \left\{ \sum_{y \in \Pi} \epsilon \, {}^*\phi_\Pi(y) \, {}^*\exp_\Pi(-2\pi i x y) \right\}$$

$$= \sum_{x \in \Pi} \epsilon F(x) \hat{\Phi}(x)$$

where

$$\hat{\Phi}(x) = \mathcal{F}_\Pi[{}^*\phi_\Pi](x) = \sum_{y \in \Pi} \epsilon \, {}^*\phi_\Pi(y) \, {}^*\exp_\Pi(-2\pi i x y) \quad .$$

The theorem will be completely proved if we show that the relation

$$\sum_{x \in \Pi} \epsilon F(x) \hat{\Phi}(x) \approx \sum_{x \in \Pi} \epsilon F(x) \, {}^*\hat{\phi}_\Pi(x) \tag{9.19}$$

where ${}^*\hat{\phi}_\Pi \equiv {}^*(\hat{\phi})_\Pi$, is satisfied for every test function $\phi \in \mathcal{S}$.

Suppose first that F belongs to $\mathbf{SC}_\Pi$; then for some number $m \in \mathbf{N}_0$ the internal function

$$\frac{F(x)}{(1 + |\lambda(x)|^2)^m}$$

is $\mathbf{S}\Pi$-integrable over Pi. Consequently

$$\sum_{x \in \Pi} \epsilon \frac{|F(x)|}{(1 + |\lambda(x)|^2)^m}$$

is a finite number and, for all $\alpha \in \Pi_\infty^+$, we have

$$\sum_{|x| > \alpha} \epsilon \frac{|F(x)|}{(1 + |\lambda(x)|^2)^m} \approx 0 \quad .$$

On the other hand, since for all $x \in \Pi_b$ we have $\mathbf{st}\hat{\Phi}(x) = \hat{\phi}(\mathbf{st}x)$, then

$$(1 + |\lambda(x)|^2)^m |\hat{\Phi}(x) - {}^*\hat{\phi}_\Pi(x)| \approx 0$$

or all $x \in \Pi$. The set defined by

$$\{z \in \Pi^+ : \forall_x [x \in \Pi \Rightarrow [|x| < z \Rightarrow (1 + |\lambda(x)|^2)^m |\hat{\Phi}(x) - {}^*\hat{\phi}_\Pi(x)| < 1/z]]\}$$

s internal and contains all finite positive numbers $z \in \Pi^+$; by overflow it will contain n infinite number $\eta \in \Pi^+_\infty$. Hence

$$\forall_x [x \in \Pi \Rightarrow |x| < \eta \Rightarrow (1 + |\lambda(x)|^2)^m |\hat{\Phi}(x) - {}^*\hat{\phi}_\Pi(x)| \approx 0]]\} \ .$$

However this fact does not imply that $(1 + |\lambda(x)|^2)^m |\hat{\Phi}(x) - {}^*\hat{\phi}_\Pi(x)|$ is infinitesimal n the whole hyperfinite line Π; it may not be infinitesimal at points of the set $\Pi \backslash [-\eta, \ldots, \eta]$. Nevertheless, from the properties of the functions of $\mathcal{S}$ we know that here exists a finite constant $C > 0$ such that

$$(1 + |\lambda(x)|^2)^m |\hat{\Phi}(x)| \le C \quad \text{and} \quad (1 + |\lambda(x)|^2)^m |{}^*\hat{\phi}_\Pi(x)| \le C$$

or all $x \in \Pi$. Hence, $(1 + |\lambda(x)|^2)^m |\hat{\Phi}(x) - {}^*\hat{\phi}_\Pi(x)|$ is a finite number for every $x \in \Pi$, nd in particular for $|x| > \eta$. Then

$$\left| \sum_{x \in \Pi} \epsilon F(x) [\hat{\Phi}(x) - {}^*\hat{\phi}_\Pi(x)] \right| \le$$

$$\left\{ \sum_{|x| \le \eta} + \sum_{|x| > \eta} \right\} \epsilon \frac{F(x)}{(1 + |\lambda(x)|^2)^m} \left\{ (1 + |\lambda(x)|^2)^m |\hat{\Phi}(x) - {}^*\hat{\phi}_\Pi(x)| \right\} \le$$

$$\max_{|x| \le \eta} \left\{ (1 + |\lambda(x)|^2)^m |\hat{\Phi}(x) - {}^*\hat{\phi}_\Pi(x)| \right\} \cdot \sum_{|x| \le \eta} \epsilon \frac{|F(x)|}{(1 + |\lambda(x)|^2)^m} +$$

$$\max_{|x| > \eta} \left\{ (1 + |\lambda(x)|^2)^m |\hat{\Phi}(x) - {}^*\hat{\phi}_\Pi(x)| \right\} \cdot \sum_{|x| > \eta} \epsilon \frac{|F(x)|}{(1 + |\lambda(x)|^2)^m} \approx 0$$

hus giving a proof of (9.19) in the case when $F \in \mathbf{SC}_{\Pi,t}$.

Now suppose that F is any function in $\mathbf{T}_\Pi$; that is that F is of the form

$$F = D^p_+ D^q_- G$$

with $G \in \mathbf{SC}_{\Pi,t}$ and $p, q \in \mathbb{N}_0$. Then

$$\sum_{x \in \Pi} \epsilon (D^p_+ D^q_- G(x)) \hat{\Phi}(x) = (-1)^{p+q} \sum_{x \in \Pi} \epsilon G(x) (D^p_+ D^q_- \hat{\Phi}(x))$$

nd

$$\sum_{x \in \Pi} \epsilon (D^p_+ D^q_- G(x)) \, {}^*\hat{\phi}(x) = (-1)^{p+q} \sum_{x \in \Pi} \epsilon G(x) (D^p_+ D^q_- \, {}^*\hat{\phi}(x)) \ .$$

The relation (9.19) is now obtained in a similar way, taking into account that for any $m \in \mathbb{N}_0$

$$(1 + |\lambda(x)|^2)^m |D^p_+ D^q_- [\hat{\Phi}(x) - {}^*\hat{\phi}_\Pi(x)]|$$

s finite for all $x \in \Pi$ (and infinitesimal for all $x \in \Pi_b$). •

9.3 Study of Π-Band-limited Functions

9.3.1 Preliminary definitions

Let I be the Π-interval $[-1/2 \ldots 1/2] = \{x \in \Pi : -1/2 \leq x \leq 1/2\}$. An internal function $\hat{F} \in \mathbf{F}_\Pi$ is said to be **finitely supported** in I if $\hat{F}(y) = 0$ for Ξ_L-almost all $y \notin I$ and

$$\mathbf{st}\left(\sum_{y \in I} \epsilon |F(y)|\right) \approx 0 \ .$$

In these circumstances the internal function of $\mathbf{F}_\Pi$ defined by $F = \bar{\mathcal{F}}_\Pi[\hat{F}]$ is said to be a **Π-band-limited** function (or, a function of **Π-limited spectrum**) on I. We shall denote by $\mathcal{B} \equiv \mathcal{B}_\Pi(I)$ the vector subspace of $\mathbf{F}_\Pi$ containing all functions which are Π-bandlimited on I. For each function $F \in \mathcal{B}_\Pi$ we will have

$$F(x) = \sum_{y \in \Pi} \epsilon \hat{F}(y) \, {}^*\exp_\Pi(2\pi i x y) \approx \sum_{-1/2 \leq y < 1/2} \epsilon \hat{F}(y) \, {}^*\exp_\Pi(2\pi i x y) \tag{9.20}$$

and therefore

$$|F(x)| \leq \|\hat{F}\|_{\Pi,1} \ .$$

Here,

$$\|\hat{F}\|_{\Pi,1} = \sum_{y \in \Pi} \epsilon |\hat{F}(y)|$$

is the Π_1-norm of $\hat{F}$, and is a non-negative hyperreal number which may be finite or infinite. The function F can be extended in a natural way to the hyperfinite complex plane $\Pi + i\Pi$, giving

$$F(x + i\eta) = \sum_{y \in \Pi} \epsilon \hat{F}(y) \, {}^*\exp_\Pi[2\pi i (x + i\eta) y]$$

$$\approx \sum_{y \in I} \epsilon \left\{\hat{F}(y) \, {}^*\exp_\Pi(-2\pi \eta y)\right\} \, {}^*\exp_\Pi(2\pi i x y)$$

from which we obtain the following upper bound

$$|F(x + i\eta)| \leq \|\hat{F}\|_{\Pi,1} \, {}^*\exp_\Pi(\pi |\eta|)$$

for all $(x + i\eta) \in \Pi + i\Pi$. Moreover, for any $j \in {}^*\mathbb{N}_0$,

$$D_+^j F(x) = \sum_{y \in \Pi} \epsilon \lambda(y)^j \hat{F}(y) \, {}^*\exp_\Pi(2\pi i x y)$$

$$\approx \sum_{y \in I} \epsilon \lambda(y)^j \hat{F}(y) \, {}^*\exp_\Pi(2\pi i x y)$$

and therefore

$$|D_+^j F(x)| \leq C_j . \|\hat{F}\|_{\Pi,1}$$

where

$$C_j = \max_{y \in I} |\lambda(y)|^j \leq 2\pi\{\max_{y \in I} |y|^j\} = \pi 2^{1-j}$$

is a constant (finite for all $j \in \mathbb{N}_0$ and infinitesimal for j infinite).

Consequently, if $\|\hat{f}\|_{\Pi,1}$ is a finite number, the function $F = \bar{\mathcal{F}}_\Pi[\hat{F}]$ is finite and all its Π-derivatives are finite on the whole hyperfinite line; moreover, F may be extended to the hyperfinite complex plane $\Pi + i\Pi$ as a finite function of finite exponential type. In general, F may not be an **S**Π-continuous function and, therefore, **st**F may not exist; however the equation

$$\mathbf{st}\left(\sum_{x\in\Pi} \epsilon F(x)\, {}^*\hat{\phi}_\Pi(x)\right) = \mathbf{st}\left(\sum_{y\in I} \epsilon \hat{F}(y)\, {}^*\phi_\Pi(y)\right)$$

is valid for every function ϕ bounded in the interval $[-1/2, 1/2] \subset \mathbb{R}$. In particular $\hat{F}$ defines a Radon measure of compact support in $[-1/2, 1/2]$ and therefore F defines the inverse Fourier transform of such a Radon measure.

The aim of this section is to obtain a Π-cardinal representation for the functions in $\mathcal{B}_\Pi$. Before doing this, however, it is necessary to develop the analogue of the (standard) Fourier series in the hyperfinite line - the so-called Fourier Π-series.

9.3.2 Π-periodic functions

Given any internal function $F \in \mathbf{F}_\Pi$ the **Π-periodic transform**[3] of F of period 1 (or, simply, the Π-periodic transform of F) is that function $\mathbf{T}_\Pi[F] \in \mathbf{F}_\Pi$ which is defined by

$$\mathbf{T}_\Pi[F](x) = \sum_{n \in {}^*\mathbb{Z}_\Pi} F(x-n), \qquad x \in \Pi$$

where ${}^*\mathbb{Z}_\Pi \equiv {}^*\mathbb{Z} \cap \Pi$. (It is necessary to suppose, as usual, that $\mathbf{T}_\Pi[F]$ is periodically extended to $\epsilon\, {}^*\mathbb{Z}$).

In particular the periodic Π-transform of the function Δ_0

$$\mathbf{T}_\Pi[\Delta_0](x) = \sum_{n \in {}^*Z_\Pi} \Delta_0(x-n)$$

is the function defined by

$$\mathbf{T}_\Pi[\Delta_0](x) = \begin{cases} \kappa, & \text{if } x \in {}^*\mathbb{Z}_\Pi \\ 0, & \text{if } x \notin {}^*\mathbb{Z}_\Pi \end{cases}.$$

For the π-convolution of any function $F \in \mathbf{F}_\Pi$ with the function $\mathbf{T}_\Pi[F]$ we have

[3] Unless otherwise stated, we always consider the Π-periodic transform to have period equal to 1.

$$F \star \mathbf{T}_\Pi[\Delta_0](x) = \sum_{y\in\Pi} \epsilon F(x-y)\mathbf{T}_\Pi[\Delta_0](y)$$

$$= \sum_{y\in\Pi} \epsilon F(x-y)\left\{\sum_{n\in {}^*\mathbb{Z}_\Pi} \Delta_0(y-n)\right\}$$

$$= \sum_{n\in {}^*\mathbb{Z}_\Pi} \left\{\sum_{y\in\Pi} \epsilon F(x-y)\Delta_0(x-n)\right\}$$

$$= \sum_{n\in {}^*\mathcal{Z}_\Pi} F(x-n)$$

and therefore

$$\mathbf{T}_\Pi[F] = F \star \mathbf{T}_\Pi[\Delta_0] \quad . \tag{9.21}$$

9.5.2.1 The Fourier Π-series

Since we have

$$\mathcal{F}_\Pi[\mathbf{T}_\Pi[\Delta_0]](y) = \sum_{x\in\Pi} \epsilon \mathbf{T}_\Pi[\Delta_0](x) \,{}^*\exp_\Pi(-2\pi ixy)$$

$$= \sum_{x\in\Pi} \epsilon \left\{\sum_{n\in {}^*\mathcal{Z}_\Pi} \Delta_0(x-n)\right\} {}^*\exp_\Pi(-2\pi ixy)$$

$$= \sum_{n=-\kappa/2}^{(\kappa/2)-1} \left\{\sum_{x\in\Pi} \epsilon\Delta_0(x-n) \,{}^*\exp_\Pi(-2\pi ixy)\right\}$$

$$= \sum_{n=-\kappa/2}^{(\kappa/2)-1} {}^*\exp_\Pi(-2\pi iny) = \mathbf{T}_\Pi[\Delta_0](y)$$

we obtain the equality

$$\mathbf{T}_\Pi[\Delta_0](y) = \sum_{n\in {}^*\mathbb{Z}_\Pi} {}^*\exp_\Pi(-2\pi iny) \tag{9.22}$$

the right-hand side of which is described as the **Π-Fourier series**[4] of the internal function $\mathbf{T}_\Pi[\Delta_0]$. For an arbitrary function $F \in \mathbf{F}_\Pi$ we have

---

[4]Similarly we obtain

$$\mathbf{T}_\Pi[\Delta_0](y) = \sum_{n\in {}^*\mathbb{Z}_\Pi} {}^*\exp_\Pi(2\pi iny) \quad \text{for all} \quad y \in \Pi.$$

$$\begin{aligned}
\mathbf{T}_\Pi[F](x) &= F \star \mathbf{T}_\Pi[\Delta_0](x)\\
&= \sum_{y\in\Pi} \epsilon F(x-y)\left\{\sum_{n\in {}^*\mathbb{Z}_\Pi} {}^*\exp_\Pi(-2\pi ixy)\right\}\\
&= \sum_{n\in {}^*\mathbb{Z}_\Pi}\left\{\sum_{y\in\Pi} \epsilon F(x-y)\,{}^*\exp_\Pi(-2\pi ixy)\right\}\\
&= \sum_{n\in {}^*\mathbb{Z}_\Pi} {}^*\exp_\Pi(-2\pi inx)\left\{\sum_{y\in\Pi} \epsilon F(y)\,{}^*\exp_\Pi(-2\pi ixy)\right\}\\
&= \sum_{n\in {}^*\mathbb{Z}_\Pi} \hat{F}(n)\,{}^*\exp_\Pi(-2\pi inx)
\end{aligned}$$

where

$$\hat{F}(n) = \sum_{y\in\Pi} \epsilon F(y)\,{}^*\exp_\Pi(-2\pi iny) \quad .$$

The right-hand side of the next equation,

$$\mathbf{T}_\Pi[F](x) = \sum_{n\in {}^*\mathbb{Z}_\Pi} \hat{F}(n)\,{}^*\exp_\Pi(-2\pi inx), \qquad x \in \Pi \tag{9.23}$$

is called the **Fourier Π-series** of the internal function $\mathbf{T}_\Pi[F]$. Writing (9.23) in the form

$$\sum_{n\in {}^*\mathbb{Z}_\Pi} F(x-n) = \sum_{n\in {}^*\mathbb{Z}_\Pi} \hat{F}(n)\,{}^*\exp_\Pi(-2\pi inx)$$

and setting $x = 0$ we get

$$\sum_{n\in {}^*\mathbb{Z}_\Pi} f(n) = \sum_{n\in {}^*\mathbb{Z}_\Pi} \hat{F}(n) \tag{9.24}$$

which is the **Π-Poisson Summation Formula.**

Definition 9.16 Any function $P \in \mathbf{F}_\Pi$ which satisfies the condition

$$P(x+1) = P(x)$$

for all $x \in \Pi$ will be called a **Π-periodic function** with period 1 (or simply a Π-periodic function). We denote by $\mathbf{P}_\Pi$ the set of all functions in $\mathbf{F}_\Pi$ which are Π-periodic.

Let F be any function in $\mathbf{F}_\Pi$: since for all $x \in \Pi$ we have

$$\begin{aligned}
\mathbf{T}_\Pi[F](x+1) &= \sum_{n\in {}^*\mathbb{Z}_\Pi} F((x+1)-n)\\
&= \sum_{n\in {}^*\mathbb{Z}_\Pi} F(x-n) = \mathbf{T}_\Pi[F](x)
\end{aligned}$$

then the Π-periodic transform of any internal function of $\mathbf{F}_\Pi$ is a function of $\mathbf{P}_\Pi$. Conversely,

Theorem 9.17 Every internal function $P \in \mathbf{P}_\Pi$ is the Π-periodic transform of a function Φ_P finitely supported on the unit Π-interval $I \in \Pi$.

Proof Let $\mathbf{H}(x)$, $x \in \mathbb{R}$ be the Heaviside unit step function defined by

$$\mathbf{H}(x) = \begin{cases} 0, & \text{if } x < 0 \\ 1, & \text{if } x \geq 0 \end{cases} .$$

Then

$${}^*\mathbf{H}_\Pi(x) = \sum_{t=-\kappa/2}^{x} \epsilon\Delta_0(t)$$

and, further,

$${}^*\mathbf{H}_\Pi(x+1/2) - {}^*\mathbf{H}_\Pi(x-1/2), \qquad x \in \Pi$$

is a function of $\mathbf{F}_\Pi$ finitely supported on the Π-interval I. The function

$$\Phi_P(x) = [\,{}^*\mathbf{H}_\Pi(x+1/2) - {}^*\mathbf{H}_\Pi(x-1/2)].P(x), \qquad x \in \Pi$$

belongs to $\mathbf{F}_\Pi$ and is finitely supported on I, being such that

$$\begin{aligned} \mathbf{T}_\Pi[\Phi_P](x) &= \sum_{n \in {}^*\mathbb{Z}_\Pi} \Phi_P(x-n) \\ &= \sum_{n \in {}^*\mathbb{Z}_\Pi} [\,{}^*\mathbf{H}_\Pi(x+1/2-n) - {}^*\mathbf{H}_\Pi(x-1/2-n)].P(x-n) \\ &= P(x) \sum_{n \in {}^*\mathbb{Z}_\Pi} [\,{}^*\mathbf{H}_\Pi(x+1/2) - {}^*\mathbf{H}_\Pi(x-1/2)] = P(x) \end{aligned}$$

which completes the proof. •

From this theorem it follows that

$$\begin{aligned} P(x) = \mathbf{T}_\Pi[\Phi_P](x) = \Phi_P \star \mathbf{T}_\Pi[\Delta_0](x) &= \sum_{y \in \Pi} \epsilon\Phi_P(x)\mathbf{T}_\Pi[\Delta_0](x-y) \\ &= \sum_{n \in {}^*\mathbb{Z}_\Pi} {}^*\exp_\Pi(-2\pi i n x) \left\{ \sum_{y \in \Pi} \epsilon\Phi_P(y)\, {}^*\exp_\Pi(-2\pi i n y) \right\} \\ &= \sum_{n \in {}^*\mathbb{Z}_\Pi} c_{P,n}\, {}^*\exp_\Pi(-2\pi i n x) \end{aligned}$$

where

$$c_{P,n} = \sum_{y \in I} \epsilon\Phi(y)\, {}^*\exp_\Pi(-2\pi i n y) = \hat{\Phi}_P(n)$$

is the **Π-Fourier coefficient** of $P \in \mathbf{P}_\Pi$. Then

$$P(x) = \sum_{n \in {}^*\mathbb{Z}_\Pi} \hat{\Phi}_P(n)\, {}^*\exp_\Pi(-2\pi i n x), \qquad x \in \Pi \tag{9.25}$$

is the **Π-Fourier series** of the Π-periodic function P. Restricting (9.25) to the Π-interval I we have

$$\Phi_P(x) = \sum_{n\in {}^*\mathbb{Z}_\Pi} \hat{\Phi}_P(n) * \exp_\Pi(-2\pi i n x), \qquad x \in I.$$

9.3.2.2 The Fourier Π-series as an extension of the classical Fourier series

Let f be a periodic function (with period 1) defined on $\mathbb{R}$. Then there exists a function $g : [-1/2, 1/2[\rightarrow C$ such that

$$f = \sum_{n=-\infty}^{+\infty} \tau_n \circ g \ .$$

The nonstandard Π-extension of g, $\Phi(x) = {}^*g_\Pi(x)$, is an **S**Π-continuous function on the unit interval $I \subset \Pi$, and moreover, $\mathbf{T}_\Pi\Phi$ is an internal function in $\mathbf{P}_\Pi$, which is **S**Π-continuous and Π-periodic. Then

$$\mathbf{T}_\Pi[\Phi](x) = \sum_{n\in {}^*\mathbb{Z}_\Pi} \hat{G}(n) * \exp_\Pi(2\pi i n x) \tag{9.26}$$

where

$$\hat{G}(n) = \sum_{x\in I} \epsilon\, {}^*g_\Pi(x) * \exp_\Pi(-2\pi i n x) \ .$$

Suppose further that g is twice differentiable over the interval $[-1/2, 1/2]$. Then $D_+^2\Phi(x)$ is finite for all $x \in I$ and therefore the sum

$$\sum_{x\in I} \epsilon |D_+^2\Phi(x)|$$

is also finite. As

$$\mathcal{F}_\Pi[D_+^2\Phi](y) = \lambda(y)^2\hat{G}(y)$$

we have, for $n \neq 0$,

$$|\hat{G}(n)| = \frac{1}{|\lambda(n)|^2}|\mathcal{F}_\Pi[D_+^2\Phi](n)| \ .$$

Since $\lambda(n) = (2\pi i n) * \exp_\Pi(i\pi\epsilon n) * \mathrm{sinc}_\Pi(\epsilon n)$, then for all $n \in {}^*\mathbb{Z}_\Pi$ we have

$$|\mathcal{F}_\Pi[D_+^2\Phi](n)| = \left|\sum_{x\in I} \epsilon D_+^2\Phi(x) * \exp_\Pi(-2\pi i n x)\right| \le \sum_{x\in I} \epsilon |D_+^2\Phi(x)|$$

where the final member is, by hypothesis, a finite number. For large values of $|n|$ it follows that

$$|\hat{G}(n)| \le C\frac{1}{|n|^2}$$

where C denotes a finite constant. Let v_1, v_2 be two infinite, positive hypernatural numbers such that $v_1 \leq v_2 < \kappa/2$. Then for any $x \in \Pi$ we have

$$\left|\sum_{|n|=v_1}^{v_2} \hat{G}(n)\, {}^*\exp_\Pi(2\pi i n x)\right| \leq \sum_{|n|=v_1}^{v_2} |\hat{G}(n)| \leq 2C \sum_{n=v_1}^{v_2} \frac{1}{n^2} \approx 0 \quad .$$

We can now state the following result:

Theorem 9.18 For any number t in the interval $[-1/2, 1/2) \subset \mathbb{R}$, the projection of (9.23) on the real universe can be given in the form

$$f(t) = \mathbf{st} \circ \mathbf{T}_\Pi[\Phi](x) = \mathbf{st}\left(\sum_{n=-\kappa/2}^{(\kappa/2)-1} \hat{G}(n)\, {}^*\exp_\Pi(2\pi i n x)\right)$$

$$= \sum_{-\infty}^{+\infty} \hat{g}(h) \exp(2\pi i n t)$$

for any $x \in mon_\Pi(t)$.

Proof For each $x \in mon_\Pi(t)$ and each real number $r > 0$, the set

$$\left\{ v \in {}^*\mathbb{Z}_\Pi \cap {}^*\mathcal{N} : \left|\sum_{|n|=v}^{(\kappa/2)-1} \hat{G}(n)\, {}^*\exp_\Pi(2\pi i n x)\right| < r \right\}$$

is internal and contains all the infinite hypernatural numbers; by underflow it contains a finite number $n_r \in \mathbb{N}$ and therefore:

$$\left|\sum_{n=-\kappa/2}^{(\kappa/2)-1} \hat{G}(n)\, {}^*\exp_\Pi(2\pi i n x) - \sum_{n=-n_r}^{n_r} \hat{G}(n)\, {}^*\exp_\Pi(2\pi i n x)\right| < r \quad .$$

Taking standard parts we get

$$\left|\mathbf{st}\left(\sum_{n=-\kappa/2}^{(\kappa/2)-1} \hat{G}(n)\, {}^*\exp_\Pi(2\pi i n x)\right) - \sum_{n=-n_r}^{n_r} \mathbf{st}(\hat{G}(n))\mathbf{st}({}^*\exp_\Pi(2\pi i n x))\right| < r \quad .$$

Since $x \in mon_\Pi(t)$ and $|n| \in \mathbb{N}_0$ we have ${}^*\exp_\Pi(2\pi i n x) \approx \exp(2\pi i n t)$ and $\mathbf{st}(\hat{G}(n)) = \hat{g}(n)$, and since r is arbitrary we get finally

$$f(t) = \mathbf{st} \circ \mathbf{T}_\Pi[\Phi](x)$$

$$= \mathbf{st}\left(\sum_{n=-\kappa/2}^{(\kappa/2)-1} \hat{G}(n)\, {}^*\exp_\Pi(2\pi i n x)\right) = \sum_{n=-\infty}^{+\infty} \hat{g}(n) \exp(2\pi i n t)$$

for any $x \in mon_\Pi(t)$. •

9.3.3 The Π-cardinal series

Let F be a function of $\mathcal{B}_\Pi$. Then it is true that

$$F(x) = \sum_{y\in\Pi} \epsilon\hat{F}(y)\, {}^*\exp_\Pi(2\pi ixy) \approx \sum_{y\in I} \epsilon\hat{F}(y)\, {}^*\exp_\Pi(2\pi ixy) \tag{9.27}$$

for any $x \in \Pi$. For each $y \in \Pi$ we have

$$\hat{F}(y) = \mathbf{T}_\Pi[\hat{F}](y) = \sum_{n=-\kappa/2}^{(\kappa/2)-1} c_n\, {}^*\exp_\Pi(-2\pi iny)$$

where, for each $n \in {}^*\mathbb{Z}_\Pi$, the coefficient c_n is given by

$$c_n = \sum_{y\in\Pi} \epsilon\hat{F}(y)\, {}^*\exp_\Pi(2\pi iny) = F(n) \quad .$$

Replacing $\hat{F}$ in (9.27) by its Π-Fourier transform we obtain

$$F(x) \approx \sum_{y\in\Pi} \left\{ \sum_{n=-\kappa/2}^{(\kappa/2)-1} F(n)\, {}^*\exp_\Pi(-2\pi iny) \right\} {}^*\exp_\Pi(2\pi ixy)$$

$$= \sum_{n=-\kappa/2}^{(\kappa/2)-1} F(n) \left\{ \sum_{y\in I} \epsilon\, {}^*\exp_\Pi[2\pi i(x-n)y] \right\} = \sum_{n=-\kappa/2}^{(\kappa/2)-1} F(n)\Gamma(x-n)$$

where $\Gamma : \Pi \to {}^*\mathbf{C}$ is the function defined by

$$\Gamma(x) = \sum_{y\in I} {}^*\exp_\Pi(2\pi ixy) = \sum_{m=0}^{\kappa-1} \epsilon\, {}^*\exp_\Pi\left[2\pi ix\left(-\frac{1}{2} + m\epsilon\right)\right]$$

$$= \epsilon\, {}^*\exp_\Pi(-i\pi x) \sum_{m=0}^{\kappa-1} {}^*\exp_\Pi(2\pi i\epsilon xm) = \epsilon \frac{1 - {}^*\exp_\Pi(2\pi ix)}{1 - {}^*\exp_\Pi(2\pi i\epsilon x)} \quad .$$

Taking into account the definition of the function λ, we get

$$\Gamma(x) = \frac{2\pi ix}{\lambda(x)}\, {}^*\mathrm{sinc}_\Pi(x) \tag{9.28}$$

and finally,

$$F(x) = \sum_{n=-\kappa/2}^{(\kappa/2)-1} F(n) \frac{2\pi ix}{\lambda(x)}\, {}^*\mathrm{sinc}_\Pi(x) \tag{9.29}$$

which is described as the Π-cardinal series of the internal function $F \in \mathcal{B}_\Pi$. The function F may be expressed in terms of its values at the hyperinteger points $n \in {}^*\mathbb{Z}_\Pi$ and of the **interpolation function** Γ.

9.3.3.1 The Π-cardinal series as extension of the Whittaker-Kotel'nikov-Shannon cardinal series for functions of compact spectrum

Let f be a continuous function with compact support contained in the interval $[-1/2, 1/2] \subset \mathbb{R}$. Then the internal function

$$\hat{F} = {}^*\hat{f}_\Pi(y), \qquad y \in I,$$

is finite, SΠ-continuous and such that $\mathbf{st}\hat{F} = \hat{f} \circ \mathbf{st}$: moreover, $\|\hat{F}\|_{\Pi,1}$ is certainly a finite number.

The inverse Π-Fourier transform of $\hat{F}$ is itself a function in $\mathbf{SC}_\Pi$. In fact for $x, \xi \in \Pi_b$ we will have

$$F(x) - F(\xi) = \sum_{y\in I} \epsilon \hat{F}(y)\, {}^*\exp_\Pi(2\pi i \xi y)\, \{ {}^*\exp_\Pi[2\pi i(x-\xi)y] - 1\}$$

$$= 2\pi i(x-\xi) \sum_{y\in\Pi} \epsilon y \hat{F}(y)\, {}^*\exp_\Pi(2\pi i \xi y)[1 + 2\pi i(x-\xi)y\mathbf{O}(1)]$$

and therefore

$$|F(x) - F(\xi)| \leq 2\pi|x-\xi| \sum_{y\in I} \epsilon |y\hat{F}(y)||1 + \epsilon\pi i(x-\xi)y\mathbf{O}(1)|$$

$$\leq 2\pi|x-\xi| \left\{ \max_{y\in I}[1 + 2\pi|x-\xi|y\mathbf{O}(1)] \right\} \sum_{y\in I} \epsilon |y\hat{F}(y)|$$

$$= \pi|x-\xi| \left\{ \max_{y\in I}[1 + 2\pi|x-\xi|\mathbf{O}(1)] \right\} \|\hat{F}\|_{\Pi,1} \ .$$

Then, if $x \approx \xi$, we will have $F(x) \approx F(\xi)$ in Π_b, as stated. The projection $f = \mathbf{st}F$ is a function well defined and continuous on $\mathbb{R}$, and is the (standard) inverse Fourier transform of the function $\hat{f}$. Moreover, since

$$D_+^j F(x) = \sum_{y\in I} \epsilon \lambda(y)^j \hat{F}(y)\, {}^*\exp_\Pi(2\pi i x y)$$

it is easy to verify that $D_+^j F$ belongs to $\mathbf{SC}_\Pi$ for every $j \in \mathbb{N}_0$ (although this is not necessarily true for infinite $j \in {}^*\mathbb{N}$). As $\mathbf{st}D_+^j F = f^{(j)}$ we can assert that $f \equiv \mathbf{st}F$ belongs to C^∞ and, furthermore, that F can be extended to the complex plane as an entire function of exponential type $\leq \pi$.

Further, suppose that

$$\|D_+\hat{F}\|_{\Pi,1} = \sum_{y\in I} \epsilon |D_+\hat{F}(y)|$$

is a finite number. For each infinite $n \in {}^*\mathbb{Z}_\Pi$ we have that

$$\bar{\lambda}(n)F(n) = \sum_{y\in I} \epsilon D_+\hat{F}(y)\, {}^*\exp_\Pi(2\pi i x y)$$

and therefore

$$|\bar{\lambda}(n)F(n)| \leq \|D_+\hat{F}\|_{\Pi,1}$$

from which, taking into consideration that $|\bar{\lambda}(n) \geq 4|n|$, (see (9.6)), we get

$$|F(n)| \leq C.\frac{1}{|n|}$$

where $C = \|D_+\hat{F}\|_{\Pi,1}/4$ is a finite constant. Since for each $x \in \Pi_b$ and infinite $n \in {}^*\mathbf{kZ}_\Pi$ we have

$$|\Gamma(x-n)| \leq \frac{\pi}{2}\left|\frac{\sin[\pi(x-n)]}{\pi(x-n)}\right| \leq \frac{1}{2}\frac{1}{|x-n|}$$

$$\leq \frac{1}{2}\frac{1}{||x|-|n||} \leq \frac{1}{2|n|}\frac{1}{1-|x/n|} \leq \frac{1}{|n|},$$

seeing that $1-|x/n| \geq 1/2$. Then for $v_1, v_2 \in {}^*\mathbb{N}_\infty$ such that $v_1 \leq v_2 < \kappa/2$ we have

$$\left|\sum_{|n|=v_1}^{v_2} F(n)\Gamma(x-n)\right| \leq 2C\sum_{|n|=v_1}^{v_2}\frac{1}{n^2} \approx 0 \ .$$

We can now derive the following result:

Theorem 9.19 Let $\hat{f}$ be a (standard) function, continuous and differentiable on the interval $[-1/2, 1/2] \subset \mathbb{R}$. If r is any point of $\mathbb{R}$ then the projection of the Π-cardinal series (9.29) can be put into the following form:

$$f(t) = \mathbf{st}F(x) = \mathbf{st}\left(\sum_{n=-\kappa/2}^{(\kappa/2)-1} F(n)\frac{2\pi ix}{\lambda(x)} * \mathrm{sinc}_\Pi(x)\right)$$

$$= \sum_{n=-\infty}^{+\infty} f(n)\,\mathrm{sinc}(t-n)$$

where x is any point in $mon_\Pi(t)$, and the convergence of the series is uniform on compacts of $\mathbb{R}$.

Proof If $\hat{f}$ is differentiable in $[-1/2, 1/2]$ then although $\mathbf{st}D_+\hat{F}$ may not be equal to $\hat{f}' \circ \mathbf{st}$, the internal function $D_+\hat{F}$ will be finite at all points of I and consequently $\|D_+\hat{F}\|_{\Pi,1}$ is a finite number. Then, for each $x \in mon_\Pi(t)$ and each real number $r > 0$, the set

$$\left\{v \in {}^*\mathbb{N}_\Pi : \left|\sum_{|n|=v}^{(\kappa/2)-1} F(n)\Gamma(x-n)\right| < \frac{r}{2}\right\}$$

is internal and contains all the infinite numbers in ${}^*\mathbb{N}_\Pi$. By underflow it will contain a finite number $n_r \in \mathbb{N}$ and therefore

$$\left|\sum_{|n|=v}^{(\kappa/2)-1} F(n)\Gamma(x-n) - \sum_{n=-n_r}^{n_r} F(n)\Gamma(x-n)\right| < \frac{r}{2}$$

from which, taking standard parts,

$$\left| \mathbf{st}\left(\sum_{|n|=v}^{(\kappa/2)-1} F(n)\Gamma(x-n) \right) - \sum_{n=-n_r}^{n_r} \mathbf{st}F(n)\mathbf{st}\Gamma(x-n) \right| < \frac{r}{2} .$$

Now, since for $x \in mon_\Pi(t)$ and $|n| \in \mathbb{N}_0$ we have $\Gamma(x-n) \approx \Gamma(t-n)$, and therefor

$$\mathbf{st}\Gamma(x-n) = \mathbf{st}\Gamma(t-n) = \text{sinc}(t-n),$$

then the result of the theorem is a consequence of the fact that $r > 0$ is an arbitrar real number. Finally, seeing that for $x, z \in \Pi_b$ such that $x \approx z$ and for all $v \in {}^*\mathbb{N}_\infty \cap$ we have

$$\sum_{|n|=v}^{(\kappa/2)-1} F(n)\Gamma(x-n) \approx \sum_{|n|=v}^{(\kappa/2)-1} F(n)\Gamma(z-n)$$

so that the series converges quasi-uniformly on $\mathbb{R}$. •

From the proof of this theorem it follows easily that it is still true under less stringe conditions:

Theorem 9.20 Let $\hat{f}$ be a function continuous in the interval $[-1/2, 1/2]$ of $\mathbb{R}$ suc that $\|D_+\hat{F}\|_{\Pi,1}$ is a finite number. If t is any point of $\mathbb{R}$ then the projection of th Π-cardinal series (9.29) can be put into the following form:

$$f(t) = \mathbf{st}F(x) = \mathbf{st}\left(\sum_{n=-\kappa/2}^{(\kappa/2)-1} F(n)\frac{2\pi i x}{\lambda(x)} {}^*\text{sinc}_\Pi(x) \right)$$

$$= \sum_{n=-\infty}^{+\infty} f(n)\,\text{sinc}(t-n),$$

where x is any point in $mon_\Pi(t)$ and the convergence of the series is uniform c compact subsets of $\mathbb{R}$.

9.3.3.2 Generalisations

It is possible to project the Π-cardinal series (9.29) into the real universe und somewhat more general conditions. Without claiming that this is the most gener approach to this question we suppose that $F \in \mathcal{B}_\Pi$ but that we only know that $\|\hat{F}\|_\Pi$ is a finite number. In particular, $\|D_+\hat{F}\|_{\pi,1}$ may not be finite. The previous sectic may not therefore apply, at least not directly.

We consider the function $\hat{G} : \Pi \to {}^*\mathbf{C}$ defined by

$$\hat{G}(y) = \sum_{t=-1/2}^{y} \epsilon\hat{F}(t) - \left\{ \sum_{t\in I} \epsilon\hat{F}(t) \right\} {}^*\mathbf{H}_\Pi(y)$$

$$= \sum_{t=-1/2}^{y} \epsilon\hat{F}(t) - F(0)\,{}^*\mathbf{H}_\pi(y) \qquad (9.3$$

where, without loss of generality, we suppose that $\hat{F}$ vanishes outside I. As $\|\hat{F}\|_{\Pi,1}$ is, by hypothesis, finite and

$$F(0) \leq \|\hat{F}\|_{\Pi,1}$$

then $F(0)$ is a finite number.

From (9.30) it follows that $\hat{G}(y) = 0$ for $y \notin I$ and $D_+\hat{G} = \hat{F} - F(0)\Delta_0$ is such that $\|D_+\hat{G}\|_{\Pi,1}$ is a finite number. Then $G = \bar{\mathcal{F}}_\Pi[\hat{G}]$ is a function which satisfies the conditions of Theorem 9.20. The Π-cardinal series

$$G(x) = \sum_{n=-\kappa/2}^{(\kappa/2)-1} G(n)\Gamma(x-n)$$

reduces to the (standard) cardinal series

$$g(t) = \sum_{n=-\infty}^{+\infty} g(n)\,\mathrm{sinc}(t-n)$$

where $g = \mathbf{st}G : \mathbb{R} \to \mathbf{C}$ is a series converging uniformly on compacts.

Since

$$\hat{F}(y) = F(0)\Delta_0(y) + D_+\hat{G}(y)$$

then, applying the inverse Π-Fourier transform we get

$$F(x) = F(0) + \bar{\lambda}(x)G(x), \qquad x \in \Pi, \tag{9.31}$$

and therefore, for $t \in \mathbb{R}$ such that $t = \mathbf{st}x$,

$$\begin{aligned} f(t) &= f(0) + \mathbf{st}\bar{\lambda}(x).g(t) \\ &= f(0) - 2\pi i t g(t) \\ &= f(0) - 2\pi i t \sum_{n=-\infty}^{+\infty} g(n)\,\mathrm{sinc}(t-n) \end{aligned} \tag{9.32}$$

where, for $n \neq 0$,

$$g(n) = \mathbf{st}G(n) = \mathbf{st}\frac{F(n)-F(0)}{\lambda(n)} = -\frac{f(n)-f(0)}{2\pi i t} \quad .$$

On the other hand, from (9.31), taking into account that $\lambda(0) = 0$, we obtain

$$F(\epsilon) - F(0) = [\bar{\lambda}(\epsilon) - \bar{\lambda}(0)]G(\epsilon)$$

and therefore that

$$g(0) = \mathbf{st}G(\epsilon) = \mathbf{st}\frac{D_+F(0)}{D_+\bar{\lambda}(0)} = -\frac{1}{2\pi i}f'(0) \quad .$$

Substituting in (9.32) we obtain, finally,

$$f(t) = f(0) + \sum_{-\infty}^{+\infty} \frac{f(n)-f(0)}{n} t\,\mathrm{sinc}(t-n) \tag{9.33}$$

where the series converges uniformly on compacts.

Appendix A

Evolution of the concept of function

The explicit appearance of the idea of function is often cited as a landmark in th history of mathematical thought (see, for example, W.L.Schaaf, [41]). It separat "classical" mathematics, originating in classical Greece and essentially geometric i content, from "modern" mathematics as developed in the last 300 years.

Introduction The term "function" did not emerge explicitly in mathematics unt about the end of the 17th century. Derived from the Latin *functio* it was introduc by Leibniz in a paper published in the *Acta Eruditorum* in order to deal with certa geometric features of curves (co-ordinates, tangents, radii of curvature, etc.). Fo Leibniz such entities were "functions" of the curve in question. The function conce was therefore conceived in what was essentially a geometric context. Neverthele its appearance was crucial with regard to the opening of quite new horizons in th development of mathematics. Although initially devoted to the resolution of problem in geometry and mechanics the *Calculo Infinitesimal* was actually concerned wi functions as such, and so the concept demanded explicit expression.

The first "formal" definition of function which we recognise was given [40] by Je Bernoulli (1677-1748) in the following form:

Definition A.1 (J.Bernoulli, 1718) A function of a variable magnitude is quantity composed, in any way, of a variable together with some constants.

The concept of function was shared at this time between
(i) the idea of a geometric curve or graph, and
(ii) the idea of a formula or algebraic expression containing variables.

Bernoulli's definition already gave a clear indication of the "de-geometrisatio process which the calculus was to bring about as it developed. However t passage from the markedly geometric character of (i) to the essentially analytic o of (ii) would be principally due to the work of Leonhard Euler (1707-1783). It w Euler who contributed most decisively to the *liberation* of mathematical analysis fr geometry. For Euler, at this level, the calculus was typically concerned with formulas, that is to say with analytic expressions. The notion of function acquired, for the first time, an explicit form in his *Introductio Analysis Infinitorum* in which there appeared the following definition:

Definition A.2 (L.Euler, 1748) A function of a variable quantity is an analytic expression composed, in any way, of the variable quantity together with numbers or constant quantities.

Although it would only be very much later that "analytic expression" would be given a formal rigorous definition, Euler does explain that an admissible analytic expression might contain the four algebraic operations, roots, exponents, logarithms, trigonometric functions, derivatives and integrals. Such is the case, for example, with the expressions

$$a + 3x, \qquad az + b\sqrt{a^2 - z^2}, \qquad c^z,$$

which emphasises the fact that, in accordance with this definition, the concept of function remains indissolubly associated with the idea of analytic expression. Euler's initial position, that of total algebraisation of the notion of function, would come to undergo important modifications in the later stages of evolution of the concept, as will be seen in what follows.

The problem of vibrating strings The discussion of the concept of function became of central importance in the middle of the 18th century with the formulation of a problem of mathematical physics known as the Problem of Vibrating Strings. This generated an enormous controversy beginning with, as principal protagonists, Leonhard Euler on one side and Jean le Rond d'Alembert (1717-17) on the other:

Consider a taut elastic thread fixed at its two extremities. Suppose that this thread, subjected to a given initial displacement, is released at the instant $t = 0$ and allowed thereafter to move freely. The problem of vibrating strings then consists of the determination of the function $y = y(x,t)$ which describes the form of the string at some later instant $t > 0$.

The equation which governs the motion of the vibrating string will have, in modern terms[1] , the following form

$$\frac{\partial^2 y}{\partial x^2} = \frac{1}{c^2}\frac{\partial^2 y}{\partial t^2} \tag{A.1}$$

where $y(x,t)$ denotes the (vertical) displacement of the point with abscissa x at the instant t and c is a physical constant. (Without loss of generality we may take $c = 1$, which corresponds to the change of variable $ct \to t$).

Applying to equation (A.1), with $c = 1$, the identities

$$d\,(\partial y \partial x) = \left(\frac{\partial^2 y}{\partial x^2}\right) dx + \left(\frac{\partial^2 y}{\partial x \partial t}\right) dt = \left(\frac{\partial^2 y}{\partial t^2}\right) dx + \left(\frac{\partial^2 y}{\partial x \partial t}\right) dt$$

$$d\left(\frac{\partial y}{\partial t}\right) = \left(\frac{\partial^2 y}{\partial x \partial t}\right) dt + \left(\frac{\partial^2 y}{\partial t^2}\right) dt$$

[1]At this time there did not even exist an entirely clear notation for the partial derivative. It was generally confused with the notation used for *growth*. For simplicity of exposition we shall always use modern notation instead.

d'Alembert, in his celebrated paper of 1747, obtained the relations

$$d\left(\frac{\partial y}{\partial t}+\frac{\partial y}{\partial x}\right)=\left(\frac{\partial^2 y}{\partial x^2}+\frac{\partial^2 y}{\partial x\partial t}\right)(dt+dx)$$

$$d\left(\frac{\partial y}{\partial t}-\frac{\partial y}{\partial x}\right)=\left(\frac{\partial^2}{\partial x^2}-\frac{\partial^2 y}{\partial x\partial t}\right)(dt-dx)$$

from which we conclude that $\frac{\partial y}{\partial t}+\frac{\partial y}{\partial x}$ depends only on $(t+x)$ and $\frac{\partial y}{\partial t}-\frac{\partial y}{\partial x}$ depends only on $(t-x)$. That is to say, we must have

$$\frac{\partial y}{\partial t}+\frac{\partial y}{\partial x}=\Phi(t+x)$$

and

$$\frac{\partial y}{\partial t}-\frac{\partial y}{\partial x}=\Psi(t-x)$$

where Φ and Ψ are two functions (of a single variable). Solving this system for $\frac{\partial y}{\partial t}$ and $\frac{\partial y}{\partial x}$ we obtain

$$\begin{aligned} dy &= \left(\frac{\partial y}{\partial t}\right)dt+\left(\frac{\partial y}{\partial x}\right)dx \\ &= \frac{1}{2}\Phi(t+x)d(t+x)+\frac{1}{2}\Psi(t-x)d(t-x). \end{aligned}$$

A further integration then gives

$$y=\phi(t+x)+\psi(t-x)\ .$$

Using the boundary conditions $y(0,t)=0=y(0,l)$ and the initial conditions $y(x,0)=f(x)$, $\frac{\partial y}{\partial x}(x,0)=0$ where $f(x), 0\le x\le l$ represents the original configuration of the string, d'Alembert obtained, as solution of the equation (A.1),

$$y(x,t)=\frac{1}{2}[F(x+ct)+F(x-ct)] \qquad (A.3)$$

where F is a function of one variable defined over $(0,l)$. For the initial form of the string

$$F(x)=y(x,0)=f(x), \quad \text{for } 0\le x\le l\ .$$

For d'Alembert the function $F(x)$, and the initial configuration $f(x)$, $0\le x\le l$, must be given by a unique formula, at least twice differentiable.

Following a similar path Euler obtained a formal solution analogous to that of d'Alembert. However the interpretation by Euler and by d'Alembert of the nature of the solutions were completely different, and even antagonistic in certain respects. In the event it was Euler who ought to be regarded as ultimately victorious in the resulting famous dispute. D'Alembert's position corresponds to what is now understood as the *classical solution*, while Euler anticipated the 20th cntury *generalised solution* of this type of equation.

For d'Alembert it was necessary to restrict the initial configuration to functions which, in modern terminology, are of class C^2. Although functions not satisfying this condition might possibly correspond to acceptable physical situations, they could ot be considered in the context of the mathematical analysis of the problem. For Euler, on the contrary, any function which described a plane curve (without breaks), owever irregular it might be, ought to be allowed as a representative of an initial onfiguration of the vibrating string.

These two positions dictated two distinct modes of mathematical interpretation f the problem. D'Alembert assessed the validity of the equation, with its partial erivatives, from the point of view of the subsequent solutions: a function which was ot continuously differentiable could not satisfy equation (A.1) and therefore could ot be considered an admissible entity in the analysis of the problem. By contrast, Euler adopted a formal approach to the same equation and accepted a solution (A.3) which constituted a mathematical expression making physical sense, whatever might e the function f representing a plane curve, free from breaks. If a function F, defined n the interval $[0, l]$, for the initial configuration f were to satisfy the equation of the ibrating string, then f would actually possess adequate regularity properties. This would imply that f would have a unique analytic expression and would not exhibit ny sharp peaks, or corners. However, an intuitive physical approach to the problem emanded that initial configurations with such sharp angular features ought to be onsidered. The following example,

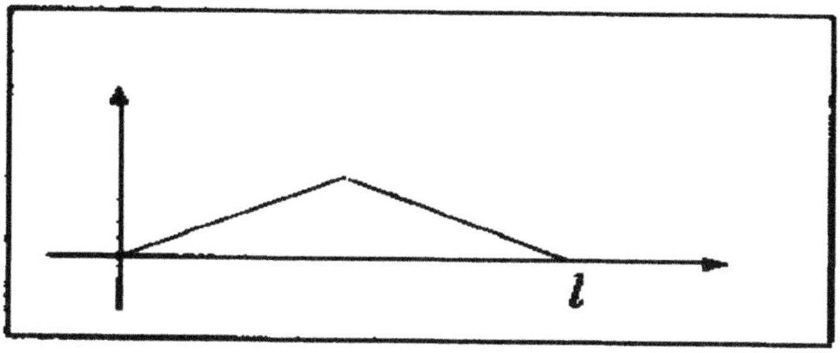

lthough it involves an angular peak nevertheless corresponds to a readily acceptable hysical situation, and one which could be seen to develop in space and time. For uler once the solution (A.3) had been obtained, all the preceding description was relevant: all that mattered was the final result.

This debate, often very heated, had the immense value of calling attention to the eed to clarify the concept of *function*, as well as the associated notions of *continuity* nd *discontinuity*. In his *Institutiones calculi differencialis* Euler reformulated the efinition of function as follows [39]:

Definition A.3 (Euler, 1755) If one quantity depends on another in such a way hat any change in the second produces a corresponding change in the first, then the rst is said to be a function of the second.

This definition already encompassed types of function much more general than those given explicitly by an analytic expression of any combination of constants and variables based on the known algebraic or transcendental operations which formed the material of practical mathematics at this time. It allowed Euler to introduce into mathematical analysis what were described as *discontinuous functions* which corresponded to curves drawn freehand, with or without breaks. Such function would generally be described today as sectionally continuous functions with sectionally continuous derivatives. Functions defined by a single analytic expression, with second derivatives similarly defined, were said by Euler to be *continuous* and these could be accepted as solutions of the problem of vibrating strings in the sense of d'Alembert.

Following Euler the concept of discontinuous function was later refined by L.F.Aarbogast (1759-1803) who distinguished those functions with breaks (*"when the different parts of the curve are not united"*), and called them *discontiguous functions.* These could appear in the course of the integration of differential equations. This distinction between discontiguous and discontinuous functions was the beginning of a process leading up to the modern concepts of the continuity and discontinuity of functions. This characterisation of discontinuous functions next led to an increasing awareness of the enormous generality of Euler's definition A.3, and new definitions of function appeared in the literature with a view to making the concept of function even more precise. In particular M.A.Condorcet (1743-1794) in a work (see [3]) which, although not published was widely circulated at the time, gave the following definition:

Definition A.4 (Condorcet, 1778-82) Given certain quantities $x, y, x, \ldots, F, .[\ldots],$ if it always happens that whenever x, y, z are determined then so also is F, though there may not be known a way of expressing F in terms of x, y, z in the form of an equation which links F, x, y and z, we say that F is a function of x, y and z.

For greater precision Condorcet additionally distinguished three types of functions:

1. those whose form is known; that is, the functions given explicitly;
2. that which is given by an equation in F, x, y, z, or equivalently, the functions defined implicitly;
3. that which is specified by means of certain conditions, such as differential equations.

Among those taking note of the work of Condorcet was S.F.Lacroix (1765-1843) who offered the following definition:

Definition A.5 (Lacroix, 1810) Every quantity whose value depends on one or more distinct quantities is designated a function of them, whether or not the operations required to pass from the latter to the former are known or not.

This development of the concept of function was the sequel to the controversy between d'Alembert and Euler, which was not fully resolved until the creation of the Schwartz theory of distributions in the 20th century - a theory of which Euler might well be considered the precursor. This dispute arose from the tension, which never totally disappeared, between the geometric and the algebraic elements co-existing within the concept of function. In the case of the vibrating string problem the first element allows us to consider any initial configuration corresponding to a function whose graph is a plane curve drawn freehand, though free from breaks, while the second requires a sufficiently well-behaved algebraic expression for its representation.

While the controversy actively involved many other great names in mathematics the most forward step in the evolution of the concept of function appears to have been made not be a mathematician but by a physicist - Daniel Bernoulli. His arguments were essentially physical: since, as was generally known, musical sounds were composed of fundamental frequencies together with their harmonics, Bernoulli[2] proposed in 1753 the following series solution for the problem of the vibrating string:

$$y(x,t) = \sum_{n=1}^{\infty} b_n \sin\frac{n\pi x}{l} \cos n\pi t l \quad .$$

He was convinced that this solution, with suitable choice of the constants b_n, must comprehend all the solutions of d'Alembert as well as those of Euler. However, it meant that the initial configuration $f(x)$ would itself have to be representable as a series of sines[3] ,

$$y(x,0) = \sum_{n=1}^{\infty} b_n \sin\frac{n\pi x}{l} \quad .$$

Both d'Alembert and Euler disagreed with Bernoulli's proposal. To understand the reasons for this summary rejection it is necessary to recall that in the 18th century it was universally held that *"two analytic expressions which coincided in an interval must coincide in all respects"*. Hence if the functions

$$f(x) \qquad \text{and} \qquad \sum_{n=1}^{\infty} b_n \sin\frac{n\pi x}{l}$$

coincided in $(0,l)$ then they would have to coincide everywhere. But this would be absurd since then *every* function $f(x)$ would be odd and periodic, which was manifestly unacceptable for the two mathematicians. The debate continued in this way for some time, but eventually died out without however reaching a completely satisfactory solution. Nevertheless it left, as an important legacy, a refinement of the concept of function together with a heightened sense of its importance.

The series of Fourier The evolution of the concept of function entered its decisive phase in the work of Joseph B. J. Fourier (1768-1830), with the introduction of Fourier series to the study of the heat equation. The (one-dimensional) equation for the diffusion of heat takes the form

$$\frac{\partial^2 T}{\partial x}\frac{1}{\kappa^2}\frac{\partial T}{\partial t} \qquad (A.4)$$

where $T \equiv T(x;t)$ is the temperature at the point of abscissa x of a uniform metallic bar at the instant t and κ is a constant depending on the material of which the bar is made. Fourier discovered a method which allowed him to obtain what he judged to

[2]Bernoulli himself gave no method for determining the b_n; this would not be achieved until much later, with the work of Fourier.

[3]A proposal showing great scientific perspicacity since, of course, nothing was known about Fourier series at the time.

be the most general possible solution of the equation (A.4). This method appeared to show that *every* function $f(x)$ defined on an interval could always be expressed as an infinite sum of sines and cosines. For example the function

$$f(x) = \begin{cases} -1, & \text{if } \; -\pi < x < 0 \\ 0, & \text{if } \; x = 0 \\ 1, & \text{if } \; 0 < x < \pi \end{cases}$$

could be expressed by means of a single formula in the following way

$$f(x) = \frac{4}{\pi}\left(\frac{\sin x}{1} + \frac{\sin 3x}{3} + \frac{\sin 5x}{5} + \dots\right) \; . \tag{A.5}$$

This result provoked enormous objections from the contemporaries of Fourier who believed the following argument to be valid:

1. the functions $\frac{\sin x}{1}, \frac{\sin 3x}{3}, \dots$ have continuous graphs, and
2. every finite sum of these functions also has a continuous graph,
3. it follows that the infinite sum (A.5) must have a continuous graph, it therefore being impossible for it to be equal to a function $f(x)$ which is discontinuous at the origin.

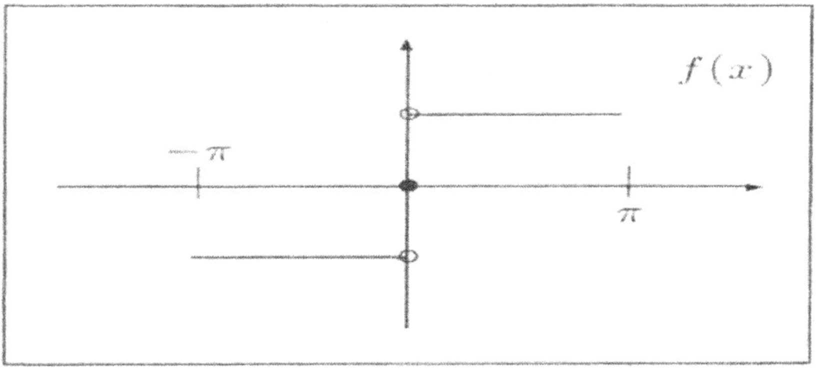

As is well known today, the conclusion (3) is false. Fourier's actual announcement can be given in the following terms:

Any function $f(x)$, defined in the interval $(-\pi, \pi)$ can always be represented in this interval by a series of sines and cosines

$$f(x) = \frac{1}{2}a_0 + \sum_{n=1}^{\infty}[a_n \cos nx + \sin nx] \tag{A.6}$$

where the coefficients are given by

$$a_n = \frac{1}{\pi} \int_{-\pi}^{\pi} f(t) \cos ntdt, \qquad n = 0, 1, 2, \ldots$$

$$b_n = \frac{1}{\pi} \int_{-\pi}^{\pi} f(t) \sin ntdt, \qquad n = 1, 2, \ldots.$$

Perhaps as a result of the controversy which his announcement generated, Fourier presented the following definition in his classic text *Théorie analytique de la chaleur*[40]:

Definition A.6 (Fourier, 1822) In general a function $f(x)$ represents a sequence of values or ordinates each one of which is arbitrary. Given an infinity of values of abscissa x there exists an equal number of values for the ordinates $f(x)$. All have numerical values, positive, negative or zero.

We do not suppose that these ordinates will be subject to a common law; they may succeed one another in any manner and each of them can be given as if it were a single quantity.

In order to understand fully the importance of Fourier's work and the reactions which it generated it should be noted that the height of controversy was not caused by the application of Fourier's result to a *particular* function: what provoked enormous reservations on the part of members of the mathematical community was the intransigent defence made by Fourier of the claim that his results applied to *all* functions.[4]

The definition of Dirichlet Although the assertion made by Fourier would have to be considered false, even in the context of his period, it did have a very positive effect. It led P.G.Lejeune Dirichlet (1805-1859) to carry out a critical analysis of Fourier's work and to establish the following result:

Theorem A.7 (Dirichlet) If a function has only finitely many discontinuities and finitely many maxima and minima in $(-\pi, \pi)$ then f can be represented by a Fourier series in $(-\pi, \pi)$. The Fourier series converges pointwise to $f(x)$ at every point where f is continuous and to $\frac{1}{2}[f(x+) + f(x-)]$ at each point of discontinuity.

The rigorous proof of this theorem first required the clarification of various important notions, including those of continuity, convergence, integral and, naturally, that of *function*. Among the first such clarifications was the suggestion by Cauchy (1789-1857) to Dirichlet to introduce a definition of function which is still used today:

Definition A.8 (Dirichlet, 1837) y is a function of a variable x, defined on the interval $a < x < b$, if to each value of the variable x in this interval there corresponds a well-defined value of the variable y. Moreover, the way in which this correspondence is established is completely irrelevant.

[4] Remember that Fourier had defined *function* in a particularly general form.

As an example to illustrate his definition, Dirichlet introduced for the first time the function which bears his name[5] today and which is defined as follows:

$$D(x) = \begin{cases} c, & \text{if } x \text{ is rational} \\ d, & \text{if } x \text{ is irrational} \end{cases}$$

where x lies between $-\pi$ and π, and c and d may be any two distinct numbers whatsoever. This function does not satisfy the hypothesis of Dirichlet's theorem on the representation of functions by Fourier series.

As Kleiner (op. cit.) affirms, this example of Dirichlet's was a true novelty in his time in that

- it was the first function not defined by one or more algebraic expressions, nor one whose graph could be drawn by hand;
- it was a function discontinuous everywhere (in the modern sense of the term):
- it illustrated, in an extreme form, a function that could be defined by an arbitrary rule of correspondence.

This function took on an enormous importance, not so much in itself but as giving an incentive for the creation of new types of function whose properties departed completely from what intuitively seemed admissible.

A celebrated example of such a so-called "pathological" function which appeared in the wake of the Dirichlet example is one provided by Weierstrass. Until about 1870 the mathematical community seemed to accept without question that geometric intuition (based on curves which could be drawn by hand) showed that any function which was continuous must be differentiable except for at most finitely many points. It was a great shock when in 1872 Weierstrass presented the function [6] defined by

$$W(x) = \sum_{n=1}^{\infty} b^n \cos(a^n \pi x)$$

where a is an odd integer, b a number belonging to the interval $(0, 1)$ and $ab > a + 3\pi/2$. This function is continuous but not differentiable.

Generalised functions The conditions for convergence of Fourier series given in Dirichlet's theorem are sufficient but not necessary. Dirichlet worked with the definition of integral given by Cauchy for continuous functions and then extended to sectionally continuous functions. It was Bernhard Riemann (1816-1866) who further extended the concept of integral to what is now known as the *Cauchy-Riemann integral* and which allows us to enlarge the class of functions which can be developed

[5]The Dirichlet function can actually be defined by means of a single analytic expression (though one which is rather complicated) as follows:

$$D(x) = (c - d) \lim_{n\to\infty} \lim_{m\to\infty} [\cos(n!\pi x)]^{2m} + d, \qquad x \in \mathbb{R}.$$

[6]Bolzano had already given an example of such a function , but it was not generally known.

n Fourier series. More powerful theories of integration have appeared since Riemann, n particular that devised by H.Lebesgue (1875-1941). But the more powerful as is he integration theory considered, the less will be the possibility that the Fourier eries obtained will converge pointwise to the function from which it originated. This eads to the consideration of new forms of convergence, namely convergence in the o-called Lebesgue spaces $L^p, (1 \leq p \leq \infty)$. The elements of L^p are not individual unctions (defined pointwise) in the sense of Dirichlet, but equivalence classes of functions equal almost everywhere - that is, except at most on a set of points of Lebesgue neasure zero: such sets may, however, be infinite and even uncountably infinite. This eneralisation of the concept of function to classes of L^p-integrable functions is particularly fruitful when $p = 2$, giving us the so-called Hilbert function space(s) and a orresponding return to a certain geometrisation of mathematical analysis.

The notion of function was generalised in other directions during the second half f the 20th century: it can even be said without much fear of contradiction that such eneralisations continue to develop in the dawn of the 21st century. The theory of listributions, created by Schwartz in the 1940s, constitutes one of the most important xtensions of the concept of function which has come to show, *a posteriori*, the justiation of the ideas put forward by Euler in connection with the nature of the solutions o the problem of vibrating strings. A brief exposition of the theory of distributions s given in chapter 8, which allows us to end this account here.

Appendix B

The Theorem of Lŏs

As remarked in Chapter 5, it is the internal objects which are the proper inheritors of the standard definitions and properties which can be transferred, term by term, modulo the ultrafilter $\mathcal{U}$, to the nonstandard universe. This process, used ad hoc in that chapter, finds its complete justification in the **Theorem of Lŏs** which we now describe briefly, together with an outline of the salient points of its proof. A more profound and more detailed study of this theorem will be found in the specialised literature. We begin with a preliminary lemma whose result is of importance in the sequel.

Lemma B.1 If $\alpha^1 \equiv [(a^1_\gamma)_{\gamma\in\Gamma}], \ldots, \alpha_k \equiv [(a^k_\gamma)_{\gamma\in\Gamma}]$ are elements of $\Pi_{\mathcal{U}}(\mathbb{R})$ then

$$^*t(\mu(\alpha^1), \ldots, \mu(\alpha^k)) = \mu([t(a^1_\gamma, \ldots, a^k_\gamma)_{\gamma\in\Gamma}])$$

where $t(x_1, \ldots, x_n)$ denotes a term of the language $\mathcal{L}(\mathbb{R})$ with the free variables $x_1, \ldots, x_k$ and $^*t(x_1, \ldots, x_n)$ is its transform in $\mathcal{L}(^*\mathbb{R})$.

Proof If t is a constant or a variable of the language the result follows immediately from the mapping $^* \; = \mu \circ \kappa$. Suppose for example that

$$t(x_1, \ldots, x_k) = a$$

where $a \in \mathcal{V}(\mathcal{R})$ is a constant. Then for each particular $\gamma \in \Gamma$ we get

$$t(a^1_\gamma, \ldots, a^k_\gamma) = a$$

and therefore

$$\begin{aligned} ^*t(\mu(\alpha^1), \ldots, \mu(\alpha^k)) &= {}^*a = \mu \circ \kappa(a) \\ &= \mu([(a)_{\gamma\in\Gamma}]) = \mu([t(a^1_\gamma, \ldots, a^k_\gamma)_{\gamma\in\Gamma}]). \end{aligned}$$

If on the other hand we have

$$t(x_1, \ldots, x_k) = x_i$$

for some index $1 < i \leq k$, then for each $\gamma \in \Gamma$ we get

$$t(a^1_\gamma, \ldots, a^k_\gamma) = a^i_\gamma$$

and therefore

$$^*t(\mu(\alpha^1),\ldots,\mu(\alpha^k)) = \mu(\alpha^i) = \mu([(a^i_\gamma)_{\gamma\in\Gamma}]) = \mu([t(a^1_\gamma,\ldots,a^k_\gamma)_{\gamma\in\Gamma}]) \quad .$$

Suppose now that the result is true for the terms $t_1,\ldots,t_m$ and let

$$t = f(t_1,\ldots,t_m)$$

where f denotes a function of m variables. Then, for each $\gamma \in \Gamma$,

$$t(a^1_\gamma,\ldots,a^k_\gamma) = f(t_1(a^1_\gamma,\ldots,a^k_\gamma),\ldots,t_m(a^1_\gamma,\ldots,a\gamma^k)$$

and therefore

$$^*t(\mu(\alpha^1),\ldots,\mu(\alpha^k)) = {^*f}({^*t_1}(\mu(\alpha^1),\ldots,\mu(\alpha^k)),\ldots, {^*t_m}(\mu(\alpha^1),\ldots,\mu(\alpha^k))$$
$$= \mu([f(t_1(a^1_\gamma,\ldots,a^k_\gamma),\ldots,t_m(a^1_\gamma,\ldots,a^k_\gamma))_{\gamma\in\Gamma}]) = \mu([(t(a^1_\gamma,\ldots,a^k_\gamma))_{\gamma\in\Gamma}])$$

which completes the proof. •

Theorem B.2 (Lǒs) Let $\phi(x_1,\ldots,x_k : b_1,\ldots,b_m)$ be a formula of $\mathcal{L}(\mathbb{R})$ in which $x_1,\ldots,x_k$ are the free variables and $b_1,\ldots,b_m$ are constants, and let

$$\Phi(x_1,\ldots,x_k) \equiv {^*\phi}(x_1,\ldots,x_k : {^*b_1},\ldots, {^*b_m}).$$

If $\alpha^1 \equiv [(\alpha^1_\gamma)_{\gamma\in\Gamma}],\ldots,\alpha^k \equiv [(\alpha^k_\gamma)_{\gamma\in\Gamma}]$ are elements of $\Pi_{\mathcal{U}}(\mathbb{R})$ then the formula of $\mathcal{L}({^*\mathbb{R}})$,

$$\Phi(\mu(\alpha^1,\ldots,\mu(\alpha^k) \equiv \Phi(\mu([(\alpha^1_\gamma)_{\gamma\in\Gamma}]),\ldots,\mu([(\alpha^k_\gamma)_{\gamma\in\Gamma}]))$$

will be true in $\mathcal{V}({^*\mathbb{R}})$ if and only if the set

$\{\gamma \in \Gamma : \mathcal{V}(\mathbb{R}) \models \phi(\alpha^1_\gamma,\ldots,\alpha^k_\gamma; b_1,\ldots,b_m]\}$ belongs to the ultrafilter $\mathcal{U}$.

The proof of this theorem uses induction over the complexity of the formulas. It is enough to consider only the two logical connectives $\neg$ and $\vee$ and the quantifier $\exists$, since all the formulas can be expressed using only these logical symbols. Thus, letting ϕ and θ be two formulas,

$\phi \Rightarrow \theta$ is equivalent to $(\neg\phi) \vee \theta$
$\phi \wedge \theta$ is equivalent to $(\neg\phi) \vee (\neg\theta)$
$\phi \Leftrightarrow \theta$ is equivalent to $\neg[[\neg[(\neg\phi) \vee \theta] \vee [\neg[(\not\theta) \vee \phi]]]$
$\forall_x\phi(x)$ is equivalent to $\neg[\exists(\not\phi)]$.

Proof If ϕ is an atomic formula the result follows from Lemma B.1. For, if ϕ is $t_1 = t_2$, where t_1 and t_2 are terms, then the equality

$$^*t_1(\mu(\alpha^1),\ldots,\mu(\alpha^k)) = {^*t_2}(\mu(\alpha^1),\ldots,\mu(\alpha^k)) \qquad (B.1)$$

gives, from Lemma B.1,

$$\mu([(t_1(a^1_\gamma,\ldots,a^k_\gamma))_{\gamma\in\Gamma}]) = \mu([(t_2(a^1_\gamma,\ldots,a^k_\gamma))_{\gamma\in\Gamma}]) \quad .$$

From the injectivity of the mapping μ it follows that

$$[(t_1(a^1_\gamma,\ldots,a^k_\gamma))_{\gamma\in\Gamma}] = [(t_2(a^1_\gamma,\ldots,a^k_\gamma))_{\gamma\in\Gamma}]$$

and therefore that

$$\{\gamma \in \Gamma : t_1(a^1_\gamma, \ldots, a^k_\gamma) = t_2(a^1_\gamma, \ldots, a^k_\gamma)\} \in \mathcal{U} \quad .$$

The converse may be proved by reversing the order of this proof.

Sinilarly, if ϕ is $t_1 \in t_2$ then from the relation

$$^*t_1(\mu(\alpha^1), \ldots, \mu(\alpha^k)) \in {}^*t_2(\mu(\alpha^1), \ldots, \mu(\alpha^k)) \tag{B.2}$$

we get

$$\mu([(t_1(a^1_\gamma, \ldots, a^k_\gamma))_{\gamma\in\Gamma}]) \in \mu([(t_2(a^1_\gamma, \ldots, a^k_\gamma))_{\gamma\in\Gamma}]) \quad .$$

whence, from the definition of μ, we obtain

$$[(t_1(a^1_\gamma, \ldots, a^k_\gamma))_{\gamma\in\Gamma}] \in_U [(t_2(a^1_\gamma, \ldots, a^k_\gamma))_{\gamma\in\Gamma}]$$

or, equivalently,

$$\{\gamma \in \Gamma : t_1(a^1_\gamma, \ldots, a^k_\gamma) t_2(a^1_\gamma, \ldots, a^k_\gamma)\} \in \mathcal{U}.$$

The converse is again easily verified, as in the previous case. The process can be generalised in a similar way for an arbitrary atomic relation containing constants and variables.

Suppose now that $\phi(x_1, \ldots, x_k)$ is a formula of the type $\neg\theta(x_1, \ldots, x_k)$. The formula θ has fewer logical symbols than ϕ and so for the induction hypothesis we take:

$$\begin{gathered} ^*\theta(\mu(\alpha_1, \ldots, \mu(\alpha_k)) \text{ is true in } \mathcal{V}({}^*\mathbb{R}) \text{ if and only if} \\ \{\gamma \in \Gamma : \mathcal{V}(\mathbb{R}) \models \theta(a^1_\gamma, \ldots, a^k_\gamma)\} \in \mathcal{U} \\ \text{and will be false if and only if} \\ \{\gamma \in \Gamma : \mathcal{V}(\mathbb{R}) \models \theta(a^1_\gamma, \ldots, a^k_\gamma)\} \notin \mathcal{U}. \end{gathered}$$

From the properties of the ultrafilter it follows that the sentence $^*\theta(\mu(\alpha_1, \ldots, \mu(\alpha_k))$ will be false if and only if $\{\gamma \in \Gamma : \mathcal{V}(\mathbb{R}) \models \theta(a^1_\gamma, \ldots, a^k_\gamma)\}^c \in \mathcal{U}$, or equivalently if and only if $\{\gamma \in \Gamma : \mathcal{V}(\mathbb{R}) \models \neg\theta(a^1_\gamma, \ldots, a^k_\gamma)\} \in \mathcal{U}$ or again if and only if $\{\gamma \in \Gamma : \mathcal{V}(\mathbb{R}) \models \phi(a^1_\gamma, \ldots, a^k_\gamma)\} \in \mathcal{U}$. Now $^*\phi$ is true if and only if $^*\theta$ is false and therefore $^*\phi$ is true if and only if

$$\{\gamma \in \Gamma : \mathcal{V}(\mathbb{R}) \models \phi(a^1_\gamma, \ldots, a^k_\gamma)\} \in \mathcal{U}$$

as was to be proved.

Again, suppose that $\phi(x_1, \ldots, x_k)$ is a formula of the type

$$\theta(x_1, \ldots, x_k) \vee \zeta(x_1, \ldots, x_k) \quad .$$

The formulas $\theta(x_1, \ldots, x_k)$ and $\zeta(x_1, \ldots, x_k)$ have fewer logical symbols than the formula $\phi(x_1, \ldots, x_k)$. For the induction hypothesis we take:

$$\begin{gathered} ^*\theta(x_1, \ldots, x_k) \text{ is true in } \mathcal{V}({}^*\mathbb{R}) \text{ if and only if} \\ \{\gamma \in \Gamma : \mathcal{V}(\mathbb{R}) \models \theta(\alpha^1_\gamma, \ldots, \alpha^k_\gamma)\} \in \mathcal{U}, \\ \text{and } {}^*\zeta(x_1, \ldots, x_k) \text{ is true in } \mathcal{V}({}^*\mathbb{R}) \text{ if and only if} \\ \{\gamma \in \Gamma : \mathcal{V}(\mathbb{R}) \models \zeta(\alpha^1_\gamma, \ldots, \alpha^k_\gamma)\} \in \mathcal{U}. \end{gathered}$$

Now ${}^*\phi(\mu(\alpha_1),\ldots,\mu(\alpha_k))$ is true in $\mathcal{V}({}^*\mathbb{R})$ if and only if ${}^*\theta(\mu(\alpha_1),\ldots,\mu(\alpha_k))$ is true in $\mathcal{V}({}^*\mathbb{R})$ or ${}^*\zeta(\mu(\alpha_1),\ldots,\mu(\alpha_k))$ is true in $\mathcal{V}({}^*\mathbb{R})$. Suppose that ${}^*\theta(\mu(\alpha_1),\ldots,\mu(\alpha_k))$ is true in $\mathcal{V}({}^*\mathbb{R})$. Then the set $\{\gamma\in\Gamma:\mathcal{V}(\mathbb{R})\models\theta(a^1_\gamma,\ldots,a^k_\gamma)\}$ belongs to $\mathcal{U}$, and so, since

$$\{\gamma\in\Gamma:\mathcal{V}(\mathbb{R})\models\theta(a^1_\gamma,\ldots,a^k_\gamma)\}\subset\{\gamma\in\Gamma:\mathcal{V}(\mathbb{R})\models\phi(a^1_\gamma,\ldots,a^k_\gamma)\}$$

we must have $\{\gamma\in\Gamma:\mathcal{V}(\mathbb{R})\models\phi(a^1_\gamma,\ldots,a^k_\gamma)\}$ belongs to $\mathcal{U}$.

Conversely suppose that $\{\gamma\in\Gamma:\mathcal{V}(\mathbb{R})\models\phi(a^1_\gamma,\ldots,a^k_\gamma)\}\in\mathcal{U}$; since

$$\{\gamma\in\Gamma:\mathcal{V}(\mathbb{R})\models\theta(a^1_\gamma,\ldots,a^k_\gamma)\}\cup\{\gamma\in\Gamma:\mathcal{V}(\mathbb{R})\models\zeta(a^1_\gamma,\ldots,a^k_\gamma)\}$$
$$=\{\gamma\in\Gamma:\mathcal{V}(\mathbb{R})\models\phi(a^1_\gamma,\ldots,a^k_\gamma)\}$$

then

$$\{\gamma\in\Gamma:\mathcal{V}(\mathbb{R})\models\theta(a^1_\gamma,\ldots,a^k_\gamma)\}\in\mathcal{U},\text{ or }\{\gamma\in\Gamma:\mathcal{V}(\mathbb{R})\models\zeta(a^1_\gamma,\ldots,a^k_\gamma)\}\in\mathcal{U},$$

and therefore the sentence ${}^*\theta(\mu(\alpha_1,\ldots,\mu(\alpha_k))$ is true in $\mathcal{V}({}^*\mathbb{R})$ or the sentence ${}^*\zeta(\mu(\alpha_1,\ldots,\mu(\alpha_k))$ is true in $\mathcal{V}({}^*\mathbb{R})$.

Finally, let $\phi(x_1,\ldots,x_k)=\exists_{y\in t}\theta(y;x_1,\ldots,x_k)$ where t is a term and the formula $\theta(x)$ has fewer logical symbols than $\phi(x)$. If the sentence ${}^*\phi(\mu(\alpha^1),\ldots,\mu(\alpha^k))$ were true in $\mathcal{V}({}^*\mathbb{R})$ there would exist an element β belonging to ${}^*t(\mu(\alpha^1),\ldots,\mu(\alpha^k))$ such that ${}^*\theta(\beta;\mu(\alpha^1,\ldots,\mu(\alpha^k))$ is true in $\mathcal{V}({}^*\mathbb{R})$. Since ${}^*t(\mu(\alpha^1),\ldots,\mu(\alpha^k))=\mu([(t(a^1_\gamma,\ldots,a^k_\gamma))_{\gamma\in\Gamma}])$ then $\beta=\mu([(b_\gamma)_{\gamma\in\Gamma}])$ and therefore $\{\gamma\in\Gamma:\mathcal{V}(\mathbb{R})\models\theta(b_\gamma;a^1_\gamma,\ldots,\alpha^k_\gamma)\}\in\mathcal{U}$. It follows that

$$\{\gamma\in\Gamma:\mathcal{V}(\mathbb{R})\models\exists_{y\in t}\theta(y;a^1_\gamma,\ldots,\alpha^k_\gamma)\}\in\mathcal{U}$$

or, equivalently, that

$$\{\gamma\in\Gamma:\mathcal{V}(\mathbb{R})\models\phi(a^1_\gamma,\ldots,\alpha^k_\gamma)\}\in\mathcal{U}\quad.$$

Conversely, suppose that $\{\gamma\in\Gamma:\mathcal{V}(\mathbb{R})\models\phi(a^1_\gamma,\ldots,\alpha^k_\gamma)\}\in\mathcal{U}$ and therefore that $\{\gamma\in\Gamma:\mathcal{V}(\mathbb{R})\models\exists_{y\in t}\theta(y;a^1_\gamma,\ldots,\alpha^k_\gamma)\}\in\mathcal{U}$. For each $\gamma\in\Gamma$ let $b_\gamma\in t(a^1_\gamma,\ldots,a^k_\gamma)$ be such that $\mathcal{V}(\mathbb{R})\models\theta(b_\gamma;a^1_\gamma,\ldots,a^k_\gamma)$. Then, since $(t(a^1_\gamma,\ldots,a^k_\gamma))_{\gamma\in\Gamma}$ is a bounded generalised sequence, the generalised sequence $(b_\gamma)_{\gamma\in\Gamma}$ defines an internal object $\beta=\mu([(b_\gamma)_{\gamma\in\Gamma}])$ such that $\beta\in{}^*t(\mu(\alpha^1),\ldots,\mu(\alpha^k))$ and $\mathcal{V}(\mathbb{R})\models{}^*\theta(\beta;\mu(\alpha^1,\ldots,\mu(\alpha^k))$, from which follows $\mathcal{V}({}^*\mathbb{R})\models{}^*\phi(\mu(\alpha^1,\ldots\mu(\alpha^k))$. •

References

[1] Bell, J. & Machover, M., *A Course in Mathematical Logic*, North-Holland, 1977.

[2] Berz, M., "Calculus and numerics on Levi-Civita fields".

[3] Bottazzini, U., *The Higher Calculus: A History of Real and Complex Analysis from Euler to Weierstrass*, Springer-Verlag, 1986.

[4] Chang, C. C. & Keisler, H. J., *Model Theory*, Studies in Logic and the Foundations of Mathematics, vol.73, North-Holland 1994.

[5] Dirac, P. A. M., *The Principles of Quantum Mechanics*, Oxford, 1930.

[6] Edwards, C. H., *The Historical Development of the Calculus*, Springer-Verlag, 1979.

[7] Eherlich, P., *Real Numbers, Generalizations ...*, Synthese Library, vol.242, Kluwer Academic.

[8] Euler, L., *Introduction to the Analysis of the Infinite*, Book I Springer-Verlag, 1988.

[9] Euler, L., *Introduction to the Analysis of the Infinite*, Book II Springer-Verlag, 1988.

[10] Ferreira, J. C., *Introdução à Teoria das Distribuições* , Fundação Calouste Gulbenkian, 1993.

[11] Fisher, D., "Extending Functions to Infinitesimals of Finite Order", *Amer. Math. Monthly*, Aug-Sept 1982, pp443-449.

[12] Gardiner, A., *Infinite Processes*, Springer-Verlag, 1982.

[13] Gillies, D., (ed.), *Revolutions in Mathematics*, Clarendon Press, Oxford, 1982.

[14] Goldblatt, R. *Lectures on the Hyperreals - An Imtroduction to Nonstandard Analysis*, Springer-Verlag, 1998.

[15] Hanqiao, F.; St. Mary, D. F.; Wattenberg, F., "Applications of Nonstandard Analysis to Partial Differential Equations - 1.The Diffusion Equation", *Math. Modelling*, vol.7, 1986, pp507-523.

[16] Hoskins, R. F., *Standard and Nonstandard Analysis*, Ellis Horwood, 1991.

[17] Hoskins, R. F. & Sousa-Pinto, J. *Distributions, Ultradistributions and other Generalised Functions*, Horwood Publishing, 2003.

[18] Hoskins, R. F. & Sousa-Pinto, J. "Nonstandard treatments of generalised functions on the line - Part I", *Cadernos de Matemática da Universidade de Aveiro* CM/I-01, Abril de 1994.

[19] Hoskins, R. F. & Sousa-Pinto, J. "Nonstandard treatments of generalised functions on the line - Part II", *Cadernos de Matemática da Universidade de Aveiro* CM/I-11, Julho de 1995.

[20] Hoskins, R. F. & Sousa-Pinto, J. "Nonstandard treatments of generalised functions on the line - Part III", *Cadernos de Matemática da Universidade de Aveiro* - CM/I-12, Julho de 1995.

[21] Hoskins, R. F., *Delta Functions*, Ellis Horwood, 1999.

[22] Hurd, A. E. & Loeb, P. A., *An Introduction to Nonstandard Real Analysis*, Academic Press, 1985.

[23] Keisler, H. J., *Foundations of Infinitesimal Calculus*, Prindle, Weber & Schmidt, Inc., 1976.

[24] Kinoshita, M., "Nonstandard Representations of Distributions I", *Osaka J. Math.*, 25, 1988, pp805-824.

[25] Kinoshita, M., "Nonstandard Representations of Distributions II", *Osaka J. Math.*, 27, 1988, pp843-861.

[26] Kleiner, I., "Evolution of the Function Concept: A Brief Survey ".

[27] Lindstrøm, T., "An Invitation to Nonstandard Analysis", *Nonstandard Analysis and its Applications*, edited by Nigel Cutland, London Math. Soc., Students Texts 10, Cambridge University Press, 1988.

[28] Lützen, J., *The Prehistory of the Theory of Distributions*, Springer-Verlag, 1982.

[29] Lützen, J., "Euler's Vision of a General Partial Differential Calculus for a Generalized Kind of Function", *Math. Mag.*,56, 1983, pp299-306.

[30] Luzin, N., "The Evolution of *Function*: Part I", edited by Abe Shenitzer, *Amer. Math. Monthly*, vol. 105, no. 1, 1998, pp59-67.

[31] Luzin, N., "The Evolution of *Function*: Part II", edited by Abe Shenitzer, *Amer. Math. Monthly*, vol. 105, no. 1, 1998, pp263-270.

[32] Maclane, S., *Mathematics, Form and Function*, Springer-Verlag, 1986.

[33] Neves, V., "An'alise Não-Standard, Uma Linguagem para o Estudo da Análise Elementar", *Cadernos de Matemática* CM/I-38 (Departamento de Matemática da Universidade de Aveiro), Marco de 1998.

[34] Pétry, A., *Analyse Infinitesimal*, Institut Supérieur Industriel Liégeois, 1977.

[35] Priestley, W. M., *Calculus:An Historical Approach*, Springer-Verlag, 1979.

[36] Ravetz, J. R., "Vibrating Strings and Arbitrary Functions", in *The Logic of Personal Knowledge: Essays Presented to M. Polanyi on his Seventieth Birthday.* The Free Press, 1961, pp71-88.

[37] Robinson, A., *Nonstandard Analysis*, North-Holland, 1980.

[38] Robinson, A., "Function Theory on some Nonarchimedian Fields ", *Amer. Math. Monthly*, Papers in the Foundations of Mathematics, 80, 1973, pp87-109.

[39] Rubio, J. E., *Optimisation and Nonstandard Analysis*, Marcel Dekker, Inc., 1994.

[40] Ruthing, D., "Some Definitions of the Concept of Function from John Bernoulli to N. Bourbaki." *The Math. Intelligencer*, vol. 6 no. 4, 1984, pp72-77.

[41] Schaaf, W. L., "Mathematics and World History", *Math. Teacher*, 23, 1930, pp496-503.

[42] Sebastião e Silva, J., "Como nasceu a teoria das distribuições. Suas relações com a fisica e a tecnica", Revista "*Ciência*", nos. 15-16, 1959, [Orgão da Associação de Estudantes da Faculdade de Ciencias de Lisboa.

[43] Sebastião e Silva, J., "Sur l'Axiomatique des Distributions et ses Possibles Modeles", *Obras de José Sebastião e Silva* - III, INIC, 1985.

[44] Sebastião e Silva, J., *Introdução a Teoria das Distribuiçãoes*, Publicação elaborada pelo Professor A. Andrade Guimarâres segundo as Lições do Professor José Sebastião e Silva em 1956/57. Publicação subsidiada pelo INIC.

[45] Sequeira, F., "Uma Introdução à Teoria das Distribuiçãoes", *Gazeta de Matemática*, 121-124, 1971, pp31-49.

[46] Sousa Pinto, J. "Introdução à Análise Não-Standard ", *Boletim de SPM*, no. 22, 1992, pp11-26.

[47] Sousa Pinto, J. "Funções Delta no Universo da Anãlise Não-Standard", *Revista Técnica* no.2, 1995, pp5-16.

[48] Sousa Pinto, J. "As funções internas do universo da anãlise não-standard", *Boletim da SPM* no. 31, 1995, pp17-27.

[49] Sousa Pinto, J. "Estudo do Principio de Leibniz", *Cadernas de Matemática da Universidade de Aveiro* - CM/D-11, Junho de 1997.

[50] Sousa Pinto, J. & Neves, V., "Monados Topologicas", *Boletim da SPM* no. 35, 1996, pp55-71.

[51] Tall, D., "Infinitesimals constructed algebraically and interpreted geometrically", *Mathematical Education for Teaching.*

[52] Tall, D. "Looking at graphs through infinitesimal microscopes, windows and telescopes", *The Math.Gazette*, pp22-48.

[53] Zemanian, A. H., *Distribution Theory and Transform Analysis*, McGraw-Hill, 1965.

Index

Index

Printed and bound by CPI Group (UK) Ltd, Croydon, CR0 4YY
03/10/2024
01040437-0005